Aquatic Entomology

Jill Lancaster and Barbara J. Downes

The University of Melbourne, Australia

T0202335

UNIVERSITY PRESS

Aquatic Entomology. First Edition. Jill Lancaster & Barbara J. Downes.
© Jill Lancaster & Barbara J. Downes 2013. Published 2013 by Oxford University Press.

OXFORD
UNIVERSITY PRESS

Great Clarendon Street, Oxford, OX2 6DP,
United Kingdom

Oxford University Press is a department of the University of Oxford.
It furthers the University's objective of excellence in research, scholarship,
and education by publishing worldwide. Oxford is a registered trade mark of
Oxford University Press in the UK and in certain other countries

First Edition published in 2013
Reprinted 2014

Published in the United States of America by Oxford University Press
198 Madison Avenue, New York, NY 10016, United States of America

British Library Cataloguing in Publication Data

Data available

ISBN 978–0–19–957322–6

Preface

After many years using freshwater ecosystems and aquatic insects as model systems to test ecological ideas, we have learnt—repeatedly—that understanding the basic biology of these fantastic animals is often pivotal to successful and insightful outcomes. Aquatic insect species are a minority of the total insect diversity, but they are no less fascinating and have long been the focus of attention for researchers, collectors, and amateurs alike. The persistent presence of aquatic insects in the Earth's fauna for over 300 million years, their numerical dominance in contemporary freshwaters (and some saline ones), and the diversity of roles they play in aquatic ecosystems speak for the importance of this group. In an academic context, aquatic insects have served as model systems for the development of understanding into many aspects of insect behaviour, biomechanics, developmental biology, ecology, epidemiology, evolution, physiology, etc., so there are many aficionados of this group.

While aquatic insects figure in diverse fields of basic research, they also feature strongly in applied research aimed at understanding and quantifying human impacts on freshwater environments. As a result of the latter, aquatic insects are widely used in monitoring programs or assessment tools that aim to detect human impacts or assess the condition of water bodies. In our opinion, a surprisingly high proportion of research on aquatic insects, both basic and applied, and commensurate, freshwater monitoring programmes and assessment tools reflect little direct knowledge of the structure and function of insects. We find this both understandable and worrying. It is understandable for two main reasons. First, there are numerous, often excellent texts on entomology (general entomology and its various sub-disciplines) and also on freshwater ecology, but

aquatic insects often get short shrift from both. Surprisingly few texts focus strongly on aquatic entomology or bring together a broad range of information on the biology of aquatic insects, how they function in water (which is markedly different from air), and how they mange to cross the aquatic–terrestrial boundary twice during their complex life cycle. There is a wealth of information on aquatic insects, but it is scattered across the scientific literature and has not, before now, been gathered together; hence it has not been easy to access. Second, some, perhaps many, of the new cohorts of freshwater researchers or those entering management agencies receive little formal training in zoology (many have followed degree programmes in engineering, hydrology, environmental science, or physical geography), and teaching entomology is largely out of fashion in tertiary-level education, with a few notable exceptions (mainly in the USA). Hence, these new generations have typically not been exposed to entomology in all its wonderful facets.

Nevertheless, the lack of entomological knowledge among many freshwater researchers and prospective managers is worrying, because, without a sound understanding of the organism, there is always a risk of misguided programmes of research and monitoring, erroneous interpretations of ecological data, and missed insights or opportunities to push the scientific frontiers. It is also worrying because the parlous status of freshwaters around the world means that the managers and researchers charged with their well-being will be best able to protect these systems only if they are well acquainted with the basic biology of aquatic insects.

Thus, our aim was to fill that conspicuous gap in the literature and produce a book that will inform, support, and strengthen the work of a diverse body of researchers and managers of freshwater systems. This book is not about the ecology of aquatic insects, but instead focuses on the basic biology of aquatic insects that underpins so much that is important to ecological research and water resources management. Nevertheless, much of this material will also be of interest to entomologists more widely and very many others. The scope of the material is very broad and draws upon research in a huge range of disciplines, but only touches the surface of some very extensive research areas. It is not intended to replace any general entomology texts, but may be a valuable companion to other texts on entomology, freshwater ecology, biomechanics, ecohydrology, water resources management, etc., depending on readers' interests. As such, this book will be a useful reference text for tertiary subjects in the aforementioned disciplinary areas, at both later-year undergraduate and postgraduate levels. It should also be a very useful text for those researching freshwater ecosystems, including research higher-degree students. Nevertheless, we have tried to include explanations of the basic terminology of entomology throughout in a way that will be accessible to anyone with a strong interest in insects, including non-academics.

This book would not have come into existence if not for the support of others. Jill is deeply indebted to the Leverhulme Trust (UK) for a Leverhulme Trust Research Fellowship which provided relief from teaching during the initial planning and writing stages. Without that support, this project would not have even started. At Oxford University Press, Ian Sherman was courageous enough to take on this project and patient as we slipped ever further behind schedule; Helen Eaton and Lucy Nash pro-

vided gentle prodding, encouragement, and technical information about the publishing world. We thank the Australian Society for Limnology for generously funding the index. Many other colleagues helped in diverse ways, including reading and commenting on chapters, suggesting references, providing original images, or supplying specimens and photographs that became figures. We are grateful to them all for their comments, insights, and generosity: Amanda Arnold, Gerald Baker, Leon Barmuta, Andrew Brooks, John Bush, Gerry Closs, Peter Cranston, Petra Ditsche-Kuru, David Dudgeon, David Hu, P.S. (Sam) Lake, Richard Marchant, Barbara Peckarsky, Stephen Rice, Belinda Robson, Andrew Sheldon, Ian Stocks, and Handoko Wahjudi. Any errors are our own.

We gratefuly acknowledge the permission granted to reproduce the copyright material in this book: the Company of Biologists, CSIRO Publishing, the Danish Natural History Society, the Ecological Society of America, Elsevier, the Entomological Society of America, the Finnish Zoological and Botanical Publishing Board, Nature Publishing Group, the North American Benthological Society, the Royal Society, Springer-Verlag, Taylor & Francis, and John Wiley & Sons Ltd. Barbara Peckarsky kindly gave permission to reproduce Figure 7.11a; Glenn Wiggins kindly gave permission to reproduce Figures 12.2 and 12.7. Every effort has been made to trace copyright holders and to obtain their permission for the use of copyright material. We apologize for any errors or omissions in the above list and would be grateful if notified of any corrections that should be incorporated in future reprints or editions of this book.

Melbourne
March 2013

Contents

Part 3 Sensory Systems, Movement, and Dispersal

Part 4 Population Dynamics and Population Persistence

PART 1

Introduction to Aquatic Insects

The first part of this book introduces the aquatic insects, some basic aspects of their biology, morphology, and evolutionary history, and discusses where these insects may be found in nature. All of this information may be considered a basic starting point to any entomological investigation. The first step in learning any discipline is to become familiar with its jargon, and entomology has numerous unique terms. Many of these words are used to describe body parts and a basic familiarity with these words is essential to comprehend later chapters. Additionally, the names of orders and some of the common families are also essential. Not only are these words used repeatedly throughout the book, familiarity with the types of insects found in different orders is necessary to appreciate some of the more detailed aspects of their biology.

We start in Chapter 1 with key characteristics that are common to all insects, starting with the main types of life cycles. Different life cycles entail different developmental stages, and which life cycle predominates varies among orders. We follow this with a description of the basic body plan and parts of insects and then, having armed the reader with some terminology and this basic information, we describe the main features of insects in the fully and partially aquatic orders. Readers with a good grasp of entomology and its associated terminology may confidently skip Chapter 1, but may still find useful material in Chapter 2. Chapter 2 briefly explores the evolutionary history of insects and tackles questions such as, when did insects become aquatic? And what kinds of aquatic insects

appeared first? The places and times in which animals evolved are important information for many types of research, but particularly investigations into the distributional ranges of species (i.e. biogeography). This is because the explanations for where some species, genera, or families are found (around the world or regionally) lie in events that happened in the past. If a taxon evolved before major episodes of continental drift or under a different climatic regime compared to now, then these historical effects can be reflected strongly in its modern distribution. If we are unaware of such historical events, there is a risk of assuming, incorrectly, that distributions are caused by proximate (i.e. modern) environmental gradients. There is another important aspect to understanding what drives distribution patterns. Aquatic insects may be found in an enormous array of different environments from rain puddles to lakes and ponds, rivers and streams, estuaries and beaches, and even the open ocean. Each of these types of environments offers diverse living places, but these entail quite specific adaptations for successful habitation. Thus, knowing the global or regional range of species is merely the starting point to understanding where they may be found. Accordingly, Chapter 2 describes some of the extraordinary range of aquatic environments and some of their characteristic insect denizens.

While appearing esoteric, issues of these sorts are critically important to investigations into the impacts of human beings and other applied research questions. Currently, there is much interest in developing models that will predict whether distributional ranges will expand or shrink under climate change, with excessive water extraction from rivers, and so forth. Without a good grasp of how history has shaped modern day distributions and how insects are distributed within habitats, often over relatively small spatial scales, such models are likely to produce misleading predictions. This is why the basic information presented in Chapters 1 and 2 is a critically important starting point.

CHAPTER 1

Insect body structure and the aquatic insect orders

1.1 Introduction

From a systematic perspective, the class Insecta lies within the phylum Arthropoda, which is a very large group of invertebrates that all have an exoskeleton and jointed limbs. The name is derived literally from the ancient Greek words *arthro*, meaning 'joints', and *pod*, meaning 'foot'. The name Insecta is derived from the Latin *insectum*, meaning 'cut into sections', but *insectum* is probably a rendering of the ancient Greek *entomos*, which also means 'cut into sections'. Along with the insects, the arthropods include spiders, mites, crustaceans (e.g. crabs, lobsters, shrimp, etc.), centipedes, millipedes, and many others, but only the insects have six legs as adults. Nevertheless, statistically speaking, most animals are insects, i.e. more than half the animals alive at any one time are insects. In addition to being numerous, insects are extremely diverse with very many extant species. Approximately 1.5 million species of insect have been described and current estimates suggest that there may be as many as 4 million species in total. While aquatic insects make up a minority of the total, the number described is still in the order of 10^5–10^6 species, but, as will become clear, many insects are semi-aquatic to varying degrees so precise estimates of exact numbers are difficult.

Importantly, the aquatic insects do not form a distinct taxonomic group within the class Insecta. Some orders contain only species that are aquatic in some life stage (e.g. mayflies, stoneflies, dragonflies), but other orders contain both aquatic and terrestrial species (e.g. beetles, bugs). Insects evolved on land and multiple orders secondarily invaded aquatic environments (Section 2.4) and, therefore,

aquatic insects are essentially terrestrial insects that have found a way to live underwater. Consequently, aquatic insects are enormously variable in morphology, development, physiology, and ecology. Accordingly, most generalizations we can make about their diets, habitats, behaviour, and so forth are weak at the coarse taxonomic level of order. While a good starting point for the novice, almost all such generalizations have many exceptions. As will become clear in later chapters, the diversity of ways in which insects have evolved an aquatic existence at some point in their life cycles is truly enormous.

Nonetheless, morphology and life cycles have features that are common to all insects. All insects follow the same basic body plan and some understanding of the basic body structure is essential to any treatise on aquatic insects. Entomology is rife with subject-specific terminology to describe different body parts and, although this can be frustrating, this vocabulary is necessary. This chapter provides a brief description of the types of life cycles of insects (Section 1.2) and then explains the external morphology of insects (Section 1.3), followed by descriptions of the insect orders that contain species that are wholly or partially aquatic (Section 1.4). The form and function of many body parts will be discussed in detail in subsequent chapters. Additionally, much more detailed descriptions and beautiful illustrations of the various taxa are available in general entomology texts and texts devoted to particular orders or identification guides. Readers wishing for more information are advised to consult such sources, while readers familiar with this material may wish to skip ahead to Chapter 2.

Aquatic Entomology. First Edition. Jill Lancaster & Barbara J. Downes.
© Jill Lancaster & Barbara J. Downes 2013. Published 2013 by Oxford University Press.

1.2 Insect life cycle

There are three main patterns of development within the insects: ametaboly, hemimetaboly, and holometaboly; the latter two are also referred to as incomplete and complete metamorphosis, respectively. During their postembryonic growth (i.e. once hatched from the egg), insects pass through a series of juvenile instars (stages of growth) until they become adults. Each instar is terminated by a moult. Ametabolous insects are wingless (the Apterygota) and comprise some of the oldest insects. Uniquely, they moult as adults, but none are aquatic and will not be discussed further. In the hemimetabolous insects (the Exopterygota), juvenile instars resemble one another and the later instars broadly resemble the adult, except that adults have wings and genitalia. The life cycle of most hemimetabolous insects has three stages (egg, juvenile, and adult), whereas most holometabolous insects (the Endopterygota) have four stages (egg, juvenile, pupa, adult). Juveniles and adults of holometabolous insects are very different from one another in morphology and habit. They undergo striking changes (complete metamorphosis), spread over two moults, in the formation of the adult. The final juvenile instar, the pupa, has become specialized to facilitate these changes. Other than the pupa, this book will refer to all juvenile stages as 'larvae'; other authors refer to only the juveniles of the holometabolous insects as larvae and use the term 'nymph' to refer to juveniles of the Hemimetabola. Arguments can be made to support both schemes and some researchers hold strong views on the subject, but we will use the simplest scheme. For virtually all aquatic insects, the transition from juvenile to adult involves a transition from aquatic to terrestrial habits, i.e. they have complex life cycles. Where both juveniles and adults are aquatic (e.g. some Coleoptera and Hemiptera), adults typically retain the terrestrial habit of breathing air rather than obtaining oxygen dissolved in water.

1.3 Insect body plan

Insects have segmented bodies, like other members of the Arthropoda, and have an exoskeleton made of cuticle. The outer layer of cuticle is hardened (tanned) to various degrees and forms the exocuti-

cle. These hardened regions (sclerites) may be fused to form apparently solid structures, such as the head capsule, or may be separated by joints or flexible regions where the exocuticle layer is missing and the cuticle remains membranous. In the majority of insects, and especially the adults, the cuticle is heavily sclerotized and forms a series of dorsal and ventral plates along the body: the terga and sterna. In many larvae, however, virtually the entire cuticle is thin and flexible, although clearly segmented. The segments of the insect body are differentiated into three main parts: the head, thorax, and abdomen. Segmentation is clear in the thorax and abdomen, but has virtually disappeared in the head.

1.3.1 Head

At the anterior end of the body, the head is typically a heavily sclerotized capsule, bearing mouthparts and some major sense organs (Figure 1.1). Thus, the head and associated organs are central to the insect's ability to acquire food resources and to collect information about the environment, which subsequently may influence myriad behaviours. The head capsule is formed from several plates or sclerites (e.g. gena, frons, clypeus) that meet at suture lines. Both adults and larvae have distinct head capsules, although in some larval Diptera the head capsule may be incomplete or reduced (e.g. Tabanidae) or withdrawn into the thoracic segments (e.g. some Tipulidae).

The most noticeable sense organs are usually a pair of compound eyes, comprising hundreds to thousands of individual facets or ommatidia, and a pair of antennae. Between the compound eyes, there may be up to three ocelli (Figure 1.1), each on a separate sclerite. Ocelli are often referred to as simple eyes, but their sensory function can be far from simple (Taylor and Krapp 2008). Larvae of the holometabolous orders lack compound eyes and ocelli, but have eyes called stemmata. These are often called 'lateral ocelli', but stemmata are morphologically and functionally different from true ocelli (Gilbert 1994). The antennae are highly variable among taxa and life stages (Schneider 1964) but they are typically segmented, with two differentiated basal segments (the scape and pedicel) and a distal, often whip-like flagellum composed of many similarly shaped elements. The scape is set in a membranous

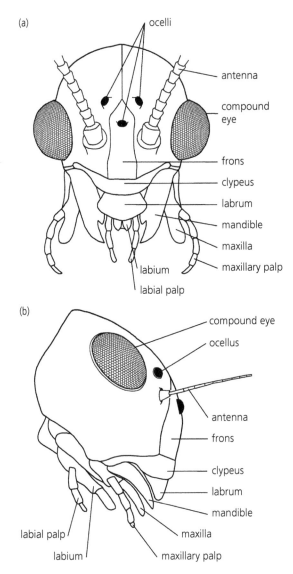

Figure 1.1 Structure of the head. Hypothetical insects viewed from (a) the front, and (b) the side.

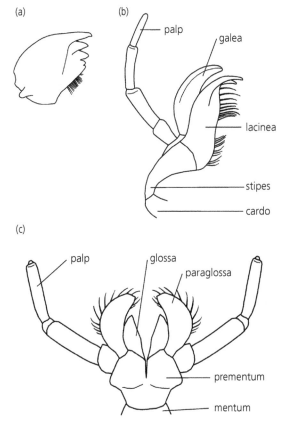

Figure 1.2 Structure of the mouthparts, as illustrated by a larval stonefly, *Cosmioperla* (Eustheniidae). (a) Mandible, (b) maxilla, and (c) labium.

socket and, in the majority of insects, movement of the whole antenna is effected by muscles inserted on the scape and attached to the head capsule.

The mouthparts and their various structures function, collectively, to acquire food, to reduce large food items into smaller bite-sized pieces (in some groups), and to move food into the mouth, which is the start of the alimentary canal. In the plesiomorphic condition (also called the primitive or ancestral form), the head is oriented so that the mouthparts lie ventrally and

the mouthparts are for chewing. Typical chewing mouthparts, with only minor modifications, occur in many larval Ephemeroptera and Trichoptera, and the adult Odonata. The primary mouthparts consist of a labrum, a pair of mandibles, a pair of maxillae, and a labium, with the mouth situated at the base of the labium. In a typical chewing insect, the labrum (Figure 1.1) is a broadly flattened plate with a membranous ventral (inner) surface, which bears chemosensilla. The mandibles (Figure 1.2) are heavily sclerotized and vary in shape according to the insect's diet: in herbivorous forms there are cutting edges and grinding surfaces; in carnivorous forms the mandible may possess sharply pointed 'teeth' for seizing prey, cutting, and tearing. The maxillae each bear a segmented maxillary palp attached to the main stipe (Figure 1.2), although this palp may be reduced or lost in some species. The labium bears a pair of segmented labial palps (Figure 1.2), arranged laterally to a pair of glossae and paraglossae. The glossae and paraglossae may be fused together to form a single ligula, as in many larval Chironomidae. There are, or course, many variations on this general scheme and the mouthparts of some groups have become highly modified and specialized for particular feeding habits (Chapter 13).

1.3.2 Thorax

The thorax is the centre for insect locomotion. In the adult, each of the three thoracic segments (pro-, meso-, and metathorax) typically bears a pair of legs, and the meso- and metathorax each have a pair of wings. All larvae lack wings, although developing wings may be visible in the wing buds of hemimetabolous orders, and many larvae have a distinct three-segmented thorax with three pairs of true legs. In some holometabolous groups, especially the Diptera, larvae lack a distinct thorax and true legs, although prolegs may be present (e.g. larval Simuliidae and Chironomidae). In cross-section, the thorax of adults is essentially a box frame made of terga, sterna, and, in some, sclerites that lie between the terga and sterna (Figure 1.3). The wings are attached to the wing processes and each leg is inserted in a coxal cavity. The thoracic sclerites have internal cuticular ridges that are attachment points for flight and walking muscles. The box-like frame of these segments is not completely rigid as wing movement is, in part, brought about by flexure of the terga, which is itself caused by muscle contraction and relaxation (Section 9.2). The thoracic segments of larvae that have true legs are analogous to the adult thorax, but much simpler and less box-like given the absence of wings.

The true legs are typically for walking, crawling, and running, but may be specialized for other functions, such as swimming or grasping prey. Each leg consists of six segments: the coxa, trochanter, femur, tibia, tarsus, and pretarsus (Figure 1.4). A narrow, annulated, and flexible membrane (the corium) lies between adjacent segments and facilitates articulation. Except for the trochanter, which is fused to the femur, articulation usually occurs between all the segments. The femur is generally the largest leg segment. The tarsus is usually subdivided into between

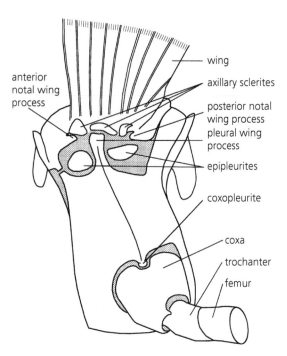

Figure 1.3 Structure of a thoracic wing-bearing segment of a hypothetical adult (lateral view). The pleural wing process acts as a fulcrum for wing movement. The anterior and posterior notal wing processes give support to small axillary sclerites on the wing base. The epipleurites are small plates for insertion of muscles attached to the wing sclerites, leg segments, and other parts of the thorax. The coxopleurite is an articulation point between the thorax and coxa.

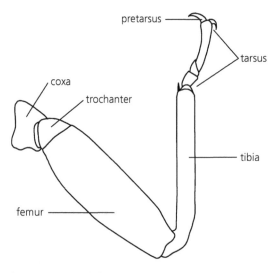

Figure 1.4 Structure of a typical walking leg, as illustrated by the third leg of a larval stonefly, *Eusthenia venosa* (Eustheniidae).

two and five tarsomeres; the pretarsus usually takes the form of a paired tarsal claw in adults and a single or paired claw in larvae. Between the tarsal claws of adult insects may be a medial lobe (arolium) which, along with other foot pads (empodium and pulvilli), play various roles in allowing insects to adhere to surfaces (Figure 1.5); this is how flies walk on the ceiling (Beutel and Gorb 2001; Gorb 2008). The thoracic prolegs (sometimes called parapods) found in some larvae (e.g. many Diptera) are not true legs; they are unsegmented, fleshy, and often bear claws or hooks.

The majority of adult, winged insects have one or two pairs of functional wings. Reduced wings or the complete absence of wings (Section 9.6) is a secondary condition. The wing is a flattened evagination of the body wall and is composed of integument. The dorsal and ventral integument layers become closely apposed and form the wing membrane. Channels through the wings, the veins, are strengthened by the surrounding cuticle and may contain nerves and trachea (tubes through which gas exchange occurs). The spaces between veins are called cells. The arrangement of veins varies markedly among the insect orders and is an important taxonomic feature (Figure 1.6). The typical wing condition may be modified in many different ways (Wootton 1992), but generally these fall into two broad categories: those that lead to improved flight (directly or indirectly), and those in which the wing takes on an entirely unrelated func-

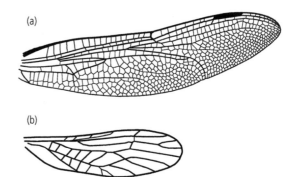

Figure 1.6 Forewings of (a) a dragonfly, *Hemianax papuensis* (Aeshnidae), that has many small cells, and (b) a stonefly, *Isoperla grammatica* (Perlodidae) that has a comparatively simple wing with few large cells. Note: images are not to scale.

tion. Aerodynamically, a two-winged condition is usually more efficient than four wings and wing-coupling mechanisms that link together the fore and hind wing have evolved in several insect orders (e.g. Trichoptera, Lepidoptera). The two-winged condition has been achieved in other orders through the loss, functionally, of the forewings (Coleoptera) or the hind wings (e.g. Diptera, some Ephemeroptera). In these insects, the wings are no longer used directly in flight, but are still present and may be modified for other functions. For example, the heavily sclerotized forewings (elytra) of Coleoptera are mainly protective in function; the modified hind wings (halteres) of Diptera are highly specialized inertial sensing systems that function according to the same principle as a vibratory gyroscope and are used in flight stabilization (Taylor and Krapp 2008) (see Chapter 9 for further discussion of flight).

1.3.3 Abdomen

The segments of the abdomen generally have a rather simple structure; they are usually distinct from one another and most lack appendages, at least in the adult stage. The plesiomorphic number of abdominal segments appears to be 12, but most insects have 10 or 11 segments, several of which may be reduced (mainly at the posterior end). Perhaps the chief differences between adults and larvae in the structure of the abdomen relates to the reproductive structures. Indeed, it is generally impossible to sex larvae as they lack reproductive structures.

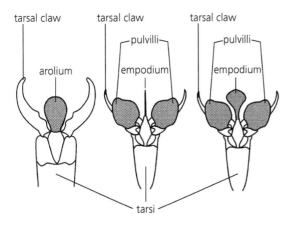

Figure 1.5 Some examples of the insect 'foot' (ventral views). Stippling indicates areas that are used for attachment.

Source: From Beutel and Gorb (2001). Reproduced with permission from John Wiley & Sons.

The external genitalia of adult insects are extremely diverse in morphology and, because of their specific form, external genitalia are widely used for taxonomic purposes. The genital opening (gonopore) is generally on or at the posterior edge of the eighth or ninth sternum in the female, and the ninth sternum in the male. The genital segments (eighth and ninth) are modified in various ways to ensure that only males and females of the same species are able to connect during sperm transfer (i.e. reproductive isolation) and for oviposition. In many orders, the terga and sterna of the genital segments remain distinct plates, and appendages on the genital segments are modified in various ways (Figure 1.7). Females of most orders have an ovipositor for egg laying; exceptions include the Ephemeroptera, most Plecoptera and Trichoptera, which release eggs directly from the gonopore. There are two kinds of ovipositor: a true or appendicular ovipositor formed from the appendages of abdominal seg-

ments 8 and 9, and a substitutional ovipositor comprising the posterior abdominal segments, which are telescoped or retracted within the more anterior segments of the abdomen (Figures 11.1, 11.2). When extended, the segments of a substitutional ovipositor form a long, narrow tube that facilitates egg-laying in inaccessible places, and sometimes the abdomen tip is sclerotized for piercing tissues (e.g. plants).

A range of other abdominal appendages can be found in both the larvae and adults of many aquatic insects. Paired cerci, appendages of the eleventh segment, are typically elongate, multi-segmented structures that often function as sense organs. A median terminal filament, which arises from the last abdominal segment, may also be present and have a tail-like appearance. External abdominal gills, often segmentally arranged along the abdomen, are conspicuous in the aquatic larvae of many orders, and remnant gills may be visible on some adults

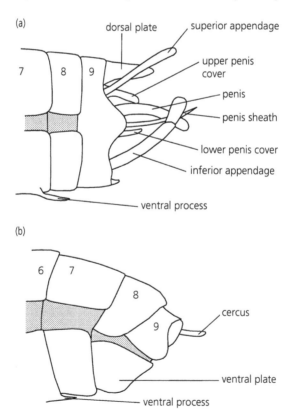

Figure 1.7 Lateral views of the abdomen and external genitalia of a hypothetical (a) male and (b) female caddisfly. Abdominal segments are numbered. The female lacks a true ovipositor.

(a)

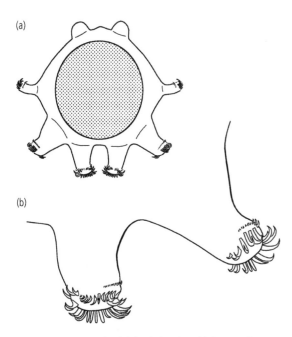

(b)

Figure 1.8 An example of abdominal prolegs. (a) Cross-section through the third abdominal segment of a tabanid larva, *Tabanus kingii*, which lives under stones in swiftly flowing streams and has a pair of ventral prolegs, plus sub-ventral and dorsal swellings analogous to prolegs. (b) An enlarged view of one ventral and one sub-ventral proleg with crotchets.

Source: Adapted from Hinton (1955). Reproduced with permission from John Wiley & Sons.

(e.g. some Plecoptera). The last abdominal segment may also bear papillae, spiracles, lobes, and hairs that function in gas exchange. Prolegs are also present on the abdominal segments of the larvae of holometabolous taxa (e.g. Lepidoptera and some Diptera such as Chironomidae, Tabanidae, Empididae), and their structure varies among taxa (Figure 1.8). They are leg-like, fleshy outgrowths of the body wall, not true legs, and they lack sclerotized segments. Used for locomotion, prolegs are usually in pairs, on the ventral surface of the abdomen, and have various crotchets, hooks, or friction pads on the tip to grasp the substrate. Unusually, a few taxa also have extra prolegs arranged dorsally or laterally, as in Figure 1.8.

1.4 Aquatic insect orders

The aquatic insects are usually split into the fully and the partially aquatic orders. Virtually all taxa within the fully aquatic orders have juvenile stages that are aquatic and adult stages that are terrestrial (Ephemeroptera, Odonata, Plecoptera, Trichoptera, Megaloptera). In the other orders, those species that are aquatic often have aquatic juveniles and some also have aquatic adults, yet the adults of many other species are largely terrestrial (Hemiptera, Lepidoptera, Diptera, Neuroptera, Coleoptera). Inevitably, there are exceptions to this scheme and some orders (e.g. Orthoptera, Hymenoptera, Mecoptera) have only a very few species that are aquatic and we will call these rarely aquatic orders. The names of most of these orders end with the suffix '-ptera', which is derived from the ancient Greek word *pteron*, meaning 'wing'.

The text that now follows is a brief introduction to the aquatic orders, with a description of the distinctive external morphology of adults and larvae, and some comments about their biology. This is not intended to be an identification guide, an exhaustive review of the literature, or a summary of everything known about the order. The reader is referred to specialist texts for each group and to alternative identification guides that often include biological information organized in a taxonomic manner.

1.4.1 Ephemeroptera—mayflies

The Ephemeroptera are hemimetabolous insects and their name is derived from the ancient Greek word *ephemeros*, meaning 'lasting for a day', and refers to the brief life of the adult. Some adults may live as long as a few days, but many live for only a few hours and adult life can be as short as 37 minutes (Taylor and Kennedy 2006). Over 3000 species of mayfly from 42 families and over 40 genera have been identified and are extant (Barber-James et al. 2008).

The adults (also called imagos) have antennae that are small (aristate) and the compound eyes are usually sexually dimorphic, with males having larger eyes than females. Adults have vestigial mouthparts and do not feed. Although they still have a gut, it serves no alimentary function (Section 14.3). The prothorax is small, the meso- and metathorax are fused and large. Two pairs of wings are present, the hind pair is always greatly reduced in size and is absent in some species. At rest, the wings are usually held vertically above the body. Wings are usually transparent, the forewings are triangular and the wing surface has a regular series of corrugations, with longitudinal veins lying on either a ridge or a furrow. Recently, brachypterous (short-winged) and flightless mayflies have been identified in the Madagascan genus *Cheirogenesia* (Ruffieux et al. 1998). For most species, much of the adult life is spent on the wing and the legs are sometimes reduced. The forelegs of most mayflies are sexually dimorphic; those of the male have very long tibiae and tarsi and are used to clasp the female during mating. At the posterior end of the adult abdomen are two or three caudal filaments: two cerci plus, in some species, a vestigial or well-developed median terminal filament. Males tend to have longer caudal filaments than females. Uniquely, male mayflies have two penes, females have two gonopores and no ovipositor.

Unique among insects, the mayflies have two, winged stages. The first stage, the subimago, emerges from the final larval stage and usually moults to the second imago stage within 24 hours. The subimago is usually duller in colour than the imago, as is most evident in the wings, which appear dull or translucent in the subimago compared to the clear wings of the imago. The subimago usually has a fringe of hairs on the posterior margin that is absent in the imago. Although subimagos are winged and superficially resemble adults much more than larvae, they are generally sexually immature. The transition from subimago to imago often involves reduction or loss of larval features (e.g. mouthparts) and maturation or expansion of adult features, such as the genitalia and male forelegs (Edmunds Jr and McCafferty 1988; Harker 1999). See Section 12.4 for further discussion.

The body forms of mayfly larvae are diverse, although similarities are apparent for species that occur in particular habitats, e.g. running versus still water, versus burrowing habits. The compound eyes, ocelli, and short antennae are similar to the adults. Mouthparts are well developed and generally of a biting form. As in the adult, the meso- and metathorax are fused, and the dorsal surface is heavily sclerotized. The legs of most species terminate in a single tarsal claw, but in *Camelobaetidius* spp. (Baetidae) the tip of the tarsal claw is divided into 17–24 denticles, which give the claw a rake-like appearance

(Salles et al. 2003 (2005)). The abdomen terminates with a pair of long cerci and usually a long median caudal filament, giving the appearance of three tails. Along the abdomen are between four and seven pairs of tracheal gills. Gill shape varies with larval habit: typically, lamellate or plate-like gills in open water species, plumose in burrowing species. Gills are important for gas exchange, although gas exchange can occur over diverse body surfaces (Chapter 3).

The adults of many mayflies are renowned for aggregating in huge mating swarms, often near water. This behaviour may be advantageous in ensuring that mates are found, especially given that adults are short-lived, or that wound-be predators are satiated before a significant proportion of the population is lost (Sweeney and Vannote 1982). A few species display obligate or facultative parthenogenesis, and the potential for parthenogenetic reproduction may be present in many mayfly species (Funk et al. 2010). The aquatic larval stage is the dominant life-history stage, with larval stages generally in the order of weeks to months, but development times may be as long as three years. Larvae moult many times (ranges of 10 to 50 moults have been reported, Ruffieux et al. 1996), the number of moults varies between species and may vary within species depending on environmental factors. The range of larval habitats is extremely wide, in running and standing waters, and varies among species. Most larvae are herbivorous or detritivorous, although a few are carnivorous (Müller-Liebenau 1978; Gattolliat and Sartori 2000).

1.4.2 Odonata—dragonflies and damselflies

The name Odonata is derived from the ancient Greek word *odous*, meaning 'tooth', and refers to the serrated mandibles. They are hemimetabolous insects with terrestrial adults and aquatic larvae; both adults and larvae are carnivorous. Approximately 5680 extant species of Odonata from 32 families have been identified (Kalkman et al. 2008). The extant Odonata are divided into two suborders: the heavy-set Anisoptera or dragonflies and the more slender Zygoptera or damselflies. The odonates are generally considered to be tropical species and, although many species do occur at high latitudes and altitudes, the highest species richness occurs at low latitudes.

Adult Odonata are remarkable for their colours on the body and the wings, arising from reflections in thin layers of cuticle and from pigment in the cells. Males are often more brightly coloured than females. The head is freely articulated with the thorax and much of the head is occupied by well-developed compound eyes; together, these characteristics facilitate incredible visual capability in many directions. The antennae are short (aristate) and, apparently, carry few sense organs. The biting and chewing mouthparts are powerful, as is often characteristic of carnivorous taxa. The prothorax is small, whereas the meso- and metathorax are large and fused together into a pterothorax. The legs are ventral (the legs of many adult insects project laterally; Section 8.2), weak, and generally unsuitable for walking, but are armed with various spines and hairs that serve to grasp prey. The Odonata are high-performance fliers (Marden 2008), with extensive flight muscles in the pterothorax (50–60 per cent of total body mass in some males, Marden 2008). Fore- and hind wings are identical in damselflies; hind wings are somewhat broader than the forewings in dragonflies. A prominent pterostigma (a pigmented patch) is often present near the tip of each wing. Other wing features such as the nodus (Section 9.4.1) are also characteristic of odonates. Perhaps the most remarkable characteristics of odonates are the suite of modified copulatory structures they possess (especially in males) and their complex reproductive behaviours (Section 10.6.1).

The heads of odonate larvae broadly resemble those of the adults and have prominent compound eyes. Their mouthparts are unique and prey are captured using a 'mask', which is an elongated labium with a pair of hinged, hook-like labial palps, and various setae, spines, and teeth for skewering prey (Chapter 13). At rest, the mask is folded and held between the legs, but it can be extended very rapidly. In contrast to the adults, the larval thoracic segments are roughly equal in size, the legs are well-developed and project laterally. The abdomen is tipped with five abdominal appendages (including two small cerci) that are small in Anisoptera, but three (the epiproct and two paraprocts) are enlarged to form caudal lamellae in Zygoptera. These caudal

lamellae serve a range of functions including gas exchange, sensory reception, and may act as tail fins during swimming (Section 8.3.2). In contrast, the Anisoptera have rectal gills and water is continually pumped in and out of the rectum (Komnick 1982).

Adult odonates typically hunt in flight. Dragonflies tend to catch prey that are also on the wing, whereas damselflies tend to catch prey that are resting. After emergence, immature or 'teneral' adults typically spend some time away from water until they become sexually mature. During this maturation period, which lasts from a few days to weeks, some Odonata migrate or travel over long distances (Chapter 9). Odonates may spend a long time attached to one another during reproduction and are often very conspicuous flying around in pairs (Chapter 10). Larvae are facultative predators and will feed on almost all appropriately sized prey (i.e. larvae are typically gape-limited). Larvae adopt primarily an ambush foraging strategy, remaining immobile until prey come within range of the mask, and prey are detected using vision and tactile stimuli (Rebora et al. 2004). Many forms burrow in the substrate or hide among detritus; others escape detection by predators through camouflage. There are normally 10–15 instars, but the duration of larval development is highly variable and can range from one or two months to several years.

1.4.3 Plecoptera—stoneflies

The Plecoptera are hemimetabolous and their name is derived from the ancient Greek words, *plektos,* meaning 'braided', and *pteron,* and refers to the prominent venation of the wings. Approximately 3500 species of Plecoptera from 16 families have been identified worldwide and are extant (Fochetti and Tierno de Figueroa 2008). The stoneflies are often considered to be cold-water species, with few species occurring at low latitudes, but there are many tropical species nevertheless, especially in the family Perlidae.

The adult stonefly body is generally dorso-ventrally flattened. The head is large and bears a pair of elongate antennae, well-developed compound eyes, and three ocelli. Mouthparts are generally for biting, but are weak, and the mandibles are vestigial in a few families. The thorax has three distinct segments and the prothorax is large. Usually present are two pairs of membranous wings, folded flat over the abdomen when at rest. Brachypterous (shortened wings) and apterous (wingless) species also occur, with shortened wings in both sexes or only in the males, and some normally fully winged species tend to become short-winged at high altitudes and latitudes (Section 9.6.1). Gills that are found in larval stages may persist in adult stages, but tend to shrivel and may leave scars. The genitalia of both sexes are diverse in form and unlike those of most other insects. The males of some northern-hemisphere families have a small lobe at the front of the subgenital plate that is used to drum during courtship (Section 10.5).

Stonefly larvae closely resemble the adults, except for the absence of wings and, in most species, the presence of several pairs of gills. The head is large, dorso-ventrally flattened, with large, laterally placed compound eyes and ocelli are often present also. The mouthparts are generally robust. Legs are always well-developed and the three-segmented tarsus ends in two claws. Developing wing-pads on the meso- and metathorax may differ noticeably between the sexes in species with brachypterous males. In the plesiomorphic condition, there are five or six pairs of abdominal gills, but in many groups these are reduced in number and secondary respiratory structures may appear anteriorly (on the submentum, neck, thorax, coxae) or may encircle the anus. The abdomen terminates in a pair of long, multi-segmented cerci that have sensory functions.

Most stoneflies have aquatic larvae and terrestrial adults, although adults of a few species are not fully terrestrial and some larvae may have terrestrial habits. For example, *Capnia lacustra* (Capniidae) is a wingless stonefly whose adults and larvae are both found at depth in Lake Tahoe, USA (Jewett 1963). Similarly, two species of *Baikaloperla* (Capniidae), endemic to Lake Baikal, Siberia, are also apterous and the adults appear to be entirely aquatic. In contrast, the larvae of a few species of Gripopterygidae in Patagonia and New Zealand have larvae that live terrestrially on cold, wet mountains far from water (Illies 1963). Indeed, there seems to be a tendency among larvae of New Zealand Gripopterygidae to leave the water (Winterbourn 1966; McLellan 1977), and several Australasian Notonemouridae also live terrestrially on wet surfaces (Hynes and Hynes 1975).

Adult stoneflies are typically weak fliers (or flightless) and this influences many aspects of their ecology as adults. Many stoneflies skim across water surfaces, often using fairly rudimentary wings and wing motions, or by using their wings as sails (Marden and Kramer 1994, 1995; Marden et al. 2000) (Section 8.5.2). At high latitudes and altitudes, adult emergence may begin when there is still snow on the ground and adults are commonly seen walking across the snow—the so-called winter stoneflies. Their dark bodies preferentially absorb solar radiation against the white snow, and this allows them to survive and be active on a cold surface (Section 4.2). Adult longevity varies among species and may last from a few days to a couple of weeks. Larval stoneflies occur in a wide range of running and standing waters, typically where temperatures are cool and dissolved oxygen concentrations are high. Development is slow, often taking more than a year, and there are many instars, with diapause common in larvae and eggs. Often the life history is timed so that larvae are not exposed to warm conditions, and diapausing eggs or larvae may be buried in the sediment during warm seasons.

1.4.4 Trichoptera—caddisflies

The name Trichoptera is derived from the ancient Greek words *tricho*, meaning 'hair', and *pteron*, and refers to the many hairs that are usually present on the wings. The Trichoptera are holometabolous, typically with terrestrial adults, aquatic larvae and pupae, and primarily aquatic eggs. Over 12,600 extant species of Trichoptera from 46 families have been identified (de Moor and Ivanov 2008), making it the most species-rich of the fully aquatic orders. Trichoptera are closely related to the order Lepidoptera (butterflies and moths) and the two groups are similar in many respects. A notable difference is that the wings of adult caddisflies are covered in hairs, whereas the wings of butterflies and moths are covered in scales.

Adult caddisflies are moth-like in appearance and often drab in colour. The antennae are long and the compound eyes usually small (but are very large in some males). The mouth parts are reduced, but the hypopharynx is well-developed and in some groups it is modified for sucking up liquids. The prothorax is small, but the meso- and metathorax are well developed. The legs are long, with spurs, setae, bristles, or spines (particularly on the tibia); the tarsi have five segments with ventral spines on each segment. Almost all adults have two pairs of membranous wings covered in hairs, although flight ability varies among taxa. Wing venation resembles some ancient Lepidoptera and, similar to moths, at rest the wings are often held like a peaked roof over the abdomen. Abdominal segment 5 may have a pair of slender filaments or raised ovoid processes that may be connected to pheromonal glands.

Larval caddisflies are perhaps best known for their ability to build cases, although many do not. Cases are built with silk and often incorporate objects from the surrounding environment (e.g. sand, twigs, pieces of leaves). All caddis are capable of producing silk and some larvae that do not build cases construct a non-portable silken web. Proteinaceous silk is produced in a pair of elongate salivary glands, which occupy much of the body cavity (Glasgow 1936; Engster 1976b). The silk is emitted as a double-stranded, single filament (Engster 1976a) from a single spinneret at the tip of the labium. The head is well sclerotized, bears a pair of very short antennae, two lateral clusters of stemmata, and the mouthparts are generally chewing. The thorax is variably sclerotized and the legs are well developed, with unsegmented tarsi and a single tarsal claw. The forelegs are often fairly short and used for holding food and for constructing cases or nets, rather than for walking. The first abdominal segment of some case-bearing families has three prominent, retractile papillae. These papillae may help the larva to maintain position in the centre of the case and allow water to flow efficiently over the abdominal surface and gills (Chapter 3). On the last abdominal segment only are two prolegs, which each have a single curved claw. The prolegs may be large, mobile, and used to anchor the larva to the substrate or net; or may be small and immobile, and function primarily to grip the case. Many species have gills, which are usually simple filamentous structures on the abdomen and occasionally on the thorax.

Adult caddis generally do not eat, but most are capable of sucking up nectar and liquids, and some may live for several weeks. Species that inhabit ephemeral systems typically emerge as adults as the water disappears, and adults of these species may be relatively long-lived. Adults of some species swarm prior to mating and, unusual among aquatic insects, many communicate via pheromones (Chapter 10). Larval development typically involves five instars and may be rapid; diapause usually occurs in the final larval or adult instar, and more rarely in the egg stage. The majority of species have aquatic larvae; only a few are semi-aquatic and they live in moist terrestrial places (e.g. the limnephilids, *Limnephilus centralis* and *Enoicyla pusilla*, of northern Europe). The larvae of most species are benthic and live on the bottom of water bodies or attached to submerged plants, etc., but some can swim with their cases (e.g. *Triaenodes*, Leptoceridae). Larval caddisflies are extremely diverse in their feeding habits and diet, and some are parasitic on other caddisflies (Wells 2005). Larvae of the family Hydropsychidae are notable for being able to produce sound underwater, a rare accomplishment among immature insects (Chapter 7).

1.4.5 Megaloptera—alderflies, dobsonflies, and fishflies

The Megaloptera is a relatively small order with only two families (Sialidae, Corydalidae) and 328 extant species have been identified (Cover and Resh 2008). The name is derived from the ancient Greek words *megas*, meaning 'big' or 'great', and *pteron*, and refers to the large, winged adult. The Sialidae are commonly known as alderflies, the Corydalidae as dobsonflies or fishflies. The Megaloptera are holometabolous with aquatic larvae, but the adults, eggs, and pupae are terrestrial.

As the prefix 'mega-' implies, adult Megaloptera are generally large-bodied insects. The Corydalidae are generally bigger than the Sialidae and wingspans of 16 cm have been recorded. At rest, the wings are held like a peaked roof over the abdomen. The head has well-developed compound eyes and long, slender antennae. Mouthparts are adapted for chewing, and the mandibles are particularly long in some male corydalids and sometimes called tusks, although their function is unknown. The legs are generally short, and the tarsi have five segments and end in two claws. The abdomen lacks cerci.

Larval corydalids, often called hellgrammites, are among the largest of the larval Megaloptera (some reach 8 cm in length). The megalopteran head is well sclerotized, with chewing mouthparts and large mandibles. The prothorax is heavily sclerotized, but the meso- and metathorax less so. The legs terminate in two claws of unequal size. The abdomen bears seven or eight pairs of abdominal filaments, or gills. Larval fishflies have a pair of dorsal respiratory tubes on abdominal segment 8 that can be used for air breathing. The abdomen of sialids ends in a single, long filament, whereas the corydalids have a pair of anal prolegs.

The adult megalopterans are poor fliers, relatively short-lived (approximately one week) and most do not feed, or feed on nectar and other sugar solutions. Larvae are ambush predators and consume a wide range of aquatic invertebrates. Most species pass through 10–12 larval instars and development in large species may take up to five years, although smaller species can complete development in one. Prior to pupation, larvae leave the water and most burrow into soil or moss, or under stones where pupation occurs (Mangan 1994).

1.4.6 Hemiptera—true bugs

The Hemiptera are primarily terrestrial taxa in which members of three infraorders may be considered to be aquatic or semi-aquatic (these are sometimes referred to as the aquatic Heteroptera). The Nepomorpha (back swimmers, water boatmen, water scorpions) have at least one life stage that is truly aquatic; the Gerromorpha (water skaters, water measurers) are semi-aquatic and live on the water surface. The Leptopodomorpha (shorebugs) also have a water dependency and are sometimes classified as semi-aquatic (Polhemus and Polhemus 2008), but this group will not be discussed here. The extant aquatic and semi-aquatic species of Hemiptera fall into 19 families and there are approximately 4430 species in 301 genera (Polhemus and Polhemus 2008). The Hemiptera are hemimetabolous insects and the order name is derived from the ancient Greek words *hemi*, meaning 'half', and *pteron*, and refers to the nature of the wings, in which the forewing (the hemielytra) is usually sclerotized basally and only the distal portion is membranous.

Body size among species of adult aquatic Hemiptera is very variable, ranging from < 1 mm in *Micronecta* (Corixidae) to > 11 cm in *Lethocerus* (Belostomatidae). In the adults, the compound eyes are usually well developed and the antennae typically have four or five segments. The heads of some species (e.g. Hydrometridae) are very elongate, with small eyes some distance away from the pronotum. All members of the Hemiptera have suctorial mouthparts and essentially a liquid diet: the mandibles and maxillae form two pairs of piercing stylets that are contained within a flexible, segmented labium. The pronotum is large, and the meso- and metanotum are generally small, with some exceptions (e.g. the Hydrometridae and Gerridae have an elongated mesothorax). Usually two pairs of wings are present, although brachyptery and aptery are common. The forewings are partially sclerotized whereas the hind wings are membranous and, at rest, are folded beneath the forewings. The structure of legs is remarkably diverse within the aquatic Heteroptera, and these include modifications for swimming, catching prey, and rowing over the water surface. Most species have five larval instars.

Most aquatic Heteroptera are carnivorous, although some, such as the Corixidae, appear to be at least partially scavengers or detritivores. Prey are subdued by injection of a venom consisting of toxins and proteolytic enzymes, and then digested fluids are sucked up the stylets. Sound production is widespread in adult aquatic Heteroptera: among the surface-dwellers, gerrids communicate by surface waves and veliids stridulate, and underwater sound production via stridulation has been documented in virtually all families of Nepomorpha (Section 7.3).

In the aquatic Nepomorpha, small instars breathe dissolved oxygen, as do all stages of the Micronectinae and Aphelocheiridae. Adults and most older instars come to the surface to pick up an air bubble that also functions as a physical gill (Section 3.4). In some species the air bubble may be held under the wings (e.g. adult water boatmen, Corixidae) or attached to hairs (e.g. backswimmers, Notonectidae). Although the Nepomorpha are fully aquatic, their ability to breath air (primarily the adults) allows them to persist out of water for relatively long periods of time, for example, when in flight, or when some belostomatids crawl out of streams to avoid imminent flash floods (Lytle 1999; Lytle and White 2007).

The semi-aquatic Gerromorpha spend most of their life on the water surface, but some adults may leave the water during dispersal or for over-wintering. Many species, and particularly juveniles, live in marginal aquatic habitats such as wet soils, floating plants, and mosses at the water's edge. Walking on water depends on a species' ability to manipulate and control the air-water interface, and there are various biological and physical means by which this can be achieved (Section 8.5). Because the Gerromorpha live primarily on the water surface, they do not face many of the osmotic stresses associated with living within water and some occur on extremely saline water, such as the well-known marine and wingless species of *Halobates* (Gerridae) that occur on the open ocean (Cheng 1985; Andersen and Cheng 2004).

1.4.7 Lepidoptera—aquatic moths

The Lepidoptera is a large and primarily terrestrial order and only a fraction of the 10^5 described species have aquatic larvae. Over 730 extant species of aquatic Lepidoptera have been identified. The vast majority are in the family Crambidae (formerly the family Pyralidae) (Mey and Speidel 2008) and a few are scattered across other families such as Arctiidae and Cosmopterigidae (Rubinoff 2008). The name is derived from the ancient Greek words *lepidos*, meaning 'scales', and *pteron*, and refers to the scales on the wings. The Lepidoptera are holometabolous, and the aquatic species generally have aquatic larvae, eggs, and pupae, but terrestrial adults.

A major portion of the adult lepidopteran head capsule is covered by compound eyes. The antennae are variable in form, but often long and slender in aquatic species. In most species, the mouthparts (primarily the maxillae) are modified as a suctorial proboscis, which is coiled beneath the thorax when not in use. The anterior region of the foregut is modified as a pharyngeal sucking pump, to draw fluids up the proboscis. The mesothorax is the largest of the pterothoracic segments and bears large tegulae (scale-like lobes overlapping the base of the forewing),

a characteristic feature of the order. Both pairs of wings are generally large and covered with scales (modified hairs).

The larva (caterpillar) has a heavily sclerotized head that typically bears a ring of stemmata, short antennae, and strong, biting mouthparts for feeding on aquatic plants. The prementum (the labium) carries a spinneret, which receives the ducts of the silk glands (modified salivary glands). Most species have three pairs of true legs on the thorax and five pairs of abdominal prolegs on segments 3 to 6 and on segment 10. Prolegs usually bear curved hooks (crochets) arranged in species-specific patterns. The only way to identify a caterpillar as truly aquatic is by the presence of filamentous gills (often branched) or the presence of a case or retreat, often made from leaf fragments or other vegetable matter. Caddisfly-like cases made of mineral substrates glued together with silk also occur in aquatic moths, such the amphibious larvae of the Hawaiian genus *Hyposmocoma* (Cosmopterigidae) (Rubinoff 2008; Rubinoff and Schmitz 2010).

Adults often form aggregations on the undersides of boulders or hanging rocks. Some adults can be found far from water, suggesting a propensity to disperse. A few brachypterous species are aquatic as adults, such as *Acentria ephemerella* (Crambidae). In common with many terrestrial Lepidoptera, most pond-dwelling species lay eggs on a larval food plant on, under, or close to the water. In contrast, stream-dwelling species typically lay eggs on submerged rock surfaces and the larvae eat lichens, algae, and mosses at or below the water surface. Many larvae of the Crambidae have been used in biological control programmes to suppress growth of aquatic weeds (Chapter 13). There are generally five to seven larval instars.

1.4.8 Diptera—true flies

The name Diptera comes from the ancient Greek word *dipteros*, meaning 'two-winged', and refers to the fact that flies have only one pair of membranous wings. As mentioned earlier, the second pair have been modified to form halteres, which are important organs for flight stabilization (Taylor and Krapp 2008). The Diptera are holometabolous and, among the aquatic species, it is almost exclusively the larvae and pupae that are aquatic in habit. The aquatic Diptera are species-rich with extant species described from 24 families. The most species-rich families, with more than 2000 species each, are the Chironomidae (Ferrington Jr 2008a), Tipulidae (de Jong et al. 2008), Simuliidae (Currie and Adler 2008), Culicidae (Rueda 2008), Psychodidae, and Ceratopogonidae (Wagner et al. 2008). Traditionally, there were two suborders of Diptera, the Nematocera (e.g. blackflies, midges, mosquitoes) and the Brachycera (e.g. horseflies, hoverflies, danceflies), and most aquatic species are in the families of Nematocera. However, the Nematocera is paraphyletic and the Brachycera is sister to only part of the Nematocera, so these are not 'good' suborders and the systematic relationships within the Diptera are still under some debate. Nevertheless, the dipteran families grouped within each of these two categories share some morphological characteristics and, for convenience, these names will be used in the discussion to follow. Because of their diverse morphology and the polyphyletic nature of aquatic Diptera, it is difficult to generalize about their morphology or biology, and the following discussion is necessarily brief.

Adult Diptera are generally soft-bodied with relatively large and mobile heads. The compound eyes are well developed, and the antennae are variable in structure (longer in the Nematocera than the Brachycera). Mouthparts are generally adapted for sucking. In some, the mandibles and maxillae are modified into a piercing proboscis, whereas in others the labium dominates the proboscis. The pro- and metathoracic segments are narrow and fused with the much larger mesothorax, which bears the single pair of membranous wings. Legs typically have tarsi with five segments.

Larvae of the aquatic Diptera are usually elongate and cylindrical, the body segments are usually distinct and the cuticle only weakly sclerotized. The head typically is a distinct sclerotized capsule (eucephalic) in the larvae of the Nematocera (e.g. Chironomidae, Simuliidae, Culicidae), but usually incomplete and retractile (hemicephalic or acephalic) in the Brachycera (e.g. Tabanidae, Empididae). In parallel, antennae and mouthparts are well developed in the Nematocera, whereas variable degrees of reduction and modification occur in the Brachycera.

Mouthparts show great diversity of form, in accordance with the diverse feeding habits. Unusually, in some groups the antennae are modified as feeding apparatuses, including the filter-feeding cephalic fans of some larval Simuliidae and the prehensile antennae for capturing prey in larval Chaoboridae (Section 13.3). The eyes of larvae are usually simple (one to three stemmata), but apparently compound eyes are present in some Culicidae and Chaoboridae. The three thoracic segments are fused in some larval Diptera (e.g. Simuliidae, Culicidae, Chaoboridae). All dipteran larvae lack jointed thoracic legs, but prolegs may be present on the thorax and/or the abdomen.

With the exception of a few that do not feed, adult Diptera feed entirely on fluids. Most feed on nectar or the fluids from decaying organic matter, sap, faeces, etc., but a few groups (e.g. Tabanidae, Culicidae, Ceratopogonidae) are adapted for feeding on body fluids (usually blood) of other animals. The majority of body-fluid feeders have a fine proboscis to pierce the skin and penetrate directly into the fluid, and the blood-sucking habit is usually confined to the females. Taxa without piercing mouthparts can usually eject digestive fluids onto solid food that is 'dissolved' before being eaten. Most adults have a muscular cibarium for sucking fluids and, for the blood-sucking species, a large pharyngeal pump is also present.

1.4.9 Neuroptera—lacewings and spongillaflies

The name Neuroptera is derived from the ancient Greek words *neuron*, meaning 'sinew' or 'tendon', and *pteron*, and refers to the pronounced veins on the wings. Of the 17 families of Neuroptera, only two (Osmylidae and Sisyridae) have aquatic larvae, and 73 aquatic species have been described (Cover and Resh 2008). The larvae of the Sisyridae (spongillaflies) are wholly aquatic, whereas the Osmylidae (lacewings) have only semi-aquatic larvae. The Neuroptera are holometabolous, with aquatic larvae, but terrestrial adults, eggs, and pupae. The Neuroptera are often described alongside the Megaloptera, with which they have evolutionary and morphological affinities.

The adult aquatic Neuroptera range in size from the small Sisyridae (5–6 mm wing length), to the larger Osmylidae (11–14 mm wing length). The head has well-developed compound eyes, long slender antennae, and generally three ocelli in the Osmylidae, but none in the Sisyridae. Mouthparts are for chewing; the mandibles are well developed and asymmetrical. The thoracic segments are well developed and bear two pairs of membranous wings, although adults are generally weak fliers. At rest, the wings are held like a peaked roof over the abdomen like many other insects.

The larval head is well sclerotized, with relatively long antennae. Of the mouthparts, the mandibles and maxillae form two tubes or suctorial jaws, which project forward in front of the head. In the Sisyridae, these mouthparts are extremely long and thin. The legs terminate in two claws of unequal size and the tarsi are generally one-segmented. Ventral, abdominal gills are present in the Sisyridae, but absent in the Osmylidae.

There are only three larval instars in the aquatic Neuroptera, perhaps the smallest number of instars for any aquatic insect. Larvae of the Sisyridae are usually found associated with freshwater sponges (typically those of the family Spongillidae) that they feed upon with piercing mouthparts. As with the Megaloptera, eggs are laid in batches on vegetation or branches overhanging the water; when neonates hatch, they drop into the water. Neonate spongillaflies drift until contact is made with a sponge and they may remain associated with that same sponge for the entire larval life. Just before pupation, larvae leave the water (sometimes travelling up to 20 m away), climb onto mosses, plants, or other objects, and spin a silk cocoon. Larval Osmylidae are largely semi-aquatic and live primarily in moss on the edge of streams, but will enter the water and feed on aquatic prey. They are carnivorous. Prey are pierced with the suctorial jaws, a salivary secretion paralyzes the prey, and then body fluids are sucked out.

1.4.10 Coleoptera—beetles

The beetles are perhaps the most species-rich of any insect order. Approximately 30 of the 170 families have aquatic or semi-aquatic representatives, but most are classified into one of six families (Dytiscidae, Hydraenidae, Hydrophilidae, Elmidae, Scirtidae, Gyrinidae) and into two of four suborders (Adephaga, Polyphaga). Approximately 18,000 extant and aquatic or semi-aquatic species of beetle have been described (Jäch and Balke 2008). The name Coleoptera is derived from the ancient Greek words *koleos*, meaning 'sheath', and *pteron*, and refers to the sheath-like nature of the hardened forewings (elytra). The Coleoptera are holometabolous, but the degree to which various life stages have aquatic habits varies among and within families. The truly aquatic beetles all have aquatic larvae and most have aquatic adults—exceptions include the Scirtidae where adults are terrestrial. Because of their diversity and the polyphyletic nature of aquatic beetles, it is difficult to generalize about their morphology or biology, and the following discussion is necessarily brief.

Adult beetles are heavily sclerotized. Compound eyes are usually present and large, and divided into upper and lower pairs in Gyrinidae (Section 6.4). The typically eleven-segmented antennae vary in form. Mouthparts are typically of the chewing type. The prothorax is the largest of the thoracic segments and, in many species, can be moved independently of the other thoracic segments. The meso- and metathorax are fused in a pterothorax, which is often invisible from above, or visible only as a small, typically triangular-shaped scutellum. Forewings are modified as hard elytra that may be smooth, pitted, striated, or variously textured. Elytra meet in the midline but are rarely fused and, except in a few taxa that lack hind wings (e.g. some Elmidae), the hind wings are membranous and typically folded beneath the elytra when not in use. In flight, the elytra are held horizontally, at right angles to the long axis of the body. Legs, especially the hind legs, are often modified for swimming, i.e. widened and often fringed with hairs. Tarsi are normally five-segmented and terminate in paired claws, although one segment may be reduced. In the Meruidae, a recently discovered family of aquatic beetle living in cascades over bedrock outcrops, the tarsal claws are pectinate (bear many teeth) and this gives rise to their common name of 'comb-clawed cascade beetles' (Spangler and Steiner Jr 2005).

The shapes of beetle larvae are highly varied, although in all species the head is well developed and sclerotized. The antennae generally have three or four segments, although some are longer (e.g. larval

Scirtidae), and six or fewer stemmata on each side of the head form the eyes. Three thoracic segments are readily distinguished and thoracic legs are usually present, although legs may be reduced (e.g. some Curculionidae). Abdominal prolegs are normally absent and some beetle larvae have filamentous abdominal gills. The tip of the abdomen may be modified or bear various structures for gas exchange, such as sclerotized respiratory horns.

The habits of aquatic beetles are as diverse as the group: most are benthic in habit; adult Gyrinidae swim at the water's surface; adults of other species can 'walk' upside down on the underside of the surface film (e.g. some Hydraenidae and Hydrophilidae; Section 8.6.2); many are adept at swimming and diving (e.g. Dytiscidae); and some live exclusively on aquatic plants. Very few species have aquatic pupae; larvae usually leave the water to pupate or construct a submerged air-filled cocoon. Exceptions include the water-pennies (Psephenidae) where pupae respire by means of spiracular gills. In adults, respiration underwater is primarily via an air bubble trapped under the elytra or attached to the ventral surface (Section 3.4). Larvae never carry external air reservoirs: some have tracheal gills and are independent of atmospheric air (e.g. Elmidae), others access air at the water's surface (e.g. many Dytiscidae), or by tapping into the air spaces in the stems of aquatic plants (e.g. Noteridae).

Feeding habits are very varied among species and often between life stages of the same species, although the majority are probably carnivores and/or scavengers. Some groups are algivorous (larval Elmidae); others are strictly phytophagous (Chrysomelidae, Curculionidae) and have been used in biological control programmes for nuisance aquatic plants. A very few species are parasitic, such as the beaver beetle *Platypsyllus castoris* (Leiondidae), where wingless and virtually eye-less adults and larvae are ectoparasites of Palaearctic and Nearctic species of beaver (*Castor fiber* and *C. canadensis*) (Peck 2006; Whitaker Jr 2006).

1.4.11 Rarely aquatic insects

Of the Orthoptera (grasshoppers and crickets), approximately 188 species are considered to be water-dependent to some degree, but only approximately 80 species would be considered aquatic (Amédégnato and Devriese 2008). The name Orthoptera comes from the ancient Greek *orthos*, meaning 'straight' or 'perpendicular', and *pteron*, and may refer to the straight wings. These are hemimetabolous insects and both larvae and adults of aquatic species occur in water. Adults generally have two pairs of wings and the forewings (tegmina) are thickened and somewhat sclerotized, although they are generally weak fliers and brachypterous and apterous forms also occur. Females typically have a well-developed, appendicular ovipositor, and males have concealed copulatory structures. Compound eyes are well developed and mouthparts are of the chewing type. The hind legs of almost all species are enlarged for jumping. In many aquatic species, the hind femora are enlarged, the tibiae are variously flattened, and equipped with spines, spurs, and sometimes fringes of hairs, that aid swimming. Like their terrestrial counterparts, aquatic Orthoptera are typically herbivorous and feed on aquatic macrophytes. Most members of the order lay eggs in soil or in burrows, but aquatic species often oviposit on or within aquatic plants, often just below the water surface. They are air breathers and dense patches of hair, typically on the abdomen and forewing, may facilitate carrying air bubbles underwater.

The Hymenoptera is a very large order of holometabolous insects, which includes the familiar bees, wasps, and ants. In total, 150 species from 11 families in the suborder Apocrita have been described and recognized as aquatic and all are parasitoids (Bennett 2008). The name Hymenoptera is derived from the ancient Greek words *hymenos*, meaning 'membrane', and *pteron*, and refers to the membranous wings. The aquatic wasps have terrestrial adults and larvae that are parasitic upon other aquatic organisms, often insects (this excludes wasps that are parasitic only on the terrestrial stages of aquatic insects). Most aquatic Hymenoptera are endoparasitoids of larvae that occur in tissues of aquatic plants, although some are ectoparasitic (e.g. *Agriotypus* spp., Ichneumonidae) and oviposit on larvae, prepupae, or pupae (Chapter 11). Adults have two pairs of membranous wings that are coupled by hamuli and have simple venation. The plesiomorphic condition is herbivory, but mouthparts of some groups have been modified for predation. As parasitoids, females generally have a long, needle-like ovipositor. Many aquatic wasps have characteristics that are uncommon in other wasps and that appear to be adaptive to life underwater: short, dense hairs on the body and wings that allow the wasp to be hydrophobic and to maintain a plastron of air around the body; and elongated, strongly curved claws that can grip the substrate and prevent the female floating to the surface or being swept in currents when searching for hosts (Bennett 2008). The larvae are generally maggot-like and their morphology is typical of internal parasites.

Mecoptera (scorpion flies) is a relatively small order of holometabolous insects. Only nine species are aquatic, all of which are in the family Nannochoristidae and all occur in the southern hemisphere (Australia, New Zealand, South America) (Ferrington Jr 2008b). The name Mecoptera is derived from the ancient Greek words *mekos*, meaning 'long', and *pteron*, and refers to the relatively long wings. The aquatic species have terrestrial adults, which only occur close to water, and aquatic larvae. The head of adult Mecoptera is typically elongate, with the clypeus and labrum prolonged ventrally into a broad rostrum with biting mouthparts. In the Nannochoristidae, however, the rostrum is narrow and needle-like. Compound eyes are well developed, the legs are long and thin, and there are usually two pairs of identical membranous wings. The abdomen has short cerci. Males have very prominent genitalia, often swollen and turned upwards, which give them a scorpion-like appearance. Larvae of most Mecoptera are caterpillar-like, but the aquatic larvae are active predators, with dorso-ventrally flattened bodies and thoracic legs, but no prolegs. The stemmata of some last-instar larvae (*Nannochorista*, Nannochoristidae) form quasi-compound eyes composed of ten or more stemmata (Melzer et al. 1994), although the stemmata are simple in larval *Microchorista* (Nannochoristidae). Normally, the thorax has only the pronotum sclerotized and the apex of the abdomen is modified into a suction disc, but in the Nannochoristidae, the abdomen ends in a pair of anal hooks. Rather little is known about the ecology of the aquatic Mecoptera.

CHAPTER 2

Evolution, biogeography, and aquatic insect distributions

2.1 Introduction

In evolutionary terms, the insects are a very old group and insects of some sorts are likely to have been in existence for over 400 million years, although the very ancient insects did not closely resemble any modern taxa. When and how insects began to occupy freshwater environments is still in debate, although at least some aquatic groups had terrestrial precursors. Unravelling the ecological roles of the earliest aquatic insects in ancient freshwater ecosystems, e.g. what they ate, is even more challenging. Insect fossils are a vital source of information in trying to unravel insect evolutionary history, but the fossil record is incomplete. Fossils are created only in certain environmental conditions and some of the extraordinarily diverse habitats occupied by modern aquatic insects are unsuitable for fossilization processes, so the evolutionary histories of these taxa and environments are likely to remain enigmatic. Another source of difficulty is that extant insects are often only distantly related to fossil specimens; many ancient lineages went extinct and it is often unclear how the various groups are related to one another (phylogenetic relationships). Interpreting the insect fossil record is therefore difficult and there are often multiple plausible interpretations. Interpretations and hypotheses about evolutionary sequences and relationships are often based on multiple lines of evidence and multiple techniques. New techniques are developing all the time so even the best explanations of insect evolution and the origins of modern distribution patterns are continually under review and subject to change.

Understanding the evolutionary origins of modern insects is inherently interesting, but is also important in understanding the origins of modern distribution patterns. At very large spatial and temporal scales, we need to consider insect evolution in the context of plate tectonics and this can explain some biogeographical distribution patterns at the level of insect orders or suborders. This becomes more difficult at smaller scales and higher taxonomic resolution, given the potential for more modern dispersal events and the enormous range of different environments that aquatic insects now inhabit.

Insect evolution is an enormous topic and our discussion will be correspondingly brief and somewhat simple. To start, we discuss the fossil record and how it can inform our understanding of evolutionary events (Section 2.2). Highlighting the aquatic and semi-aquatic groups, Section 2.3 outlines what we know about insect evolution, at the coarse taxonomic level of insect orders. Readers are referred to other sources for in-depth discussions of the underlying evidence (e.g. Grimaldi and Engel 2005). In a similarly superficial and somewhat speculative manner, we consider the history and evolution of aquatic habits (Section 2.4) and the biogeography of the major aquatic insect groups as revealed by plate tectonics (Section 2.5). The final Section 2.6 describes some of the many different environments occupied by aquatic insects and which groups characterize their faunas. This is a difficult task and not intended to be an exhaustive review; the aim is simply to illustrate the extraordinary range of environments that aquatic insects inhabit.

2.2 The fossil record and establishing phylogenies

Most of what we know about the timing and sequence of insect evolution is based on fossils, including whole insects, insect sclerites or other body parts, especially wings or impressions of wings. Dating the fossils themselves is difficult (although sometimes possible with isotope methods), so dating specimens often relies on dating the surrounding layers in which the fossils are embedded. Many of the major events in insect evolution occurred millions of years ago, so dating is often in terms of geological periods rather than precise numbers of years (Table 2.1). Insect fossils often occur in layers of sedimentary rock formed by the accumulation of sediments at the bottoms of lakes or in peat bogs. Insects that lived in the lake, died, drowned, or got swept into the lake (via streams or wind) may become preserved in these sediments, especially if the latter are cold, acidic, and anoxic, which retards decay processes. Some insects also become trapped and preserved in tree resin (ancient tree resin is called amber), which is also an important source of fossils, including some winged adult aquatic insects, but few aquatic larvae. Indeed, aquatic larvae are unlikely to encounter tree resin, except in unusual circumstances. Most amber deposits come from the northern hemisphere and former parts of Laurasia (see Section 2.3), and rarely pre-date the Cretaceous

period (Schmidt et al. 2010), which is when flowering plants (the source of resin) began to dominate terrestrial environments. The formation of amber also requires that resin is entrapped in a watery environment with anoxic sediments to prevent the resin deteriorating. Less common, but not requiring water for preservation, are asphalt and tarpit traps. For example, the La Brea tarpits of California are famous for the impressive prehistoric mammals that were trapped and preserved there, but the tarpits also trapped many insects, including aquatic insects (Miller 1983). Dispersing adult aquatic insects often identify water bodies (and potential oviposition sites) by the way in which reflected light is polarized by the water surface (Chapter 6). Unfortunately, tarpits polarize light in the same way and, to the eyes of an insect, may be indistinguishable from water.

Establishing phylogenetic relationships and reconstructing plausible evolutionary events are often based on a combination of information from fossil and modern specimens. The systematic methods commonly used in phylogenetic analyses include phenetics (morphological, behavioural, ecological, and genetic similarities), cladistics (shared ancestral characteristics), and the occurrence of insects over evolutionary time scales, e.g. presence in the fossil record. Traditionally, inferences about insect relationships were based primarily on morphological information, especially external structures, but

Table 2.1 Timeline showing approximate dates and names for different geological periods.

Era	Period	Approx. age (10^6 years ago)
Caenozoic—Quaternary		0–1.2
Caenozoic—Tertiary		1.2–66
Mesozoic	Cretaceous	66–135
	Jurassic	135–181
	Triassic	181–250
Palaeozoic	Permian	250–280
	Carboniferous	280–345
	Devonian	345–405
	Silurian	405–425
	Ordovician	425–500
	Cambrian	500–600

sometimes also based on behaviours or ecological traits. However, groups that are similar morphologically, behaviourally, or ecologically, may be only distantly related in an evolutionary context, because convergent or parallel evolution can lead to similar morphological structures in largely unrelated taxa. More recent developments in genetic techniques, especially molecular sequence information from nuclear and mitochondrial genomes, have provided many insights into phylogenetic relationships and genetic tools are now used widely. As with morphological characters, however, similarities in genetic material can also arise through convergent or parallel evolution, so genetic techniques do not necessarily provide definitive answers. Cladistic patterns are based on the presence of shared characteristics that are unique in an evolutionary context. For example, modified salivary glands that can produce silk are found in the Leipdoptera and Trichoptera, indicating a close evolutionary relationship of these two sister groups. All these methods have strengths and weaknesses, and a mixture of techniques is often used to build phylogenies. Of course, we can never be 100 per cent confident that proposed phylogenies are correct and many current interpretations are certain to be revised with further research.

2.3 Evolution of the insects

Insect evolution began during the late Silurian period of geological history (405–425 Mya, million years ago), when the insects arose from a group of six-legged arthropods, the Hexapoda (Grimaldi and Engel 2005). Not all hexapods became insects and the modern, non-insect hexapods include springtails (Collembola) and the soil-dwelling Protura and Diplura. The earliest insects present during the Silurian and Devonian (345–405 Mya) probably did not closely resemble any of the forms we recognize today. It was during the next geological period, the Carboniferous (280–345 Mya), that extensive radiation began and recognizable precursors or representatives of the extant orders first appeared (Figure 2.1). These earliest insects included the Ephemeroptera and Odonatoptera (a group which includes the modern Odonata), and also some basal forms of orthopteroid and hemipteroid groups. Thus, the mayflies are the most basal, extant lineage of winged insects,

although many lineages within the Ephemeroptera have gone extinct and modern species comprise only one recently evolved lineage. Radiation of the insects continued during the Permian (250–280 Mya), with roughly 30 distinct insect orders represented by the end of the Palaeozoic era, including the Plecoptera, Megaloptera, Hemiptera, and Coleoptera, or at least stem-group lineages of these orders. The end of the Permian (the Permian–Triassic boundary) saw many extinctions, especially in marine taxa, and there was a corresponding decline in insect diversity and a loss of several major lineages.

Gigantism was common in the late Palaeozoic (i.e. the Carboniferous and Permian periods) and insects with wing spans of up to 71 cm have been recorded for some fossil griffenflies (dragonfly-like insects in the now extinct order Protodonata) and 45 cm in some mayflies. Not all the griffenflies had such large wings, and much smaller species also occurred. In terrestrial environments, the maximum size of insects is set by various factors, including strength (it is difficult to support the weight of a large, heavy exoskeleton) and gas exchange, which is partly via diffusion and becomes inefficient at large body size (Chapter 3). How such large-bodied insects could exist in the past is uncertain, but some apparently credible arguments support a link with atmospheric oxygen concentrations (Harrison et al. 2010; Verberk and Bilton 2011), which were higher during the Palaeozoic. The general argument is that higher O_2 concentrations would permit larger bodies with the same gas exchange system. Note, however, that not all late palaeozoan insects were giants and most were of similar sizes to the common insects we see today. On a similarly cautious note, some modern insects are also quite large, if not gigantic, such as some damselflies in the family Pseudostigmatidae with wingspans up to 19 cm.

The dominance of dinosaurs and the origin of the mammals are often the most celebrated events of the Triassic (181–250 Mya) and Jurassic (135–181 Mya) periods, but insect radiation also continued after the Permian. Some of the major, modern orders were established during these periods, including the Trichoptera and their sister group the Lepidoptera, plus the Hymenoptera and Diptera. A few more orders, mainly terrestrial insects, came into being during the Cretaceous (66–135 Mya), but all the major insect

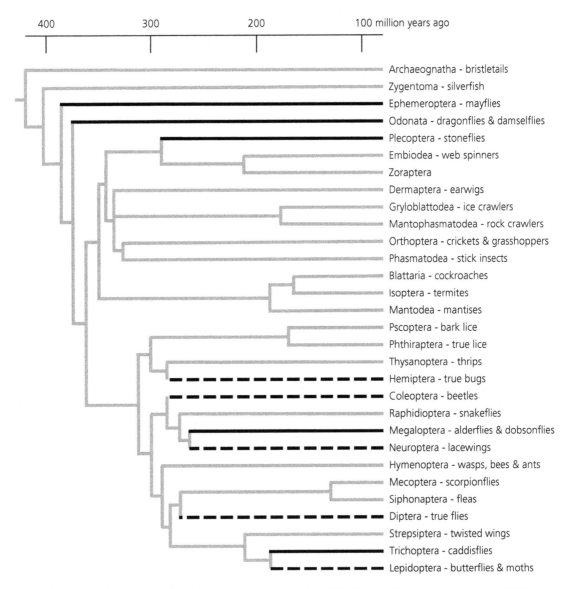

Figure 2.1 Phylogeny of extant insect orders. Fully aquatic orders are shown in solid black lines; partially aquatic orders in dashed black lines; other orders in grey.

Source: Based on information in Grimaldi and Engel (2005).

orders we see today were probably present by 66 Mya. Insect evolution continued throughout the Cretaceous and subsequent Tertiary periods, with many further radiations and extinctions within orders (i.e. at the level of families, genera, and species). There was also a massive radiation of the angiosperms (flowering plants) during these periods, and, coincidentally, the primarily phytophagous

orders (e.g. Lepidoptera and Coleoptera) also diversified. The Cretaceous–Tertiary boundary was marked by a major extinction event of disparate animal groups (the K-T extinction event), including the demise of the dinosaurs, which is linked to the impact of an extra-terrestrial object on the Yucatán Peninsula of Mexico. For the insects, there appears to have been no noticeable loss at the family level, possibly attributable to the high species richness within families, and their diverse and flexible geographic ranges. There is evidence of a diversity bottleneck for terrestrial insects with specialized plant associations and this has been linked to a marked decline in many plants at this time (Labandeira et al. 2002), but the aquatic insects appear to have been unaffected.

2.4 History and evolution of aquatic habits

Were the first insects aquatic or terrestrial? Did any terrestrial insects secondarily invade aquatic environments and, if so, when and which groups? Did the first aquatic insects live in running or still water? When did the modern ecological habits (e.g. feeding habits) of aquatic insects first arise? Were aquatic insects present in ancient freshwater systems and what functional roles did they play? These questions are fascinating, but most are very difficult to answer and opinions are often divided about what the correct answers are likely to be. The biggest problem is the shortage of convincing or complete fossil evidence.

The existence of an aquatic group can be based on three main kinds of fossil evidence, as identified by Wootton (1988). The first and strongest kind of evidence comprises fossils with visible aquatic adaptations (usually relating to gas exchange or locomotion), provided that the structures are interpreted correctly. The second kind of evidence comprises fossils belonging to two or more subdivisions of a single group that have modern aquatic representatives. This is less reliable than the first type of evidence, unless it can be proved that the fossil forms were also aquatic. Third are those fossils representing a single modern group that is aquatic. This is the weakest evidence because it assumes that the fossil forms were aquatic also. The biggest difficulty is that the strongest kind of evidence is also the hardest to acquire because most fossils are of adult insects and, for most aquatic insects, the larval stage occupies a freshwater environment, i.e. they are ontogenetically amphibious (ontogenetic relates to an organism's development). Fossilization did not occur equally in all aquatic environments and some groups may have been very abundant in some environments, but left few fossils. Another problem is that some insects could live in water but show few

obviously aquatic adaptations. For example, fossil Plecoptera first appear in the early Permian, and many Palaeozoic and Mesozoic larvae are known, but few show aquatic adaptations, even though they could have lived in water.

2.4.1 The first aquatic insects

The conventional view is that the basal forms of modern insects were terrestrial and that aquatic insects are derived from terrestrial precursors that secondarily invaded aquatic environments. This view, however, has repeatedly come under attack, with some arguing that an aquatic state, or at least an aquatic juvenile stage, was present very early in the evolutionary history of all or many insects. These arguments often revolve around the origins and evolution of wings and of the gas exchange system; these are complex arguments and we will not review them here. Whichever view is correct, and the debate may continue for some time, it has profound implications for the ecological roles of insects in ancient freshwater ecosystems (Wootton 1988). Briefly, if the conventional view is correct, then it is likely that the earliest freshwater ecosystems had no insects and that aquatic groups with terrestrial precursors appeared progressively over time. It is generally agreed that an aquatic habit is secondary for some orders, namely the Hemiptera, Diptera, and Coleoptera. These orders or suborders invaded water multiple times (at least 14 times in the semi-aquatic Hemiptera, Gerromorpha; Andersen 1995) and some quite recently, such as some lepidopteran species in the genus *Hyposmocoma* that began to invade water only approximately 6 Mya (Rubinoff and Schmitz 2010). However, the aquatic or terrestrial origins of other, older orders are often debated (e.g. the Ephemeroptera). Many authors argue that the different gill types in the Ephemeroptera, Odonata, and Plecoptera constitute strong evidence for the independent origins of aquatic lives among these very old lineages (Pritchard et al. 1993; Grimaldi and Engel 2005). In contrast, if an aquatic or ontogenetically amphibious state is ancestral for many insects, then insects of some sorts were likely to have been present in freshwater ecosystems for around 400 million years (since the Silurian or Devonian). Also, if this alternative view is correct, then all fully terrestrial insects had aquatic precursors. Unfortunately, the fossil record does not allow us to discriminate between these different views, and the arguments to support different hypotheses are often based on indirect evidence.

What the fossil record can tell us about when aquatic habits arose in the extant aquatic orders is shown in Figure 2.2, and discussed in detail by Wootton (1988). Fossils of any aquatic juveniles are rare before the Early Permian, but a variety of

unquestionably aquatic insects existed from then on. It is plausible that most Permian and Triassic aquatic insects lived in running water, given that their closest modern relatives (e.g. Ephemeroptera, Odonata) also live in streams and rivers (Wootton 1988). Detailed treatments of some groups are consistent with this proposal of an initial running-water existence, including the Trichoptera (Wiggins and Wichard 1989), and shifts to standing waters happened multiple times in many groups, including in

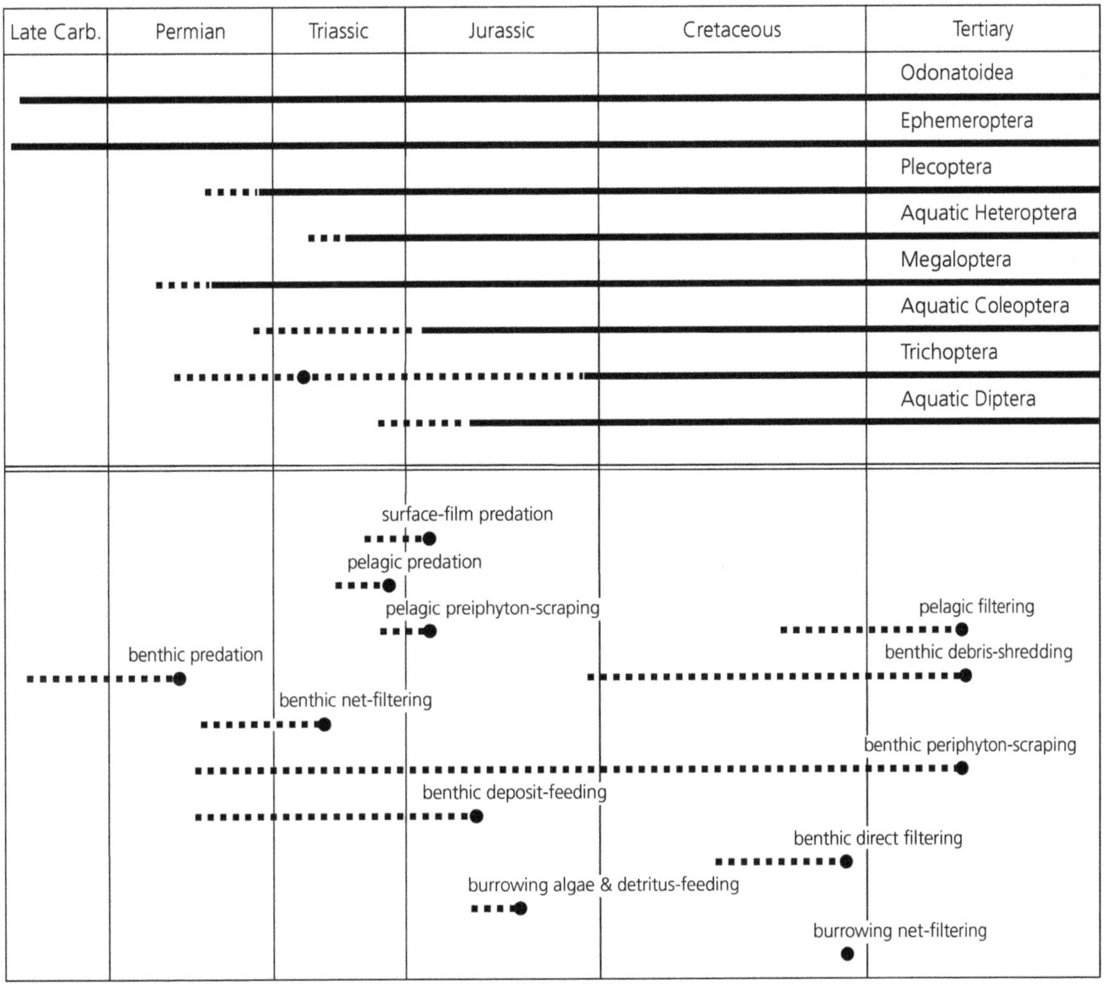

Figure 2.2 Evolution of aquatic habits. Top (above the double line): distribution over time of the major aquatic and semi-aquatic insect orders. The transition from dotted to solid lines indicates the approximate age of the earliest unequivocal fossil evidence for an aquatic habit. The circle on the trichopteran line indicates the approximate age of earlier, indirect fossil evidence for an aquatic habit. Bottom (below the double line): the origins of modern insect feeding strategies. Circles indicate the approximate age of the earliest fossil evidence for a particular strategy; dotted lines the time in which that strategy probably arose.

Source: Adapted from Wootton (1988).

the Coleoptera (Ribera 2008). Definitive evidence for an abundant still-water fauna does not appear until the Late Triassic and Early Jurassic, when there appears to have been a rapid diversification of aquatic Hemiptera, Coleoptera, and early aquatic Diptera. Ancient standing-water ecosystems were not devoid of life, but were likely inhabited by non-insect taxa, including various crustaceans, molluscs, and worms. Aquatic Coleoptera (swimming adults and aquatic larvae) are known from the Early Jurassic, but they may have been around much longer. The first aquatic hemipterans of the Triassic were the truly aquatic Nepomorpha (e.g. Belostomatidae, Naucoridae, Notonectidae); the water-walking Gerroidea appeared a little later and were derived from ancestors that occupied humid terrestrial or marginal aquatic habitats (Andersen 1982, 1995). Other older groups that initially occupied running waters, such as the Ephemeroptera, may have diversified their habitat use and begun to occupy standing waters during the Late Triassic or Early Jurassic. The Trichoptera present a dilemma because the oldest (Permian) forms are known only as adults and mostly from wings (Wiggins 2004; Grimaldi and Engel 2005). The first larvae and irrefutable evidence of an aquatic life appears in the late Jurassic, although there is weaker but plausible evidence for the middle Triassic. The earliest caddisfly taxa were probably caseless and case-building behaviours may have come later, but were well established by the Early Cretaceous. The now species-rich Chironomidae were among the first of the aquatic Diptera to appear in the late Triassic (Cranston 1995a, b; Grimaldi and Engel 2005). As mentioned above, there was a radiation of many insect orders and families during the Jurassic and Cretaceous, and it seems reasonable to assume there was an insect component in most Cretaceous freshwater ecosystems. Aquatic insects occupy an enormous range of different aquatic environments (Section 2.6) although we know little about when these habitats were first occupied by insects.

2.4.2 Early freshwater feeding habits

What did the earliest aquatic insects eat? Unsurprisingly, some guesswork is involved in tracing the development of food exploitation patterns through evolutionary time. One possible scenario, compiled by Wootton (1988) is shown in Figure 2.2. The scarcity of strong evidence of the ecological habits of ancient insects means that this scenario, and the description to follow, are necessarily simplistic and speculative. It is, nevertheless, fascinating speculation and provides a framework against which to test new discoveries.

The earliest aquatic insects appear to have been predatory and this habit is still apparent in some groups, such as the Odonata and Megaloptera, that have presumably always been predatory. The early Ephemeroptera appeared to be predatory also, as suggested by their large mandibles, even though modern mayflies are predominantly herbivorous or detritivorous. Predation in the mid-water column was underway by the Early Jurassic, with the diversification of the aquatic Hemiptera and Coleoptera, and the chaoborid flies. Surface-film predation appeared shortly afterwards with the appearance of water-walkers and gyrinid beetles.

It is unclear how many of the very early aquatic insects had non-predatory feeding habits and when other food resources were exploited by aquatic insects. We should first ask: what other food resources were available in ancient freshwater ecosystems? Aquatic macrophytes were unknown before the Triassic, but they are not a very significant food source for many modern aquatic insects (Chapter 13). Phytoplankton, attached and filamentous algae, other denizens of the biofilm (e.g. bacteria, fungi), and waterside vascular plants all likely had the potential to contribute to freshwater food webs in the Permian (and earlier) in much the same way as they do today. The relative abundance of these putative food sources is less certain, but they would have been available. Fine particles of various origins, but suitable for filter-feeding, would have been available in streams very early on. By the late Triassic, therefore, it is quite possible that aquatic insects exploited most of the broad categories of food resources they exploit today. An important exception, allochthonous inputs and coarse detritus-feeding habits, common in many small streams today, may have been delayed until catchments were well vegetated (the earliest angiosperms, flowering plants, only appeared approximately 140 Mya). The Jurassic period saw the appearance of burrowing, fine-detritus feeders (e.g. some mayflies). The cased caddisflies first appeared at the end of the Jurassic or early in the Cretaceous, and most of these were probably shredders of coarse plant material, as are many modern species. The last major feeding groups to appear, in the Cretaceous, were the fine-detritus feeders characteristic of fine sediment substrates of lakes and slow-moving rivers (e.g. tube-dwelling chironomids), and the phytoplankton filtering insects (e.g. mosquito larvae). Although phytoplankton were present long before the Cretaceous, this food resource was exploited by Crustacea well before the insects.

2.5 Historical biogeography of aquatic insects

The geographical distribution patterns we see in most modern insects reflects their evolutionary origins (at least over large spatial and temporal scales), because the spatial arrangement of land masses has changed dramatically over the periods of insect evolution. As we have seen, the sequence of events in the evolutionary origins of the different insect groups is subject to debate and, accordingly, so are their biogeographic origins. The theory of plate tectonics provides the mechanism underlying the movement of the Earth's lithospheric plates and associated land masses (continental drift). Understanding the sequence and timing of continental drift events (especially the fragmentation or joining of land masses) helps biogeographers explain some modern distribution patterns, so we begin with some brief material on plate tectonics.

All the present continental landmasses were once assembled in one supercontinent called Pangaea, which began to break up in the late Jurassic and the continents began to drift apart (Figure 2.3). Many of the modern insect orders were present by the late Jurassic and when Pangaea began to fragment, but there was still considerable radiation within insect orders over the next 180 million years. The megacon-

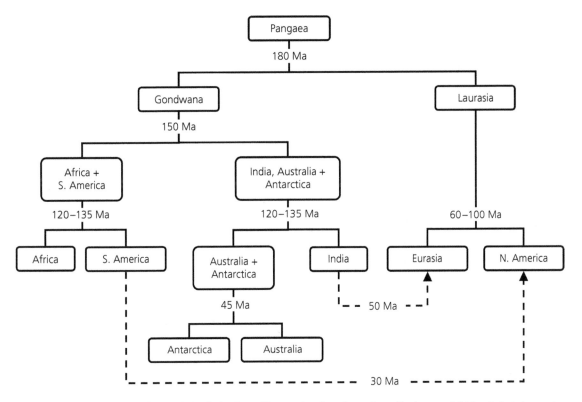

Figure 2.3 Approximate timing of events during the break up of Pangaea into the major continental land masses. Solid lines indicate fragmentation of land masses, dashed lines indicate land masses joining.

tinent of Pangea was actually rather short-lived (on geological timescales!), but the configuration and movement of continents before that are more difficult to reconstruct. Approximately 150 Mya, Pangea broke into two major landmasses, Laurasia (comprising what are now the northern continents: North America + Eurasia) and Gondwana (comprising the southern continents and India: South America + Africa + India + Antarctica + Australia). Note that the timing of these events is only approximate and there are multiple models to describe possible event sequences (Upchurch 2008). Gondwana then split into western and eastern land masses (150 Mya). The subsequent phases in the break-up of Gondwana (120–135 Mya) were more complex due to simultaneous rifting between Africa and South America, and between Antarctica + Australia and India. Antarctica and Australia only split 40 Mya. Laurasia also fragmented into Eurasia and North America during the Cretaceous (100 Mya). Drifting continents not only broke apart into smaller fragments, but some fragments also collided, such as India with Eurasia (50 Mya), and North with South America (30 Mya).

Importantly, this sequence of continental rifting is widely used by historical biogeographers to explain the modern distributions of many plants and animals, especially those on the Gondwanan plates. The basic idea is that these geological events produced 'vicariance patterns' where vicariance is defined as a distribution pattern common to several groups and that is produced by the impact of geographical barriers on speciation events. In the case of Gondwana, these barriers are oceans that separated the land masses, and vicariance is revealed by phylogenetic relationships that correspond to the continental drift sequence. Although associating patterns of geographic distribution, continental drift events, and evolution may sound quite straightforward, it is in fact a complicated process, because many vicariance patterns have become subsequently modified by more recent dispersal, extinction, and species radiation. The frequency and timing of such events have varied geographically and between taxonomic groups, and so the biogeographic history of many organisms is still being debated (Upchurch 2008).

Once the theories of plate tectonics and continental drift gained wide acceptance (circa 1960s) some of the first entomological studies of historical biogeography were on beetles, but perhaps the best early example involved chironomid midges (Brundin 1967). Focusing on the austral landmasses of Gondwana (New Zealand, Australia, South America, Africa), Brundin found that closely related groups of chironomids lived on separate continents, which argues for a Gondwanan origin. Brundin's interpretations have largely withstood the tests of further research on Gondwanan chironomids, but not without some debate (Cranston 1995b; Cranston et al. 2002). Among the aquatic insects, other austral disjunctive distribution patterns (disjuncts are closely related taxa that are widely separated geographically) that most likely arose through continental drift are well known within the Ephemeroptera, Odonata, Plecoptera, and some Coleoptera (e.g. Psphenidae), groups that were well established before the break up of Gondwana. For example, the initial split of Pangea into Gondwana and Laurasia coincided with a split of the Plecoptera into the suborders Arctoperlaria and Antarctoperlaria, and the Antarctoperlaria families became disjunct with the further break-up of Gondwana (Zwick 2000). Continental drift, however, does not explain the modern distribution patterns completely, as evidenced by the presence of some Arctoperlarian families in the southern hemisphere that may have dispersed much later on.

Beyond this general overview, detailed explanations of the historical drivers behind modern-day distributions of each of the aquatic insect orders are beyond the scope of this book. While there are interesting differences in species diversities in different regions or with latitude (Vinson and Hawkins 2003), each order has had something of a unique history that reflects differences in the effects of vicariance, dispersal, extinction, and radiation. Detailed treatments of biogeographic patterns for various orders, suborders, and families are provided by many different authors.

2.6 Environments inhabited by aquatic insects

Anywhere there is water, there are likely to be aquatic insects, from small puddles to large lakes and rivers, and even the open ocean. Even water that is highly saline, acidic, anoxic, above the boiling point, or regularly frozen may contain insects. Some species are restricted to living in particular environments, whereas congeners may occupy a wide range of habitats. Thus, while there are some weak generalizations about where particular groups of aquatic insects may be found, there are always exceptions. As a result, most higher-order taxonomic groups of aquatic insects, such as orders, suborders, and families, collectively span a wide variety of aquatic environments, and some, such as the Diptera and Coleoptera, do so to an extraordinary degree. Additionally, within the broad categories of well-known aquatic habitats, such as lakes and rivers, there are many different types of environment, which offer diverse living spaces for aquatic insects.

In the following sections, we discuss the many environments where aquatic insects occur and the predominant types of insects found in them. In many cases, variation between aquatic ecosystems is so great or insect diversity so high that only a simplistic description of the fauna is possible. This is not intended to be an exhaustive review. Instead, our aim is to illustrate the extraordinary range of environments that aquatic insects inhabit.

2.6.1 Flowing surface water: rivers and streams

Most flowing water is fresh, and limnologists refer to running waters as lotic environments. Aquatic insects occur in every type of lotic system from the tiniest streamlet, only centimetres wide, to the largest river. Multiple species of all the fully aquatic and semi-aquatic orders are routinely found in flowing waters. Stoneflies are the most strongly associated with flowing waters, but not standing waters, and the highest diversities are typically in cool, temperate streams (Vinson and Hawkins 2003). All the other orders are also common and diverse in many other kinds of aquatic environments.

Few insects living in flowing waters live predominantly in the water column and most live on the benthos (bottom). Different species may be found in swiftly flowing, rocky environments compared to slow-flowing, muddy or sandy environments. Living in fast-moving water requires certain adaptations of either body shape or equipment to avoid displacement, and living on mud or sand can present different challenges, such as to species with delicate external gills, whereas others are clearly adapted for burying through sediments (Chapters 5, 8). While some species are restricted to living in fast water (e.g. filter-feeding caddisflies and blackflies) or slow water, others may occur in a wide range of flow conditions. Some species are found only in headwater streams (which are often fast-flowing, cool, shallow, and heavily shaded) whereas others occur predominantly in the warmer, deeper, slower-moving, and turbid waters of lowland systems. Rivers that freeze during the winter or bake in the summer also present significant challenges that may be reflected in the species composition, but again, many aquatic insects span many stream and river types and many locations along them. Substrate type is important to benthic insects. One distinction is between rocky and sandy/muddy surfaces (as above), but bedrock, wood, in-stream macroalgae, and plant detritus (leaves, bark, twigs) are inhabited by species specially adapted to those places as well as many generalists. Some insects rarely occur in the main flow (thalweg) and live predominantly along stream margins, where they may live among the protruding roots of riparian trees or on emergent aquatic macrophytes. Insects also live on the water surface, particularly close to stream banks or in backwaters, where they are less exposed to the current. Common along many river margins are semi-aquatic Hemiptera, which possess a fantastic array of adaptations that permit them to run around on the water surface without getting wet (Chapters 5, 8).

Outside the obvious boundaries of river channels are many aquatic or semi-aquatic environments. Some aquatic insects live alongside channels in the spray zone of waterfalls or live in waterfalls themselves (Kalkman et al. 2008). Others live in the deep sediments below the surface of the stream bed, a habitat called the hyporheos. These individuals can be found many metres below the surface and even multiple kilometres from the main river channel (Stanford and Ward 1988). Many hyporheic species also live on the surface, thus neatly illustrating the versatility of many taxa, but some others are unique to groundwater (Section 2.6.4).

In many parts of the world, streams dry up seasonally, occasionally, or for multiple years at a time. In these environments, insects may take refuge in moist cracks within the stream bed, under leaves, barks, or rocks to await the return of water (Boulton and Lake 2008; Chester and Robson 2011). Other species move rapidly through their life cycles and disperse as adults before drying is complete, and there may be substantial exchange of individuals between temporary or ephemeral streams and nearby permanent waters. Nevertheless, stoneflies and odonates tend to be scarce in temporary streams, which are more likely to contain mayflies, bugs, beetles, caddisflies, and dipterans (Williams 2006).

Finally, variability in water chemistry is associated with differences in insect fauna, typically owing to species-specific physiological tolerances (Chapter 4). Naturally acidic or basic streams, for example, often have characteristic sets of species, with some species being restricted to those environments. While relatively rare, naturally salty streams do occur, particularly in arid or semi-arid zones. They have a reduced diversity compared to freshwaters, but beetles, dipterans, mayflies, and odonates may be found (Moreno et al. 2010), and some naturally saline streams have high endemism, especially among beetles (Abellán et al. 2007).

2.6.2 Standing surface water: lakes and ponds

Lakes and ponds are common the world over and aquatic insect inhabitants are abundant. These standing waters (lentic environments) are typically contained in a distinct basin; some may have outlet or inlet streams, whereas others are completely closed. There is no strong ecological distinction between ponds and lakes, but water bodies that are sufficiently shallow that rooted vegetation can be present throughout the basin are usually called ponds. However, we should consider also that words to describe lakes and ponds vary enormously around the world (e.g. meres, tarns, lochs, lochans) in part depending on geography and language, and in part depending on the origin of the water body. For example, oxbow lakes are derived from old river channels that have become cut off from the main channel (called billabongs, in Australia); kettle hole lakes are formed by melting blocks of ice stranded after glacial retreat.

As for running waters, many species from all the fully aquatic and semi-aquatic insect orders are rou-tinely found in lakes and ponds. As also in rivers, few insects live permanently in the water column because few are able to control buoyancy (Chapter 5). Only a few Diptera are able to hold position effortlessly in the water column (e.g. phantom midges, Chaoboridae). Other insects are good swimmers (beetles and bugs) and move actively through the water column, but remaining submerged without sinking constantly requires continual propulsion (insects are negatively buoyant) and this is energetically expensive. Consequently, most active swimmers stay near the shore or where there is emergent vegetation (e.g. macrophytes) to rest and hide.

Water-surface dwellers (semi-aquatic Hemiptera) are often more abundant and diverse in standing than running waters, but most of the lentic insects are benthic. As described in Section 2.6.1, substrate type (e.g. sand, mud, pebbles, rocks, bedrock) has a strong effect on the types of species present. Additionally, many aquatic insects live on and around macrophytes, which are restricted to relatively shallow water. In ponds, insects may be found throughout the water body, but most lake insects live close to the lake margins (the littoral zone), where the water is relatively shallow and vegetation may be present. Lake depth creates a number of effects. Deep lakes typically have strong temperature gradients with depth (because of stratification), and dissolved oxygen concentrations may be low at the bottom. This profundal zone is typically a fairly homogenous habitat. Some chironomid genera are characteristic of the profundal zone and most live buried in the surface sediments. There are some deep-water mayflies and megalopterans, but otherwise few species of benthic insects live in the profundal.

As with all water bodies, standing waters can vary greatly in water permanence and physico-chemistry: water chemistry, water transparency (e.g. due to nutrient status), turbidity, colour (e.g. water may be brown from humic compounds), the presence of inlet and outlet streams, whether the lake freezes, whether the lake is large enough for wind-generated waves, and so forth. Ponds also vary in similar respects, but also in dissolved oxygen, water volume and depth, depending on location, size, and exposure to sun. All these factors affect the kinds of aquatic insects that may be present. Particularly striking and characteristic faunal assemblages occur in lakes and ponds that dry out seasonally, in some years, or are completely dry much of the time. The insect fauna of such temporary waters (typically ponds not lakes) includes odonates, dipterans (chironomids, mosquitoes, ceratopogonids), and many bugs and beetles. Certain genera are widespread, such as *Gerris*, *Hydrometra*, and *Microvelia* (Hemiptera); several beetles (e.g. *Berosa*, *Agabus*); and the chironomid, *Chironomus* (Williams 2006).

Some lakes are quite acidic (e.g. bog lakes of the boreal zone) whereas others are saline and, although many taxa may find these environments physiologically stressful, some are well adapted. Salt lakes are common in some parts of the world (e.g. Australia, North America, Africa). They vary in salinity, with some being hypersaline (saltier than seawater) whereas others are much less salty. In Australia, where salt lakes are common, salt lakes may range up to 350 parts per thousand (ten times saltier than sea water) and there are other extremely salty lakes elsewhere (e.g. Great Salt Lake in Utah). Salinity often varies seasonally or between years (depending on rainfall or stream inflows). Some lakes are salty because of historical connections to the ocean, whereas the athalassic saline lakes have no marine connections (although they may be near the coast), and the dominant ions vary accordingly. In Australian salt lakes, the dominant ions are sodium and chloride, but in other places (e.g. Rift Valley in Africa, parts of North America), carbonate or sulphate ions predominate (Bayly and Williams 1973; Hammer 1978). The fauna of such systems can comprise species that are tolerant of very high salinities (halobionts), those able to tolerate moderate salinities, and freshwater taxa that have some salt tolerance. Typically, the saltier the lake the fewer species that are present but exceptions occur, for example, where lakes are coastal and include marine species (Timms 2009).Which taxonomic groups dominate hypersaline lakes varies geographically, but they are often Diptera. In Australia, a chironomid (*Tanytarsus barbitarsis*) typically dominates the saltiest lakes, whereas ephydrids dominate some North American lakes, such as Great Salt Lake and Mono Lake (Foote 1995). Chironomids also dominate some salt lakes in northern Africa (Ramdani et al.

2001). Other highly salt-tolerant insects in Australian lakes include beetles, such *Berosus* and *Megaporus* and some ceratopogonid dipterans (Halse et al. 1998). Freshwater taxa with some salt tolerance include a wider variety of taxa (e.g. caddisflies and mayflies as well as multiple species of dipterans and beetles, Timms 1983). Ion composition can affect the fauna, and thus different species of Ephydridae occur in chloride, carbonate and sulphate lakes, whereas other taxa (e.g. corixids) show no anionic preference (Foote 1995; Herbst 2001).

Some salt lakes and ponds are dry for some or much of the time. Salinities can fluctuate widely during drying phases, and water temperatures may exceed boiling point in arid-zone systems (Timms 1998). Intermittent salty waters typically have fewer species than permanent salt lakes (Timms 2001), but nevertheless can contain insects from many orders and including some more often associated with freshwaters. Many of these freshwater species may colonize saline lakes when they later become wet, rather than being permanent residents (Timms 1998; Herbst 2001). Special mention should be made of soda lakes, which are highly saline as well as strongly alkaline, and which also may dry up. Fauna in such lakes must tolerate not only high salinity and high pH, but also sometimes high temperatures and low dissolved oxygen. Not surprisingly the diversity of fauna is low; Diptera (e.g. Ceratopogonidae, Chironomidae) are common, but Hemiptera and Coleoptera may also occur (Verschuren et al. 1999; Alcocer et al. 2001).

2.6.3 Wetlands, springs, pools, puddles, and phytotelmata

Wetlands are difficult to define and include a wide range of environments where water is present either permanently or occasionally, and often where vegetation dominates all or most of the water body. Such places are often called marshes, swamps, fens, or bogs, depending on whether water arrives from rainfall only or also via runoff and groundwater, and whether water is still or flowing. They can cover thousands of km² in area, e.g. the Okefenokee Swamp in southern North America, but can also be very small. For example, open pools found within peat bogs can be < 1 m² yet fairly permanent.

Many of these wetlands contain fresh water and the majority (if not all!) contain aquatic insects. Few aquatic insect species are unique to particular kinds of wetlands, but some wetlands contain a high diversity of particular taxa, predominantly Diptera, which live in sediments, rotting vegetation, mud flats, sand shores, or on emergent vegetation (Batzer and Wissinger 1996; Keiper et al. 2002). Diversity of the Diptera in wetlands will often be much higher than nearby streams. Worthy of special mention are peat bogs, not least because of the enormous area they cover at high latitudes. Peat bogs typically occur in cold places, are oligotrophic or even dystrophic (very nutrient-poor), and the water tends to be naturally acidic. As such, these are relatively challenging places to live, but insects are nevertheless common in bogs, and some species are found nowhere else. There are dragonflies, bugs, beetles, and some dipterans that occur only in the acidic pools of boreal peat bogs (Spitzer and Danks 2006). Many other inhabitants, however, are not restricted to these environments and have wide distributions. Wetlands also include coastal ecosystems, such as intertidal marshes, swamps, and mangroves (see Section 2.6.5), but some inland wetlands are also naturally salty. As described for saline lakes, the degree of saltiness can determine which species predominate in wetlands (Hart and Lovvorn 2005).

Springs are places where water arises predominantly from groundwater rather than rainfall runoff. Groundwater usually has more consistent discharge, water temperatures, and water chemistry than rainfall runoff. Consequently, springs can produce some fairly distinct aquatic environments with distinct faunas, including some characteristic Diptera and Trichoptera (Cantonati et al. 2012). Important physico-chemical factors influencing the insect fauna include water temperature, hardness, and conductivity, but many inhabitants of springs also live in other types of water bodies. Mound springs are unusual in that chemical deposits from the water (e.g. calcium carbonate) produce a mounded rim around the spring. Geothermal springs may be exceedingly hot and yet still support aquatic insects. For example, some dipterans live in the hot springs in Yellowstone National Park where water temperatures are routinely over 30 °C and can reach 45 °C (Foote 1995). In the Eyre Peninsula in South Australia, marine springs contribute salty water to nearby salt lakes and contain a variety of unusual insects including a leptocerid caddisfly, *Symphitoneuria wheeleri* (Timms 2009).

Very small water bodies include puddles and rock pools, which may form only after rains or may be permanently wet. Insect inhabitants include beetles, mosquitoes and other dipterans (chironomids, ceratopogonids), and pool longevity can play a strong role in determining species composition (Batzer and Wissinger 1996; Anusa et al. 2012). Some species rely on dispersing before pools dry, whereas others can withstand drying. Some Australian chironomids survive drying by building tubes into which they retreat until water returns, and they can withstand prolonged drying and temperatures up to 56 °C (Bayly and Williams 1973). The African rock-pool chironomid *Polypedilum vanderplanki* is so hardy it can tolerate almost complete dehydration coupled with exposure to extremely high temperatures (100–200 °C) repeatedly and yet still metamorphose. This tough species can also survive submergence in liquid helium (−270 °C) and storage in a dry state for 10 years (Hinton 1960a), although they are unlikely to encounter these conditions naturally.

Phytotelmata are small bodies of water collected within parts of plants, such as tree holes, leaf axils, flowers, modified leaves, fallen leaves, bracts, fruit husks, and stem rots (where plants rot after trauma to the stem). Many animals live in phytotelmata and some species are unique. The Diptera once again reign supreme, with 13 families found in all kinds of phytotelmata, particularly mosquitoes, but also ceratopogonids, chironomids, and tipulids. Beetles are also common (scirtids, dytiscids, and hydrophilids). Six families of odonates breed regularly in phytotelmata, including the giant tropical damselflies (Pseudostigmatidae, Megapodagrionidae), and some species of odonate occur nowhere else. Finally there are some truly odd occurrences—several species of veliid (Hemiptera), caddisflies (a calamoceratid and a limnophilid), and perlid stoneflies found living in tropical *Heliconia* flowers and bromeliads (Greeney 2001).

2.6.4 Groundwater

Animals that live only in groundwater are called stygofauna. Most of the insect representatives are beetles, but many aquifers remain to be examined and this is an area where information is rapidly increasing. Recent work, for example, discovered 99 new species of dytiscid beetles living in Australian groundwater aquifers in the last few years alone (Watts and Humphreys 2009), and it is likely that many more insect species remain to be discovered, given the large volumes of groundwater yet to be examined (Guzik et al. 2010).

Subterranean estuaries arise where salty water abuts fresh groundwater (Humphreys et al. 2009). These can occur where groundwater meets the ocean at coasts, but also where groundwater is linked to the sea through subterranean channels, which can be extensive and distant from the ocean. Additionally, salt lakes may have subterranean estuaries if they overlay relatively porous substrate that permits exchange between ground and surface waters. Such groundwater estuaries can be similar to river estuaries in having distinct layers of saltwater and freshwater with some mixing, but with changes in layering and mixing dictated by the often slow recharge of the groundwater. These interesting ecosystems have unique stygofauna and, although many are not insects, dytiscid beetles appear to be common and diverse. In Western Australia, such aquifers contain different species that may have been isolated from each other for about 5–8 million years.

2.6.5 Marine environments, including estuaries

The highest diversities of aquatic insects occur in fresh, inland waters, but insects have colonized marine environments as well. Many are semi-aquatic insects that are rarely or never immersed in water, but some are fully aquatic. Intertidal mud flats and sandy beaches often contain insects, particularly where there is rotting vegetation (seaweed), but insects also live in intertidal rock pools and in crevices and under rocks. Most intertidal insects are dipterans, particularly chironomids, some of which have quite extraordinary life cycles, but there are other Diptera (e.g. ceratopogonids, mosquitoes), Hemiptera, Lepidoptera, and Coleoptera that inhabit intertidal zones. For example, carabid beetles in the genus *Aepus* can be found living under stones and crevices in the intertidal zone where they are completely submerged during high water (King et al. 1982). Similarly, the bug *Aepophilus bonnairei*

lives in the low intertidal zone associated with the seaweed *Fucus* (Leston 1956), where it can survive days of submergence (King and Fordy 1984). Corixids live in rock pools and can be significant predators upon marine invertebrates (Schuh and Diesel 1995). Perhaps the most amazing examples of intertidal insects are caddisflies in the family Chathamiidae. Recorded in Australia, New Zealand, and some offshore islands, the larvae live in intertidal rock pools and construct cases from fragments of coral, coralline algae, shell, or seaweed (Cowley 1978). Even more startling, the eggs of at least one species (*Philanisus plebeius*) are laid and develop within the coelom of the starfish *Patiriella regularis* before being released as first instar larvae (Winterbourn and Anderson 1980).

Along coasts, insects inhabit intertidal salt-marshes and mangrove forests. Most insects in intertidal marshes are not aquatic, although some supposedly terrestrial species remain in marshes during high tide and can survive inundation (Cameron 1976). However, aquatic or semi-aquatic species can be abundant in saltmarshes, such as chironomids, biting midges and mosquitoes (Breidenbaugh et al. 2009; Dale and Breitfuss 2009). There are even marine butterflies, *Coenonympha tullia nipisiquit*, that live in saltmarshes and their larvae can tolerate complete submergence for up to 24 hours (Sei 2004). Similarly, the hemipteran family Saldidae has saltmarsh species able to tolerate submergence at high tide (Brown 1948).

A surprisingly wide diversity of insects inhabits hypersaline coastal wetlands, brackish lagoons, and estuaries. Common taxa include beetles, bugs, and chironomids, but caddisflies, mayflies, stoneflies, odonates, and even blackflies have all been recorded living in estuaries or other brackish waters (e.g. Denis and Malicky 1985; Timms 2001; Boix et al. 2007; Dimitriadis and Cranston 2007; Keats and Osher 2007; Day 2010). Individuals in estuaries are not necessarily accidental dislocations from freshwater locations upstream, because many species can survive immersion in brackish water long enough to survive high tides (Williams and Williams 1998). Insects may also occupy brackish water in the hyporheos of estuaries (Williams 2003).

Impressively, even the open ocean has been colonized by insects. The champions of this habitat are water striders in the genus *Halobates*. While most *Halobates* are coastal and associated with either mangroves or other vegetation, at least one species is completely oceanic (Andersen and Cheng 2004). There are also records of corixids living out on the open ocean (Gunter and Christmas 1959), although their biology is largely unknown. The marine chironomids *Clunio* and *Pontomyia* have been found amongst the epibiota attached to the shells of hawksbill sea turtles living out at sea, and *Pontomyia* can occur at depths of 30 m (Schärer and Epler 2007). Another marine chironomid, *Telmatogeton japonicus*, has been found living on the sides of windmills and buoys out in the open ocean, suggesting either substantial dispersal ability or transport via shipping (Brodin and Andersson 2009).

2.6.6 Artificial human-made environments

Human beings create many artificial aquatic environments, some of which are deliberately polluted (e.g. sewage lagoons, urban runoff wetlands, mine tailings) or devoid of natural sediments or habitat structure (e.g. canalized streams). Typically, insects will still be present, but usually in lower densities or diversities compared to comparable, natural aquatic systems nearby. Unpolluted, unnatural environments may have faunas comparable to their natural analogues. Some truly bizarre places created by humans have become habitat for insects and are worth mentioning simply to illustrate the extraordinary flexibility of insects.

Fluid but not truly aquatic environments are puddles of crude petroleum in Californian oilfields, which have been colonized by an ephydrid dipteran. The larvae of *Helaeomyia petrolei* feed on insects that fall into the puddle and become trapped in the petroleum. The larvae rely on atmospheric oxygen for gas exchange and have 'float hairs' surrounding the posterior spiracles, which ensure the spiracles stay above the surface so larvae can move around without the spiracles blocking. Another interesting example is ephydrids that colonize wood that has been heavily contaminated with human urine in ancient European buildings. The larvae of *Scatella (Teichomyza) fusca* graze on the microflora that develops on the presumably nitrogen-rich surface (Foote 1995).

PART 2

Environmental Constraints on Distribution

Part 2 examines how and why insects live in different types of environments. In Chapter 2, we described the extraordinary range of aquatic habitats where insects may be found. Each habitat presents a range of physical and chemical conditions, and these environmental factors affect survivorship, growth, development, and reproduction, that is, they influence fitness. As such, physical and chemical gradients in the environment form important niche axes. Quantifying niches requires an understanding of how insects can tolerate and thrive along physico-chemical gradients, particularly in extreme environments. Investigating the niches of species is an important core area of basic ecology. Such research underpins questions about how and why species live where they do, and the geographical range limits that are set by proximal—as opposed to historical—factors. It follows that such research is pivotal also to investigations into the effects of natural and anthropogenic environmental change, such as climate change. For example, thermal limits may provide a starting point for investigations into whether species are likely to expand or contract their ranges in response to climate change. The three chapters comprising Part 2 examine the major environmental gradients affecting aquatic insects and the ranges of adaptations allowing them to survive and thrive in particular aquatic environments.

Important physical and chemical gradients include temperature, dissolved oxygen concentration, salinity, and turbidity. The substrate (bottom or benthos) of environments can vary from fine clays to sand, gravel, cobbles, or bedrock. Water may be still or flowing. All of these gradients affect basic life-functions, such as respiration and gas exchange, metabolism, water balance and desiccation, and excretion and osmotic balance. Aquatic insects must also deal with the basic properties of living in a fluid that is denser and more viscous than air, and that creates substantial challenges with maintaining position or dealing with drag in a sometimes fast-moving medium. Any and each gradient may set limits to growth and survivorship, development, or reproduction. Additionally, insects have complex life cycles with different stages that often inhabit different environments. To understand any species' niche means we must consider how environmental gradients place limits on fitness at each life-cycle stage. Because different stages, such as eggs, larvae, pupae, and adults, live in different environments we should expect that they are exposed to different physical and chemical conditions. As a result, niche widths may be limited in some environments and at some life-cycle stages much more strongly than in others.

In the applied context, stream insects are routinely used to assess the ecological condition of many water bodies, but particularly streams and rivers. Such methods date back to the early 1900s, when limnologists observed that rivers polluted with sewage and other forms of organic pollution had a fairly distinctive fauna comprising particular insects, such as the midge *Chironomus* (Chironomidae), as well as worms and other species. Today, a huge number of indices of stream and river condition have been constructed using the presence/

absence or relative abundance of aquatic insects (and other invertebrates). Nevertheless, the underlying connections between index values and the responses of insects is less well understood and often not acknowledged. Indeed, some papers using insects to draw conclusions about river condition often appear to have little factual knowledge of the basic biology of the invertebrate fauna of rivers and streams. For example, a common assumption is that entire families of insects, some comprising hundreds of species, will respond in a fairly uniform way to environmental gradients associated with pollution, water abstraction, or other human impacts. While a handful of families may respond in a somewhat predictable way, most contain species with highly variable responses. Often such individual variation is not measured and is therefore unappreciated. The information in the next three chapters should go a long way to highlighting the enormous variation between different insects in how they overcome obstacles to life in different aquatic habitats.

CHAPTER 3

Gas exchange

3.1 Introduction

Like most animals, insects must obtain oxygen from the environment for metabolic processes, such as those associated with growth, development, foraging, and movement, all of which may influence an insect's survival and fitness. At a cellular level, this oxygen is used in the biochemical process of respiration to convert 'fuel' (e.g. carbohydrates, proteins, and fats from food) into a molecular form of energy. In the absence of oxygen (O_2), insects typically have poorly developed capabilities for anaerobic metabolism, although some adaptations may allow brief exposure to anoxic or hypoxic conditions (Hoback and Stanley 2001). In the process of consuming oxygen, carbon dioxide (CO_2) is produced as a by-product of respiration. If too much CO_2 accumulates in tissues, it can become toxic and therefore must be removed. Thus, gas exchange has two interrelated components: transporting O_2 from the environment to the tissues, and expelling CO_2 from the tissues back to the environment.

Movement of O_2 and CO_2 inside the insect body occurs in the tracheal system, a series of internal, air-filled tubes (tracheae). Terminally, the tubes are highly branched, forming tracheoles, and this provides an enormous surface area over which diffusion occurs. Tracheoles are so numerous that oxygen readily reaches most tissues and, similarly, carbon dioxide can be transported away. Thus, unlike many other animals, gas transport in insects is largely independent of the circulatory system. Insects have an open circulatory system in which the blood (haemolymph) occupies the general body cavity (the haemocoel). Blood-based gas delivery via oxygen-binding respiratory pigments (e.g. haemoglobin, haemocyanin) moving through the haemolymph

does occur in at least a few insect groups some of the time, but these are the exception rather than the rule.

Despite their aquatic habits, a gas-filled tracheal system occurs in virtually all aquatic insects, perhaps reflecting an evolutionary origin in the terrestrial environment (Section 2.4). Only rarely is the tracheal system filled with liquid (e.g. some early instar Diptera), and hence most of this chapter is devoted to explaining the diverse ways these animals nevertheless manage to respire underwater. There are two main kinds of tracheal systems depending on how insects exchange gases between the tracheal system and the environment. In open tracheal systems, aquatic insects exchange gases—in a gaseous state—directly with atmospheric air (e.g. by visiting the surface periodically) or through the use of a gas store (e.g. an air bubble), which they carry underwater. In closed tracheal systems, gas diffuses directly across the body surface where the integument is thin, and into the tracheal system (cutaneous gas exchange). For aquatic insects, this typically involves converting gases dissolved in the water into a gaseous state in the tracheae (and vice versa). It is important to note that net gas exchange in any individual may occur via several different routes, and the relative importance of these various routes may vary with environmental factors. For example, larval fishflies are bimodal breathers, normally exchanging gases cutaneously but also using respiratory tubes to air breath (Hayashi 1989).

Before discussing the various gas exchange systems of aquatic insects in any detail, we will first review some of the basic principles of diffusion and the physical properties of gases (Section 3.2), because this is instrumental to understanding how the diverse range of gas exchange systems work. With

Aquatic Entomology. First Edition. Jill Lancaster & Barbara J. Downes.
© Jill Lancaster & Barbara J. Downes 2013. Published 2013 by Oxford University Press.

a good grasp of the physical processes, we will next describe the general structure of the tracheal system and how gases move within it (Section 3.3). Subsequent sections will consider open tracheal systems and how insects have achieved various degrees of independence from the atmosphere (Section 3.4), followed by a discussion of the various morphological adaptations and behaviours insects may exploit to maximize cutaneous gas exchange in closed tracheal systems (Section 3.5). Finally, we will briefly discuss what happens when environmental oxygen concentrations are very low (Section 3.6) and the few aquatic insects that can use respiratory pigments for blood-based gas exchange (Section 3.7).

3.2 Diffusion and the physical properties of gases

Ultimately, gas exchange in insects depends on diffusion, so it is important to understand the basic principles of this physical process and how it is influenced by environmental factors. Diffusion of gases occurs at multiple places within the tracheal system and often across membranes or cuticular surfaces, including between tracheoles and tissue cells, along tracheal tubes, between water and the body wall, and between water and the air of gas stores. In all cases, the diffusion rate is proportional to the surface area over which diffusion occurs and to the diffusion gradient. Diffusion as the sole means of gas exchange can be employed only by organisms with a high surface area to volume ratio (i.e. all cells are close to the body surface) and organisms with low metabolic rates. Otherwise, organisms must increase the rate at which gas moves by increasing the surface area for diffusion (e.g. with specialized respiratory structures) and/or maximize the diffusion gradient (e.g. with systems that quickly move large quantities of gases to and from the diffusion surface). For most terrestrial insects, avoiding water loss is also important (i.e. water vapour often moves in association with O_2 and CO_2), and this has had a significant influence on the structure and function of gas exchange systems. In the terrestrial environment particularly (e.g. adult aquatic insects), gas exchange is restricted to certain areas of the body and the remaining areas are impervious.

The greater the concentration difference, the faster the gases diffuse. When O_2 is used by the tissues in metabolic processes, O_2 concentration within the cell drops and new O_2 will diffuse along the concentration gradient from the tracheoles into the tissues. Conversely, CO_2 produced by the tissues during metabolism will increase CO_2 concentration within the cell relative to the adjacent tracheoles, and CO_2 will diffuse along the gradient from the tissues into the tracheoles. Diffusion rate can be increased by increasing the surface area over which diffusion occurs or by creating a transport system that increases the flow of gases over the diffusion surface. Although it is common to discuss diffusion in terms of concentration gradients, it is perhaps better to consider the partial pressures of gases. The total amount or pressure of gas in solution (in air or liquid) is equal to the sum of all the partial pressures of all the constituent gases (this is known as Dalton's Law). Atmospheric air is primarily nitrogen, oxygen, and water vapour (78 per cent N_2, 21 per cent O_2, 1 per cent H_2O). The relative proportions of these three gases in air remains the same over large geographical areas, but their partial pressures may vary because, for example, total air pressure varies. At sea level, total atmospheric pressure is approximately 1.25 times higher than at an altitude of 2000 m and, consequently, the partial pressure of all the gases is 1.25 times higher at sea level than at 2000 m. The opposite phenomenon occurs deep under water where total pressure is much higher than at the water's surface. Thus, the gases maintain their relative proportions, but their partial pressures differ with altitude and depth. Importantly, the potential for gases to diffuse between air and water is closely associated with their partial pressures, rather than their concentration per se.

Finally, we must consider differences in the ways that certain gases behave in air versus water, which have important implications for how gas stores function. In particular, gases differ in solubility so their relative proportions in air are different from the proportions dissolved in water (as predicted by Henry's Law). Gases differ also in their ability to diffuse in air and water. Air contains approximately 25 times as much O_2 per unit volume as does water, and it diffuses approximately 3 million times faster in air than in water. In contrast, CO_2 diffuses faster

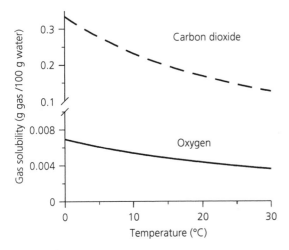

Figure 3.1 Relationship between temperature and gas solubility of oxygen and carbon dioxide in freshwater at sea level, as predicted by Henry's Law.

than O_2 and is more soluble in water (Figure 3.1). Both these important processes of diffusion and solubility decrease with temperature, but increase with pressure. Thus, at a given temperature, O_2 solubility (and hence partial pressure) in water will be higher at sea level than at higher altitudes. Although altitudinal gradients may be associated with variations in the morphology of gas exchange surfaces and O_2 consumption rates of aquatic insects, the relationships and causal mechanisms are complex because temperature and stream gradient (and thus velocity) are often correlated with altitude (Jacobsen 2000; Jacobsen and Brodersen 2008).

3.3 The tracheal system

As mentioned in Section 3.1, in most insects gases move through a branched network of internal, air-filled tubes or tracheae. The branching pattern differs among species in relation to the various routes by which air enters the system and the metabolic demands of different tissues. The finest branches, the tracheoles, contact all internal organs and tissues, and they are especially abundant in tissues with high O_2 requirements (e.g. flight muscles). Air usually enters the tracheal system via spiracular openings positioned laterally on the body. In the plesiomorphic condition, there is one pair of spiracles per thoracic and abdominal segment, giving a total

of 12 pairs. Extant insects, however, never have more than ten pairs of spiracles (one on each of the meso- and metathorax, and eight on the abdomen), and the number of functional spiracles is often reduced further. Entomologists have classified and named the various arrangements of functional and non-functional spiracles along the body (e.g. holopneustic, metapneustic, apneustic, etc.), and these classifications may provide insight into the evolutionary relationships between insect groups. For our purposes, however, the most important distinction is between open tracheal systems, which have at least one pair of functional spiracles, and closed tracheal systems where there are no functional spiracles. Both open and closed systems are common in aquatic insects, and these will be discussed in more detail below.

3.3.1 Tracheae and tracheoles

Generally, a pair of large-diameter longitudinal tracheae (the lateral trunks) runs along the length of the insect just internal to the spiracles (Figure 3.2). Other longitudinal trunks may be associated with various organs, such as the gut, heart, and ventral nerve chord. Various transverse tubes connect these longitudinal trunks. Tracheae are often dilated to form air sacs, especially in adults, which can be important in ventilation. Numerous smaller tracheae branch off the main tracts and repeatedly subdivide until the finest branches, the tracheoles, have a diameter of 1 µm or less. The tracheoles ramify throughout most tissues and, in metabolically active tissues (e.g. muscles) the tracheoles indent individual cells so that gaseous O_2 is brought very close to the energy-producing mitochondria within the cell.

Tracheae are derived from the integument and have both cuticular and epidermal components (see Section 12.3). Like the exoskeleton, the cuticular lining of the wider tracheae (but not the tracheoles) is shed at moulting, and these are often visible as fine thread-like structures on larval or pupal skins. Tracheae close to the spiracles and the tracheal trunks are the thickest; they comprise epicuticle, exocuticle, and endocuticle, and there may be little gas movement across these thick walls. Only epicuticle is present in the tracheoles. The tracheae are given strength yet flexibility in much the same way as a high-pressure hose or vacuum-cleaner hose (Figure 3.3a)—the

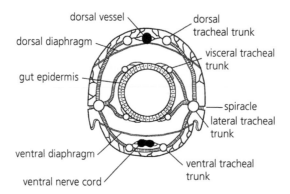

Figure 3.2 Schematic cross-section of a hypothetical insect showing arrangement of the tracheal system. In practice, details of the tracheal system differ between taxa and along different parts of the body (e.g. abdomen versus thorax).
Source: Adapted from Snodgrass (1935).

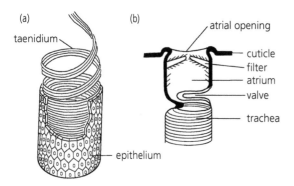

Figure 3.3 Schematic illustrations of (a) a tracheal tube with the taenidium separated to show a continuous, coiled structure, and (b) an atriate spiracle with a closing valve at the inner end of the atrium.
Source: Adapted from Snodgrass (1935).

tracheal cuticle has internal ridges which may be separate rings (annuli) or form a continuous helical fold (taenidium), although the term taenidium often refers to both arrangements. The epicuticle of the tracheae has the same layers as the integument. In the tracheoles, however, the wax layer is absent and the cuticulin layer has fine pores, but fine taenidia are present. These two features are associated with the movement of liquid into and out of the tracheoles in association with gas exchange.

3.3.2 Spiracles

Normally, the opening of a trachea (the tracheal pore) is just below the body surface, not at the surface. A small cavity, the atrium, separates the trachea from

the surface, and the spiracle generally includes the atrium and the tracheal pore (Figure 3.3b). The spiracles of most insects are equipped with valves to prevent water loss into the air. Valves may be opened and closed by muscles, or by the natural elasticity of the surrounding cuticle. Spiracles of adult Odonata, for example, are closed by muscles and opened by the natural elasticity of the region surrounding the spiracle. Often, CO_2 concentration within the tracheal network is the stimulus for closer muscles to relax and the spiracle to open. The atrium may be lined with hairs or have a sieve plate (a cuticular pad with many fine pores) that covers the atrial opening. These structures may serve several functions, including preventing waterlogging of the tracheal system during rain, preventing entry of parasites (especially mites), and reducing evaporative water loss through bulk flow of gases. This last function would be disadvantageous in insects that ventilate the tracheal system and, notably, spiracles that are important for ventilation commonly lack a sieve plate or have a divided sieve plate.

3.3.3 Movement of gases within the tracheal system

Gaseous exchange between tissues and the tracheal system occurs almost exclusively across the tracheole walls, as these are the only parts of the tracheal system that are sufficiently thin for effective diffusion. It is therefore necessary that O_2 pressure in the tracheoles is high enough to meet tissue requirements and, at the same time, that CO_2 produced during metabolism is removed quickly. In small or inactive insects, diffusion of gases in the air between the spiracle and the tracheoles may be sufficiently rapid to meet these requirements. In all insects, however, movements of the body, muscles, and gut also pump the tracheal system and move gases within it. If spiracles are kept open permanently to maximize gas movement in and out of the tracheal system, water loss via the tracheae may be significant. To reduce this loss, many insects use passive (suction) ventilation. Active ventilation movements may be used by large, active insects to shorten the distance over which diffusion must occur, although some diffusion still occurs between the primary (ventilated) tubes and the tracheoles.

In passive or suction ventilation, the spiracular valves are kept nearly closed. Only some of the CO_2 produced during metabolism is returned to the tracheal system in a gaseous state and the rest is stored in the haemolymph and the tissues. In this way a slight vacuum is created in the tracheal system that sucks in more air and effectively prevents the outward flow of water vapour.

During active ventilation, the volume of the tracheal system is alternately increased and decreased by compression and expansion of the larger air tubes. Thus, a proportion of the air in the system is constantly renewed and the diffusion gradient between tracheae and tissues is maximized. These volume changes are usually brought about by muscle contractions, which increase haemolymph pressure and cause the tracheae to flatten or collapse. Both inspiration and expiration are under muscular control in some insects; in others, only expiration is by muscle contractions and inspiration occurs as a result of the natural elasticity of the body wall.

Air sacs in tracheal systems can be expanded and compressed like bellows because their taenidia are reduced, allowing the air sac walls to stretch. This movement increases the diffusion gradient at the tracheoles by increasing the volume of air in the system and the rate at which it passes through the system. Air sacs are often large in insects that are adept fliers. However, simple tidal flow (pumping air in and out of all spiracles) is somewhat ineffective because a considerable volume of air (dead air space) remains within the system at all times. The situation can be improved markedly by directional ventila-

tion in which air is made to flow in one direction. This is achieved by synchronizing the opening and closing of spiracular valves so that air enters via one or more anterior spiracles, is pumped through the body by muscular ventilatory movements, and is expelled through one or more posterior spiracles.

During flight, O_2 consumption increases enormously, almost entirely due to the metabolic activity of the flight muscles. Oxygen use by flight muscles may be 30–50 times the maximum rate for leg muscles of active vertebrates. Various other modifications can increase air exchange in the pterothorax during flight, and these may be pronounced in high-performance fliers, such as dragonflies. The tracheal system is branched in a way that delivers oxygen to the muscles efficiently and the tracheoles become intracellular (they do not actually penetrate the muscle cell membrane, but they lie within invaginations of the muscle cells). The tracheal system can be compartmentalized so that the tracheae of the pterothorax are effectively isolated from the rest of the body by reducing the diameter of the main longitudinal tracheae. In addition, there is autoventilation of the flight muscle tracheae. This results from movements of the nota and pleura of the thorax during wing beating, which produces a considerable flow of air into and out of the thoracic tracheae. In these situations, normal directional flow can be masked by the sheer volume of air movements in and out of the thorax.

The spiracles of many insects are kept closed for long periods (hours or days) and flutter almost imperceptibly or open only periodically, especially during periods of inactivity. This is called the discontinuous gas exchange cycle (DGC) and is seen, for example, in some semi-aquatic waterstriders, the Gerridae (Contreras and Bradley 2011). The open phase is often initiated when carbon dioxide levels within the tracheae become critically high, and a burst of CO_2 is released once the spiracles open. DGC may serve several functions, although debate continues over its adaptive significance and evolutionary origins (Chown et al. 2006; Matthews and White 2011).

3.4 Open tracheal systems

In open tracheal systems, air enters the system via functional spiracles, and such insects must exchange gas with the environment either by visiting the surface periodically, by accessing air-filled spaces in aquatic plants, or through use of a 'gas gill' (a bubble or film of air that covers the spiracles). Therefore, aquatic insects with open tracheal systems vary in their dependence on atmospheric air. Some are completely dependent, some achieve partial independence by using compressible gas gills, whereas others use incompressible gas stores and are completely independent.

Being partially or completely independent of atmospheric air and having an open tracheal system requires that insects hold a gas store about the body. There are primarily three locations where gas may be stored: below the wings, as in some Coleoptera (e.g. Dytiscidae) and Hemiptera (e.g Corixidae); held in place over parts of the body by hydrofuge hairs, as in some other Coleoptera (e.g. Elmidae) and Hemiptera (e.g. Notonectidae); or held in spiracular gills (cuticular extensions of the body wall adjacent to the spiracles), as in some pupal Diptera (e.g. Simuliidae, Blephariceridae). The location of a gas store must coincide with the location of functional spiracles. Gas stores also give the insect buoyancy and serve a hydrostatic function (topics that will be discussed further in Chapters 5 and 8). For example, when returning to the surface to renew the gas store, the insect will rise passively (i.e. swimming to the surface may not be required) and the part of the body containing the bubble will reach the surface first, so air is renewed with a minimum of exposure. Three main factors determine the degree of independence provided by a gas store (e.g. time that an insect is able to remain submerged), including the insect's metabolic rate, the surface area of the gas store in contact with the water, and the concentration of dissolved oxygen in the water.

3.4.1 Complete dependence on atmospheric air

Many aquatic insects that are completely dependent on atmospheric air spend a lot of time at the water surface where atmospheric air is readily accessible. They face two important challenges. First, they must prevent waterlogging of the tracheal system whilst submerged and, second, they must be able to overcome the surface tension force at the air–water interface. Hydrofuge (water-repellent) structures around

the spiracles can solve both problems. Special epidermal glands secrete an oily material at the spiracle entrance of many Diptera. Hydrofuge hairs surround the spiracles of other aquatic insects and, when submerged, the hairs close over the spiracular opening, but spread out again upon contact with the water surface to permit exchange of air. The tracheal system may be modified in other ways, such as a reduction in the number of functional spiracles and restriction of the functional spiracles to certain sites. Functional spiracles are often restricted to the tip of the abdomen and form a post-abdominal respiratory siphon, as occurs in some larval beetles and mosquitoes (Figure 3.4). Such insects occur most often at the water surface, suspended from the water meniscus, with spiracles open to the air. During submergence, the spiracles are closed and a small vacuum is probably created within the tracheal system due to O_2 use (as in suction ventilation), and this aids in rapid air intake when the insect resurfaces.

Instead of rising to the water surface, some species (mainly Diptera and Coleoptera) are able to exploit

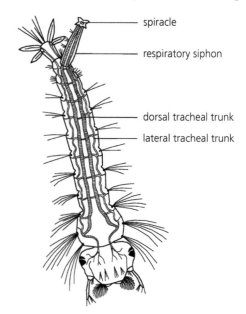

Figure 3.4 Schematic illustration of the tracheal system of a larval mosquito (Culicidae) showing the dorsal tracheal trunks opening through the posterior spiracle in the respiratory siphon. The dorsal tracheae are the primary respiratory passages. The lateral trunks are reduced and lie along the line of the adult spiracles, which are closed in the larvae.

Source: Adapted from Snodgrass (1935).

gas stores in aquatic plants. The roots and stems of many aquatic macrophytes contain a high proportion of aerenchyma, which is full of oxygen-rich air from the atmosphere that the plant transports down from the air above the water to facilitate metabolic activities within the plant roots. The insect inserts body parts with functional spiracles into the aerenchyma or open gas spaces, and effectively uses the macrophyte as a snorkel (Figure 3.5). There are various mechanisms of penetration and attachment, including saw or blade-like structures on the respiratory siphons of various Diptera: larval Culicidae (e.g. *Mansonia, Coquillettidia*), Ephydridae (*Notiphila riparia*), and Limoniidae (*Erioptera*) (Houlihan 1969a). For at least some mosquitoes, carbon dioxide released from plant roots is an attractive cue for larvae to locate potential attachment sites (Sérandour et al. 2006). Among the Coleoptera, some larval weevils and Chrysomelidae (e.g. *Donacia*) have special hooks associated with their spine-like spiracles (Houlihan 1969b; May 1970). In the scirtid beetles, the late instar larva bite into emergent macrophytes and the pupa pushes its anterior end into the hole to access air (Zwick and Zwick 2008). In general, insects that pierce plant roots and stems to access air have

rigid, sharply pointed spiracles that are long enough to penetrate the plant epidermis.

3.4.2 Compressible gas gills

Temporary gas stores (compressible gas gills) are bubbles of air typically held under wings or attached to hydrofuge hairs (Rahn and Paganelli 1968). At the most basic level, these compressible gills provide a supply of air that the insect can use while diving, much like a SCUBA-diver's air tank. Unlike a SCUBA diver, however, a certain amount of O_2 can be extracted from the surrounding water, effectively 'topping up' the O_2 supply without returning to the surface and thereby prolonging the maximum submergence time. Diffusion of O_2 from the water into the bubble can, however, only partially replace O_2 used by the insect because the gill volume decreases, i.e. it is compressible. This strategy of temporary gas stores is widely exploited by aquatic insects, but also by many terrestrial insects that may be vulnerable to soaking by rain or occasional submergence, including adult aquatic insects that oviposit under water (Chapter 11). In these examples, hairs or bristles on the body and wings may be adequate to hold a temporary gas store

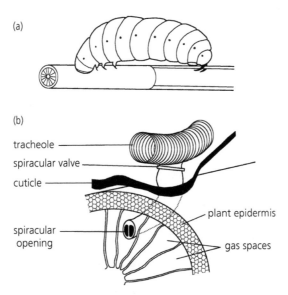

Figure 3.5 (a) Larva of the aquatic beetle *Donacia* (Chrysomelidae) with a pair of root-piercing, posterior spiracles inserted into gas spaces of a plant root (*Typha*). Part of the root epidermis has been removed to show the posterior spiracles inserted into root. (b) Enlargement of one posterior spiracle inserted into root, to show tracheal system of larva and spiracular opening in the gas spaces of the root.

Source: Adapted from Houlihan (1969a). Reproduced with permission from Elsevier.

that allows much longer submergence times than is possible with gases held within the tracheal system alone (Tsubaki et al. 2006).

How do compressible gills work? Because carbon dioxide produced in metabolism dissolves readily in water (Figure 3.1), the gas in the store is principally oxygen and nitrogen. A gas store that has just been replenished with atmospheric air will be approximately 20 per cent O_2 and 80 per cent N_2, and will be at equilibrium with the surrounding water. As the insect uses O_2, the relative proportion and partial pressure of N_2 in the bubble increases. Thus, diffusion gradients are established such that O_2 tends to diffuse from the water to the bubble, and N_2 tends to diffuse from the bubble into the water. This loss of N_2 is ultimately responsible for the gas store shrinking (Figure 3.6). However, O_2 diffuses into the gill about three times as fast as N_2 diffuses out and this prolongs the life of the gill considerably (over eight times longer in some cases), compared with a completely closed gas store. Maintaining that diffusion of O_2 into the gill requires, however, that the insect ventilates the gill, i.e. maintains a flow of water over the surface of the gas bubble (Matthews and Seymour 2010). Eventu-

ally, however, the proportion of O_2 in the gas store will be reduced below a critical threshold (usually before the gas store volume reaches zero), and the insect will return to the water surface. Adult diving beetles with subelytral gas stores (i.e. under the wings) are often seen with the tip of the abdomen at the water surface while the gas store is replenished. In contrast, hydrophilid beetles extend their pubescent antennae above the water to collect air, which is then passed to the gas store held by hairs on the prothorax. The surface area of the gas bubble exposed to the water, and over which diffusion occurs, strongly influences the longevity of the store (Rahn and Paganelli 1968). Subelytral stores have relatively little contact with the water and tend to be shorter lived than gas stores held by hydrofuge hairs (for insects of similar body mass, activity, and metabolism). At any particular concentration of dissolved gases in the water, the lifespan of a compressible gill is determined by temperature and the insect's metabolic rate. Thus, when temperatures are high and insects are active,

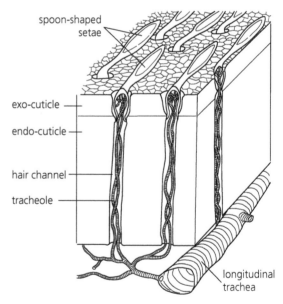

Figure 3.7 Schematic illustration of the spoon-shaped, setal tracheal gills on the elytra of *Dytiscus aubei* (Dytiscidae). The highly branched tracheoles in the basal part of the seta are connected with one of the longitudinal tracheae on the lower side of the elytra, via the hair channels that penetrate the cuticle.

Source: From Kehl and Dettner (2009). Reproduced with permission from John Wiley & Sons.

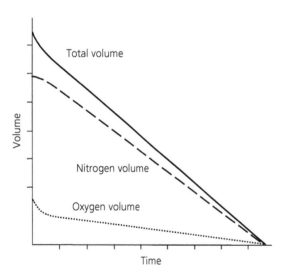

Figure 3.6 General relationship between total and partial (O_2 and N_2) gas volumes in a compressible gill as a function of time, based on a mathematical model of compressible gas gills (Rahn and Paganelli 1968). For an empirical example, see Matthews and Seymour (2010). If O_2 consumption or dive depth increase, then the slopes of the lines become steeper and gill duration becomes shorter.

the gas gill will be replenished more frequently than at colder temperatures when insects are comparatively inactive (Calosi et al. 2007).

Paradoxically some beetles with compressible gas gills can remain submerged for extraordinarily long periods—sometimes days or even weeks—yet they have no capacity for cuticular gas exchange. This paradox was resolve by the recent discovery of tracheated setae mainly on the elytral surface (Figure 3.7), which act as tracheal gills and can provide adequate gas exchange in the absence of a subelytral gas store (Kehl and Dettner 2009). Such tracheated setae may be common in diving beetles, especially those inhabiting running waters (where these gills may work particularly well), the hyporheos, and in subterranean groundwaters (stygofauna) where atmospheric air may be totally absent.

3.4.3 Plastrons and spiracular gills

Total independence of atmospheric air is possible only if insects have a permanent gas store or incompressible gas gill, called a plastron. Unlike compressible gas stores, the volume of a plastron remains constant and it is incompressible. Thus, the insect's oxygen requirements can be fully satisfied by O_2 diffusion from the water into the gas store (Thorpe 1950; Hinton 1976a; Flynn and Bush 2008). In many aquatic insects, the plastron forms around a dense mat of hydrofuge hairs, and the density of hairs is often in the region of 3–4 million per square millimetre (Thorpe 1950; Hinton 1976a). These hairs are typically bent at the tip and slightly thickened at the base, and this maximizes water repellency and resistance to becoming flattened (and the plastron shrinking or being destroyed) by the high pressures that may arise when the insect moves into deep water or depletes its O_2 supply. The density and length of plastron hairs, and their curvature strongly influence plastron efficiency. Because the plastron volume remains constant, N_2 does not diffuse out and O_2 diffuses in as long as there is a concentration gradient. The plastron typically forms a thin layer over a large surface area of the insect (Figure 3.8) and must connect with functional spiracles, which may be modified to expedite gas transport to the spiracular opening. In the water bug *Aphelocheirus* (Aphelochiridae), for example, the abdominal

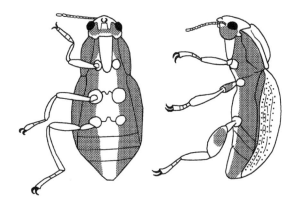

Figure 3.8 Extent of the plastron, indicated by stippling, on the ventral (left), and lateral (right) sides of the beetle, *Elmis maugei* (Elmidae).

Source: Thorpe and Crisp (1949). Reproduced with permission from the *Journal of Experimental Biology.*

spiracles associated with the plastron form a rosette structure with branching arms that radiate outwards and are lined with hairs (Thorpe and Crisp 1947b). Similarly, in the adult riffle beetle *Elmis maugei* (Elmidae) the spiracles open into a lateral groove, well protected with hydrofuge hairs (Thorpe and Crisp 1949). Periodically, elmids may refresh air in the plastron by picking up air bubbles from plants in a way that bubbles adhere to the head. The bubbles are then pushed backwards and onto the plastron surfaces using brushes on the legs, a process often described as grooming behaviour.

A successful plastron requires high rates of O_2 diffusion into it and, therefore, insects with plastrons occur most commonly in well-oxygenated waters, such as high-gradient streams. If oxygen pressure is low, an insect using a plastron will effectively suffocate because plastrons work in reverse, such that O_2 diffuses from the plastron into the water. Viable plastrons must be of sufficient surface area to accommodate the exchange of O_2 and CO_2 necessary to meet the insect's metabolic demands. The surface area to volume ratio is more important than volume per se. The plastron responds poorly to high O_2 demands, so many plastron-bearing insects are relatively sedentary or slow (e.g. adult beetles in the families Elmidae, Curculionidae, and Dryopidae). Insects with plastrons and that are active, such as the water bug *Aphelocheirus*, tend to live in fast-flowing, oxygen-rich water where the flow of water helps maintain a high

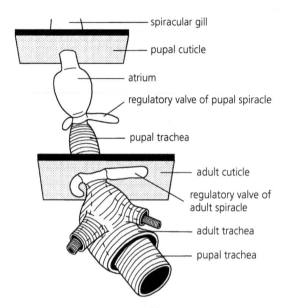

Figure 3.9 Schematic illustration of relationship between the spiracular gill and spiracle of the pupa, and spiracle of the adult developing within the pupa, as seen in a psephenid beetle.

Source: Adapted from Hinton (1966).

chemical gradient and diffusion across the plastron gradient (Thorpe and Crisp 1947a).

Spiracular gills also provide total independence of atmospheric air and typically involve a plastron. Spiracular gills are always pupal structures, but they are also the main gas exchange organ of the adult before it has shed its pupal skin (Figure 3.9) and analogous structures also occur in some insect eggs (Chapter 11). In general, spiracular gills consist of an elongate or branched, air-filled cuticular struc-

ture, which is connected at its base with a spiracle. These gill structures are either modifications of the spiracle or of the body wall adjacent to the spiracle, or both, and they form an air-filled projection away from the body (Figure 3.10). The plastron forms in the air spaces of this structure and O_2 that diffuses from the water into the plastron air, then diffuses through the air channels directly into the spiracles and the tracheal system. Pupae of many aquatic Diptera and Coleoptera respire via spiracular gills. Within the Diptera, the exceptions are chironomid pupae, which have spiracular gills but no plastron, and a continuous layer of haemolymph lies between the water and the spiracular gill. The three-dimensional arrangement of the gill structure varies considerably among species (Hinton 1968b), but generally provide a large surface area for gas diffusion. Many spiracular gills consist of air-filled, vertical struts that are branched at the tips in a plane perpendicular to the strut (Figure 3.10b, c). These horizontal branches form a hydrofuge, an open lattice or honeycomb-like structure that holds the air bubble of the plastron and also provides the water–air interface for gas exchange. The plastron network (Figure 3.10c) is overlain by a thin, laminate outer epicuticle (Hinton 1976b). In gills formed from the spiracle itself, such as some Tipulidae, the atrium often extends up into the gill and connects with the surface via a series of minute pores called aeropyles. Spiracular gills generally function as well in air as they do in water, and they are often found on insects living in the splash zone of streams or environments subject to fluctuating water levels.

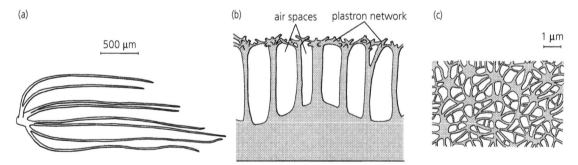

Figure 3.10 Spiracular gills of the blackfly pupa, *Simulium ornatum*, showing (a) external structure of the branched gill, (b) a cross-section through the plastron at the base of one gill branch, and (c) a close up of the plastron network on the gill surface. Note (b) and (c) are drawn to the same scale, (a) is different.

Source: From Hinton (1964). Reproduced with permission from Elsevier.

The hydrofuge hairs associated with the various gas exchange systems described above typically require some sort of behavioural maintenance or grooming by the insect in order to maintain their hydrofuge properties. Agents that lower the surface tension of the water (e.g. soaps) will cause wetting of the plastron and an inability to retain the air space. Hydrofuge hairs can also become contaminated with epizoic microorganisms, potentially resulting in the reduction or loss of the gas store. Secretions from metathoracic glands of some aquatic Hemiptera contain antimicrobial agents that the insects applies to the hairs during 'secretion-grooming' behaviours, thereby maintaining functional gas stores (Kovac and Maschwitz 1989, 1991). Analogous grooming of respiratory structures has also been recorded for a range of adult aquatic beetles (Kovac and Maschwitz 1999).

3.5 Closed tracheal systems

In a closed tracheal system, there is no direct connection between the air-filled tracheae and the external environment. Instead, there is a rich development of fine tracheal branches immediately beneath the thin cuticle (Figure 3.11), which facilitate rapid diffusion of O_2 into the tracheae over the body surface, and diffusion of CO_2 in the reverse direction—called cutaneous gas exchange. Gases also move within the tracheal system (usually by diffusion along the air-filled tubes) to permit gas exchange with tissues that are distant from the body surface. Thus, diffusion occurs across 'barriers' at two different kinds of site: at the body surface where gases diffuse between the external environment (e.g. water) and the tracheal system, and at internal tissues where gases diffuse between tracheoles and body tissues. Insects with closed tracheal systems can often survive out of water for short periods (or very long periods in some terrestrial species), but they typically require moist air to prevent skin surfaces drying out and to facilitate gas exchange.

Body surfaces used for cutaneous gas exchange may be modified to have large surface areas to maximize diffusion rates. These richly tracheated outgrowths of the body wall are known collectively as tracheal gills, and the tracheoles typically run parallel to the gill surface and lie just beneath the cuticle. Tracheation can often attain high structural order

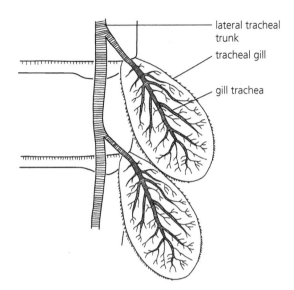

Figure 3.11 Lamellar (plate-like) gills of the middle abdominal segments of a hypothetical mayfly larva, similar to *Baetis*. Oxygen diffuses from the water, across the gill cuticle, into the gill tracheae, and finally into the lateral tracheal trunk, which then transports the oxygen to metabolically active tissues.

such that tracheae are all the same size (diameter) and evenly spaced, which provides functional efficiency with the minimum amount of tracheolar tissues (Wichard and Komnick 1974). Tracheal gills vary in form and occur externally on many parts of the body (usually the abdomen and thorax), and include plate, feather strap-like abdominal gills (in Ephemeroptera, some Zygoptera, Plecoptera, Megaloptera, Neuroptera, Coleoptera), rectal gills (Odonata), finger-like structures, often in tufts (some Plecoptera and Trichoptera), and caudal lamellae (Zygoptera). For species with heavily sclerotized bodies, gills are usually located on intersegmental membranes between plates. Many good examples occur in the stoneflies and include gills that lie between the head and thorax (some Nemouridae), at the base of the legs (some Perlidae), between abdominal segments (some Eustheniidae), and at the tip of the abdomen (some Gripopterygidae). In a few species, external gills are housed within a kind of branchial chamber, such as mayfly larvae in the family Prosopistomatidae where the thoracic terga and wing sheaths are greatly developed to form a kind of carapace over the gills, and give the mayfly the superficial appearance of a beetle.

Species vary in how much gas exchange occurs over special tracheal gills or other body surfaces, and the loss of gills may not prove fatal. Although gill removal in experiments is somewhat artificial, gills may be lost under natural circumstances, such as after close encounters with would-be predators or during moulting (Robinson et al. 1991; Apodaca and Chapman 2004). In some insects, gills are essential for gas exchange and experimental removal of gills results in death, as in the stonefly *Paragnetina media* (Kapoor 1974). Among the mayflies and over a range of O_2 concentrations, gill removal had no impact on O_2 consumption by *Baetis* sp., affected *Chloeon dipterum* only at low O_2 concentration, and affected *Ephemera vulgata* at all concentrations (Wingfield 1939). The functional role of gills in the Ephemeroptera is clearly varied. Caudal lamellae of damselflies of *Lestes disjunctus* are normally responsible for 20–30 per cent of O_2 uptake and this can increase to 60–80 per cent at high temperatures and low dissolved O_2 concentration but, when the caudal lamellae are removed, larvae are still able to respire (at least for a short time) (Eriksen 1986). In order to minimize any metabolic depression that might accompany the loss of caudal lamellae, damselfly larvae may behave in ways that increase cutaneous respiration (Apodaca and Chapman 2004) or simply move to areas of higher O_2 concentration or current speed (Robinson et al. 1991).

3.5.1 Currents and ventilatory movements

For insects with closed tracheal systems, movement of gases within the tracheal system has the potential to increase diffusion rates only a little. Instead, increasing the movement of water over gills and body surfaces used in gas exchange is a more effective and widely exploited strategy for increasing diffusion rates. The simplest way to achieve this is to live in flowing water (i.e. streams and rivers) and, indeed, rates of cutaneous gas exchange tend to be higher for insects living in running than in still waters. Alternatively, insects may move body parts, or the entire body, in such a way that currents are created over gas exchange surfaces. From a physical perspective, there are two primary components that explain why water currents increase diffusion. First, currents simply deliver 'fresh' water, i.e. a continuous supply of water with slightly higher O_2 and

lower CO_2 concentrations than in water the insect was exposed to previously. Second, as the velocity of water over a surface increases, the thickness of the boundary layer decreases (i.e. a layer of slowly flowing water close to the body surface) and O_2 diffuses more quickly across this thinner layer (see Chapter 5 for a detailed discussion of boundary layers).

By moving their bodies, aquatic insects with closed tracheal systems can create currents across gas exchange surfaces. Some species that live in burrows, tubes, cases, or net retreats (e.g. Chironomidae, Trichoptera, Lepidoptera, Megaloptera) often show rhythmic movements of the body or abdomen, which create currents over the body so that water in the tube is periodically renewed. Oxygen concentrations are often especially low within sediments, especially those with a high organic content, and animals living in sediments often construct tubes that can be ventilated to draw in relatively oxygen-rich water from above the sediment surface. Generally, ventilatory behaviour increases as O_2 concentration decreases. These movements require energy, however, and movements will cease once some minimum threshold of dissolved O_2 is exceeded, because the cost of ventilation (energy expended) is greater than the benefit.

Many larval chironomids live in silk-lined tubes constructed within soft sediments and, while these tubes may take on a variety of shapes (e.g. McLachlan and Cantrell 1976), they generally have an entrance and an exit, and undulations of the body drive water through the tube. Low O_2 concentrations in sediments results from respiration of the very small organisms that live in sediments (algae, bacteria, fungi, etc.) and the slow movement of water through sediments. Consequently, the water layer at the sediment surface is typically lower in O_2 than water layers higher in the water column. The tubes of sediment-dwellers many extend above the sediment surface forming chimneys, especially when O_2 concentrations are low (Kon and Hidaka 1983; Stief et al. 2005), and this avoids drawing in the most oxygen-depleted water at the sediment surface. In low O_2 conditions, tubes of *Chironomus anthracinus* may anastomose and allow for a degree of mutualism, whereby some larvae are free to feed while others ventilate the network of interlinked tubes (Jónasson and Kristiansen 1967). In some, but not all, tube-dwelling chironomid species, the frequency of undulations increases with decreasing O_2 concentrations, often at the expense of other activities such as feeding (Heinis and Crommentuijn 1992), until some critical point where movements cease. Undulatory activity is not necessarily continuous, and larvae of *Micropsectra* spent 10–20 minutes irrigating tubes, interspersed with 20–40 minutes resting (Frenzel 1990). Some chironomid pupae are also enclosed in silken tubes and respiratory movements, analogous to those of chironomid larvae, drive water through the tube (Langton 1995). The anal lobes of the pupa are often fringed with setae that make the respiratory movements more effective.

Some caddisflies (both larvae and pupae) with fixed or portable retreats or cases make dorso-ventral undulating movements of the abdomen, which draws water through the anterior case opening and out the posterior opening. Some larvae can also reverse the direction of flow. Multiple lines of evidence indicate that these body movements are indeed important for respiration. When deprived of their cases, O_2 consumption and survival decreases in many caddis species and families (Jaag and Ambühl 1964; Williams et al. 1987). Similarly, a rise in temperature or a decrease in dissolved O_2 is often accompanied by an increase in the rate or amplitude of ventilatory movements (Fox and Sidney 1953; Becker 1987; van der Geest 2007). For stream-dwelling caddis, stream currents can create flow through the tube, provided that they are oriented in an upstream direction. Abdominal ventilations are, nevertheless, common in stream-dwelling caddis, and ventilation rates may increase when velocity of the surrounding stream water decreases (Philipson 1954; Philipson and Moorhouse 1976). Inevitably, there are exceptions and O_2 consumption of some cased caddis is unaffected by case presence or absence (Williams et al. 1987), and others may even abandon their cases at very low O_2 concentrations (Otto 1983). Variations within and among species in their responses to respiratory stress may be explained by the range and magnitude of environmental conditions experienced (in experiments and natural situations), and also the innate characteristics of different species. Among species of Polycentropodidae, for example, the weakest response to experimental manipulations of O_2 occurred in species most likely to experience temperature and O_2 fluctuations in their natural environment (Philipson and Moorhouse 1976).

Aquatic insects that do not live in tubes also may exhibit ventilatory movements involving all or most of the body. Ordinary locomotory activity (swimming, walking) alone will ensure that water moves across gas exchange surfaces and this may be sufficient to maximize diffusion. Indeed, a common response to respiratory stress (e.g. when dissolved O_2 levels become too low) is to move, presumably away from the area of stress and in search of less stressful conditions. In extreme cases, as with the onset of drying in ephemeral habitats, this may involve animals leaving the water entirely and moving across land (Otto 1983). Within water and when walking or standing still, ventilation may occur via leg extension movements, which give the appearance that the animals are doing push-ups (or pull-downs). Such behaviours occur in some Odonata (Eriksen 1986) and Plecoptera (Hynes 1976; Genkai-Kato et al. 2000), particularly when under respiratory stress.

Some insects have internal tracheal gills, as exemplified by the richly tracheated hind gut of many Odonata, and these may be ventilated by pumping water in and out of the hind gut through the anus. In some Anisoptera, a portion of the hind gut is modified to form a branchial chamber (Figure 3.12). There are six longitudinal folds in the branchial chamber and the tracheal gills are borne directly on these folds, or indirectly via a series of cross folds. The shape and attachment of gills varies among species, but the total surface area of the gills is generally large. In a large larva of *Aeshna cyanea* (271 mg), for example, the total branchial surface area is ≈ 12 cm^2 and tracheoles make up 6 per cent of the epithelial volume of the gills (Kohnert et al. 2004). Such statistics alone suggest an important function of this structure in gas exchange. Members of both odonate suborders ventilate the rectum, even though distinct rectal gills are present only in the Anisoptera, and only some Zygoptera have a rectal epithelium that is rich in tracheae. Note, however, that the rectum has functions other than just respiration, and rectal ventilation may serve other functions, such as osmoregulation, locomotion, and ejection of faecal pellets (Corbet 2004). Rectal ventilation is under muscular control and in larval *Aeshna*, and probably most Anisoptera, involves rhythmic movements of opening

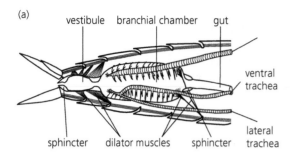

(a)

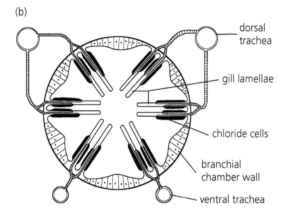

(b)

Figure 3.12 Schematic illustration of the rectal gills and branchial basket of a dragonfly larva with duplex gills (i.e. each row of rectal gills is double): (a) in longitudinal section, and (b) cross-section through the middle of the branchial chamber. The muscular vestibule pumps water in and out of the branchial chamber. Gas exchange occurs through the densely tracheated gill lamellae. Chloride cells are used in osmoregulation (Chapter 4).

Source: (a) from Hughes and Mills (1966). Reproduced with permission from the *Journal of Experimental Biology*.

and closing the anal valve (Mill and Pickard 1972; Pickard and Mill 1972, 1974). The frequency of rectal ventilation increases as dissolved oxygen concentration decreases and as temperature increases. Among damselflies, rectal ventilation may function primarily in osmoregulation but appears to be related to oxygen supply in at least some species (Miller 1994).

Where tracheal gills project beyond the main body, the gills themselves may beat rhythmically to move water across the gill and body surfaces. The frequency at which gills beat depends on body size and the O_2 content of the water. Typically, the frequency of gill beats increases as O_2 concentration decreases to a point where the energetic cost of beating outweighs the benefits of ventilation (Bäumer et al. 2000). Species that characteristically inhabit fast-

flowing water, often have fixed gills (e.g. mayfly nymphs in the genus *Baetis*). However, removal of fixed gills does not necessarily affect O_2 consumption (Wingfield 1939), indicating that most gas exchange is cutaneous. Similarly, removal of gills that typically beat and create currents can also have no effect on O_2 consumption, or effects that are dependent upon O_2 concentration (Wingfield 1939). Thus, although gas exchange does occur over the surfaces of tracheal gills, their primary function may often be to create water movements over other body surfaces where cutaneous respiration occurs. Indeed, the flow patterns created by gills are complex (Figures 3.13, 3.14) and differ among species, with individual gills pivoting and flexing in different ways to create distinctive flow patterns (Eastham 1934, 1936, 1937, 1939). In *Ecydonurus* (Heptageniidae) (Eastham 1937), for example, gill movements create currents that are symmetrical with the body axis, and that flow from the sides and in front of the body, upwards towards the dorsal surface and backwards (Figure 3.13). Although gills within each pair move simultaneously and synchronously, different pairs move sequentially (metachronal rhythm) such that any gill begins its backwards beat before the gill behind it. The time difference, however, between adjacent pairs of gills is so small that gills may appear to beat in phase. In contrast, gill movements of *Caenis* (Caenidae) (Eastham 1934; Notestine 1994) create currents that flow laterally across the body, and the direction of flow can be reversed (Figure 3.14). Pairs of gills move metachronally, but gills within a pair are out of phase with one another.

3.6 Respiration when oxygen is scarce

Oxygen is essential for life in insects and species differ in their ability to tolerate periods when oxygen concentrations may be low. The concentration or partial pressure at which oxygen becomes 'too low' will vary enormously between and within species (e.g. depending on body size, developmental stage, and metabolic activity), and with environmental factors, especially temperature (see also Chapter 4). As we have seen, many species have behavioural strategies to maintain high O_2 diffusion, e.g. gill movement, and here we consider—very briefly—some physiological strategies.

(a)

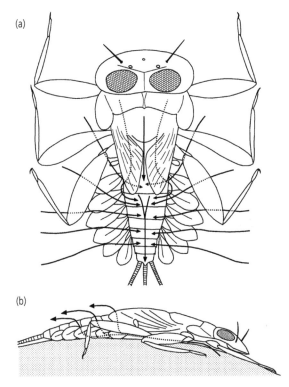

(b)

Figure 3.13 Schematic illustration of gill beating in the larval mayfly *Ecdyonurus venosus* (Heptageniidae) at rest and facing upstream. (a) Dorsal view, (b) side view. In this species, pairs of gills beat together. Arrows show direction of water currents created by beating gills.

Source: From Eastham (1937). Reproduced with permission from the Journal of Experimental Biology.

(a)

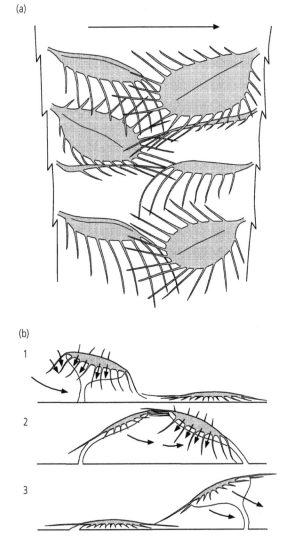

(b)

1

2

3

Figure 3.14 Schematic illustration of gill beating in the mayfly *Caenis horaria* (Caenidae) when producing a current from left to right. Note that gills on the right and left side of the abdomen are out of phase with each other. Arrows indicate the direction of flow. (a) Dorsal view of the four pairs of oscillating gills. (b) Posterior view of a pair of gills in three successive movement phases: 1, 2, and 3. In 1, the left gill is rising; in 2, the right gill is rising and the left gill is falling; in 3, the right gill is falling. In producing left to right flow, each gill turns its lower surface to the left when rising and to the right when falling. The left gill is one-third of a complete oscillation in advance of the right gill. Pivoting movements of left and right gills are nearly identical.

Source: L. E. S. Eastham (1934). Metachronal rhythms and gill movements of the nymph of *Caenis horaria* (Ephemeroptera) in relation to water flow. *Proceedings of the Royal Society of London. Series B, Containing Papers of a Biological Character*, 115 (791), 30–48, by permission of the Royal Society.

Many species are able to regulate or maintain their oxygen consumption (i.e. metabolic respiration) in a way that is independent (to some degree) of oxygen concentrations in the surrounding water and, typically, such species can tolerate lower environmental O_2 levels than species that are unable to regulate. Species are classified as oxygen conformers, regulators, or stressors, depending on how respiration varies with environmental O_2 (Figure 3.15). For all species, there is a critical point or threshold oxygen concentration (in the environment) below which their ability to regulate and maintain sufficient oxygen uptake fails, and respiration is closely correlated with environmental O_2 (e.g. the extreme left-hand parts of the graphs in Figure 3.15). The interesting part of the relationship lies above that low environmental O_2 threshold (i.e. the right hand

---

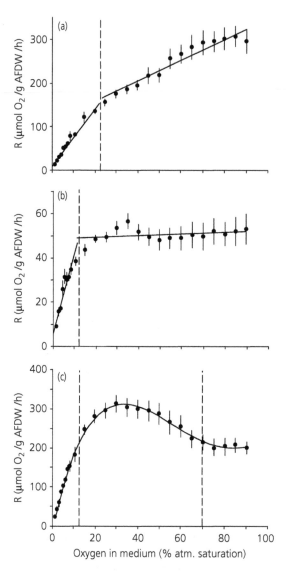

Figure 3.15 Respiration curves (oxygen consumption with respect to dissolved oxygen concentration), for three species of larval Chironomidae at 20 °C. (a) An oxy-conformer, *Cricotopus* sp., (b) an oxy-regulator, *Chironomous hyperboreas*, and (c) an oxy-stressor *Micropsectra* sp. Vertical, dashed lines indicate critical points or thresholds (i.e. breakpoints in the slope of the line). See text for further explanation.

Source: From Brodersen et al. (2008). Reproduced with permission from John Wiley & Sons.

parts of the graphs in Figure 3.15). Above the threshold, respiration rates of oxy-conformers are correlated with environmental O_2, although the slope of the line may be lower than below the threshold (Figure 3.15a). In contrast, oxy-regulators are able to maintain and regulate their respiration over a wide range of environmental conditions above the threshold (Figure 3.15b). Oxy-stressors have a second, higher threshold and can actually increase respiration as environmental O_2 approaches the low threshold, before regulation begins to fail (Figure 3.15c). To make it more interesting, some species may change from oxy-conformers to oxy-regulators as they develop. For example, small *Baetis* appear to be oxy-conformers whereas large *Baetis* are oxy-regulators (Fox et al. 1947). In general, low O_2 thresholds are lower for oxy-regulators than oxy-conformers (Brodersen et al. 2008), oxy-conformers tend to live in environments where environmental O_2 levels are usually high (e.g. high-gradient, highly oxygenated streams), and oxy-regulators tend to occur in places where fluctuations in environmental O_2 may be common (e.g. small ponds).

3.7 Blood-based gas exchange

For the overwhelming majority of aquatic insects, gas exchange occurs cutaneously or via the tracheal system, as discussed in the previous sections. A few, however, can exploit a blood-based gas exchange system, at least in some circumstances. Typically, oxygen enters the body and diffuses into the haemolymph (instead of the tracheal system) where it becomes bound to transport proteins, which are moved around the body via the haemolymph. Perhaps best known for this phenomenon are chironomid larvae, especially those in the genus *Chironomus* (the so-called blood worms), whose bright-red colour results from the presence of the respiratory pigment, haemoglobin. However, several respiratory pigments occur in aquatic insects (haemoglobins and haemocyanins) and occur in several aquatic insect groups, including some perlodid stoneflies, Chironomidae and Notonectidae. The function of these pigments, however, is not always clear and they may be involved in oxygen transport, oxygen storage, or are simply non-functional evolutionary 'leftovers' from an ancestral species in which they did have a function (Hagner-Holler et al. 2004).

Respiratory pigments (blood transport proteins) have an affinity for oxygen, with which they can combine reversibly. At the core of the proteins is a metal, iron in haemoglobin and copper in haemocyanin, and it is these atoms of metal to which the O_2 molecules bind. All insects have copper-based storage molecules called hexamerins in the haemolymph, but these do not bind O_2 and are thought to act as storage proteins, which are used as a source of energy and amino acids during non-feeding periods, among other functions (Telfer and Kunkel 1991; Burmester 1999; Hagner-Holler et al. 2007). Haemocyanin has recently been discovered in the haemolymph of a stonefly (Hagner-Holler et al. 2004). Groups of respiratory pigments are homologous; the haemoglobin of mammals and insects have the same derivation, but the molecular structure of hae-

moglobin is highly variable between species and so too are their O_2 affinities (Weber and Vinogradov 2001). For example, the O_2 affinity of haemoglobin is high in *Chironomus* (Chironomidae), higher than that of *Anisops* (Notonectidae), and also higher than the haemoglobin in humans.

Among the Chironomidae, blood-based O_2 storage and transport is most common in tube-dwelling species that live in sediments subject to very low O_2 concentrations, and haemoglobin allows larvae of *Chironomus* to maintain aerobic metabolism under hypoxic conditions. When larvae ventilate the tubes with water, O_2 is taken up cutaneously and enters the tracheal system (as discussed in Section 3.5), but O_2 also diffuses into the haemolymph where it becomes bound to haemoglobin. Oxygen remains bound as long as the tissues receive adequate O_2 from the tracheal system. If the larva ceases to ventilate, and environmental O_2 pressure decreases, then this bound O_2 will be released and enters tissues directly from the haemolymph. In *Chironomus*, circulation of the haemolymph is structured to facilitate O_2 transport to all tissues. Larvae generally do not ventilate continuously and must spend some time feeding, and haemoglobin would appear to allow the larva to maintain active feeding when O_2 concentrations are low (Walshe 1950). Some *Chironomus* can also excrete lactic acid (a toxic by-product of anaerobic respiration), which further enhances their ability to tolerate anoxic conditions (Walshe 1947).

Haemoglobin also occurs in chironomid larvae that occur in well-oxygenated environments, such as some tube-building *Tanytarsus*, but it may not necessarily serve any useful function. Although it has been demonstrated experimentally that their haemoglobin does function as an O_2 store and bound O_2 is distributed to tissues in hypoxic conditions, these experimental conditions may be more extreme than those experienced by larvae in natural situations. Further, the blood-based O_2 transport in these extreme conditions is inadequate and may just delay death a little rather than facilitate survival (Walshe 1948). Is this a case of ancestral, functional haemoglobin with little or no contemporary purpose? Similarly, the haemocyanin found in some stoneflies can bind O_2, but it is likely to be ancestral and probably does not serve a functional role in modern animals (Hagner-Holler et al. 2004).

CHAPTER 4

Physico-chemical gradients and extremes

4.1 Introduction

The distribution and abundance of species in different environments is often related to abiotic factors and the capabilities of different species to tolerate, and maintain viable populations, in the face of abiotic extremes. For aquatic organisms, the amount of oxygen dissolved in water can be critical, and this was discussed in Section 3.2. Two other, very important abiotic variables influencing the distribution of insects are temperature and water. Obviously, water bodies in which to live are absolutely essential for aquatic insects in at least some life stages. However, retaining water within the body, i.e. preventing excess water loss and desiccation, is a major stress for both terrestrial and aquatic insects. Water is the major component of living organisms, but body fluids are not pure water and typically contain many solutes. For insects living in water, the concentration of solutes in the surrounding medium relative to internal body fluids will determine whether the insect is at risk of desiccation, as in highly saline waters, or of 'drowning' from excess water uptake in very dilute water. There are many other potential stressors, especially chemicals of anthropogenic origin (e.g. insecticides), but we will focus on the major abiotic gradients that occur commonly in the natural environment and leave the rest for toxicology texts. Insects are subject simultaneously to multiple environmental gradients and many environmental factors covary. For example, high temperatures are often associated with low concentrations of dissolved oxygen in water and with high desiccation potential in air. Disentangling such combined effects is difficult, even at small scales (e.g. Ybarrondo 1995; Lancaster et al. 2009). Remember

also that aquatic insects have complex life cycles so environmental factors act on all life stages, limiting factors may differ between stages, and distribution patterns could be determined by any life stage.

What constitutes an extreme environment is difficult to define and is best viewed from the insect's perspective, not a human perspective. Thus, temperatures that might damage human tissues and disrupt metabolic processes, could present few problems for taxa that have suitable adaptations. There are, however, upper and lower limits to the abiotic conditions within which most biological or metabolic processes will work. Few species, and possibly none, will be able to tolerate the entire range, most will be adapted to a subset of conditions, and some will have special adaptations that allow them to live near the limits or thresholds. Ironically, species that can function near the biological limits may be unable to cope with apparently more benign environmental conditions. Insects can respond to abiotic extremes in several ways. One general strategy is to avoid the physiological stress, and this could be achieved by migration or long-distance movement (leave the area), dormancy or diapause (Section 12.7.3), and behavioural avoidance (e.g. small-scale movements). Avoidance is impossible in many situations and there are various physiological and morphological adaptations that allow some taxa to remain *in situ* and cope with the physiological stresses.

In this chapter we will first discuss temperature (Section 4.2), with examination of how insects are affected by temperature variations, with emphasis on the different adaptations that allow some independence of ambient temperatures (i.e. thermoregulation), and that facilitate survival at extremely

Aquatic Entomology. First Edition. Jill Lancaster & Barbara J. Downes.

high or extremely low temperatures. The effects of temperature on growth and development will be discussed in Section 12.7. Stresses from water loss are discussed in the context of water balance (Section 4.3), and the focus is on the mechanisms that allow some aquatic insects to survive, and even thrive, in environments where water loss may be very high. Finally, Section 4.4 considers some really extreme examples of aquatic life stages that can survive almost total dehydration, a form of cryptobiosis. By their very nature, these topics draw strongly on research in the fields of physiology and biochemistry. We will avoid, however, any detailed discussions in those areas and instead provide a general description of the stresses aquatic insects may experience, and the strategies that allow survival. The intention is to provide an overview of why aquatic insects respond to these environmental gradients. Readers interested in the physiology should consult insect physiology texts.

4.2 Temperature

Temperature has an enormous impact on the activity, metabolism, and developmental rates of all insect life stages. The relationship between insect performance (i.e. any metabolic activity, including locomotion, development, reproduction, and ultimately survival) and body temperature, T_b, is described by a right-skewed, humpbacked curve that is bounded by the upper and lower critical thermal limits, CT_{max} and CT_{min}, also called the critical thermal endpoints (Figure 4.1). Being poikilotherms, T_b of insects is closely correlated with ambient temperature, T_a. (Please note, T_a and T_w are often used to refer to air and water temperature respectively; we will use T_a to refer to ambient temperature of either medium.) Therefore, performance is correlated with the temperature of the environment, and the x-axis in Figure 4.1 is often correlated with T_a. The shape of this performance–temperature curve typically varies for different activities and, thus, the threshold that constrains foraging may be quite different from the threshold that results in death. Similarly, thermal limits for any particular response may differ between taxa (Dallas and Rivers-Moore 2012), between populations of the same species (Garten Jr and Gentry 1976; McKie et al. 2004), or the same population at different times of year (Heiman and Knight

1972). Finally, the ability of an insect to survive as temperatures become more extreme will depend on the duration of exposure, the rate of temperature change (acclimation), and whether an insect has previously been exposed to extreme temperatures (hardening).

How insects respond to increasingly high or increasingly low temperatures is somewhat similar. First, they enter a stage of stupor or knockdown, then a prolonged coma, and eventually irreversible trauma and death. In practice, most measures of thermal stress focus on mortality thresholds. Determining sub-lethal thresholds is difficult, but may be more relevant to ecological investigations because fecundity and fitness may be compromised well before the lethal threshold is reached. There are few mechanisms by which insects can survive very high temperatures, but a wider variety of ways in which they can prepare for sub-lethal and potentially lethal temperature lows.

Although T_a and T_b are correlated, the situation is actually more complex because T_b is also a function of several physical factors, including radiation, convection, conductance, metabolism, and evaporation. Radiative heat gain is typically from the sun, but radiative heat loss occurs if $T_b > T_a$. Convective heat gain from, or loss to, the surrounding medium (air or water) depends on movement of the medium (water velocity, wind speed, turbulence) and how or whether the animal moves through the medium. Metabolic processes generate heat, some of which may be lost to the environment, and evaporative

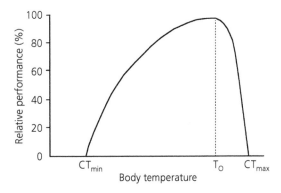

Figure 4.1 General shape of the relationship between body temperature (T_b) and performance of various metabolic activities in ectotherms. T_O is the optimum temperature for performance, CT_{min} and CT_{max} are the lower and upper limits of performance.

cooling also results in heat loss for terrestrial stages. Given this complexity, many studies measure operative environmental temperature (T_e), rather than T_a, because T_e can integrate some of these different factors and provide a more realistic reference temperature. In an analogous way, weather forecasts in cold climates often report wind-chill temperatures that are lower than air temperatures because, at sub-zero temperatures, skin will freeze faster if there is any wind. Thus, wind-chill temperature is a better indicator of how unpleasant it could be outside.

Temperature can vary enormously over different spatial and temporal scales, and many insects have behavioural or physiological adaptations to alter body temperature relative to the environment (thermoregulation, Section 4.2.1), and have adaptations to resist the lethal effects of extreme temperatures (e.g. heat, Section 4.2.2, and cold, Section 4.2.3). Therefore, how temperature variations in the environment 'play out' to influence distribution patterns and behaviours can be complex.

At large scales, species distribution patterns along gradients of latitude and altitude are likely to be underpinned by associated temperature gradients and species-specific thermal tolerances. Many such patterns are well-documented. For example, the Odonata are considered to be primarily tropical insects and most species occur at low latitudes, although some species occur only at high latitudes. In contrast, the Plecoptera are generally considered to be 'cold-climate' species, with comparatively few species in tropical regions compared with higher latitudes. Interestingly, inter-specific variation in lower thermal tolerance limits also varies with latitude, with higher variation at high latitude (Addo-Bediako et al. 2000). The presence or absence of species varies along large-scale gradients, but within species, there may also be variations in life-history traits along environmental gradients, such as voltinism (Chapter 12). Climatic conditions may be suitable year-round in some places, such as in tropical areas, and there may be continuous development and turnover of generations. Seasonal temperature cycles in other areas may preclude growth and development in some seasons. In these cases, the life history of many species is arranged so that insects are in diapause or an immobile, non-feeding life stage (egg, pupa) that is temperature-resistant during the stressful

season. Alternatively, the timing of aquatic and terrestrial stages may be arranged so that the insects are aquatic larvae during winter and terrestrial adults during summer, or vice versa depending on thermal tolerances of the different life stages and local temperature differences between air and water.

The specific heat capacity of water thermally buffers aquatic environments, resulting in more gradual changes in water temperature than air temperature, over space and time. Thus, aquatic life stages may not be exposed to the same temperature extremes or rates of temperature change experienced by terrestrial insects. Consequently, aquatic insects may continue to grow, develop, and carry out normal activities in water bodies at the same time that this is impossible for their terrestrial counterparts. Ice and snow are good insulators, so water temperatures under surface ice can be much higher than air temperatures, and aquatic stages may be abundant and active in apparently extreme winter or polar conditions. Water temperature can also vary with colour of the water and of the substrate, especially in shallow water bodies. For example, the shallow pools of peat bogs have brown, peaty substrates and brown water coloured by humic substances, both of which absorb radiant light and radiant heat. Consequently, water temperature is typically higher in bog pools than non-bog pools in the same general location and, for example, this may provide a thermal advantage to aquatic insects that inhabit bog pools (Sternberg 1994). However, most aquatic insects do spend part of their life in the terrestrial environment as adults, pupae, or eggs, and even immobile life stages may be exposed to lethal air temperatures. Accordingly, aquatic insects living in cold climates must have some strategies to avoid or tolerate extremely low temperatures or to survive temperatures well below freezing (Danks 2007, 2008).

4.2.1 Thermoregulation

Body temperature tends to track T_a or T_e, but many insects can alter T_b, at least for short periods. Such thermoregulation processes may involve behaviours that allow the individual to exploit external sources of heat (ectothermy), or physiological mechanisms that generate heat (endothermy). The obvious advantage of thermoregulation is that it allows the individual to be somewhat independent of ambient temperatures and to maintain activity or performance under otherwise unfavourable conditions. (Note: many studies specifically measure thoracic temperature T_{th} rather than the more general T_b, because muscles for locomotion are in the thorax.) Insect body size is also an important factor in thermoregulation, because heat exchange rates are strongly influenced by body mass and volume, or more specifically, surface area to volume ratio. Heat is dissipated more rapidly from small objects than large ones but, conversely, small objects can warm up more quickly than large ones when using an external heat source. These fairly straightforward physical relationships have been demonstrated in many insects, including various species of adult dragonflies (Sformo and Doak 2006). The important consequence, however, is that the effectiveness of ectothermy relative to endothermy, as a thermoregulatory strategy, increases as body size decreases. This explains why endothermy is rare in small-bodied species.

The most common forms of behavioural thermoregulation (ectothermy) involve moving between microenvironments that differ in temperature and absorbing radiant energy from the sun or from warm substrates (review: Casey 1988). Of course, this requires that individuals have sensory structures that can detect temperature (Section 7.4.3). During ectothermy, radiative heat gain and convective heat loss are the major avenues for heat exchange. Moving between thermal microenvironments is virtually the only mechanism exploited by aquatic life stages. There are even a few examples of insects moving between air and water, such as the terrestrial adults of the stonefly *Zapada cinctipes* that enter the water at night to avoid subzero air temperatures (Tozer 1979). Similarly, some water striders (*Gerris* spp.) that normally live on the water surface, crawl into the water along pieces of vegetation, emergent logs, or floating debris, when water temperatures are higher than air temperatures, a behaviour that appears to speed up gonad maturation and the rate of egg production (Spence et al. 1980). Radiative heat gain can be maximized if the insect moves into a location where incident radiation is high and also

adopts particular postures that maximize the surface area of the body exposed to the heat source, i.e. basking or heliothermy. The reverse process, cooling, may be achieved by changing body posture to minimize heat uptake and/or by moving to shady locations. Thermoregulatory behaviours, and their consequences, are particularly well documented for adult odonates (review: Corbet 2004) and many body postures for heating and cooling are easily recognized (Figure 4.2). At high T_a, postures minimize exposure to the sun and minimize warming, typically by orienting the long axis of the body parallel with the sun's rays (Figure 4.2a, b). Orienting the long axis of the body perpendicular to the sun's rays (Figure 4.2c, d) will have the opposite effect and maximize warming. Postural thermoregulation can be enhanced by the position of the wings, such as by reflecting radiation onto the abdomen, forming a greenhouse or tent-like structure over a warm substrate, sheltering the body from convective cooling, or simply by avoiding shading the body. Basking postures are very effective and dragonflies that spend much of their active period perching can have body temperatures up to 15 °C higher than ambient (e.g. May 1976).

The function of basking in odonates is generally to maintain the thermal values necessary for flight, which is essential for foraging, courtship, mating, mate guarding, territorial defence, etc. The more time an adult can spend flying, or being able to fly, the greater its potential fitness. Basking on emergent stream rocks has been reported for some adult stoneflies, but the function of this behaviour is unclear (adult stoneflies are typically poor fliers). One hypothesis is that higher body temperatures achieved through basking increases the development rate of reproductive organs, thereby reducing the time to oviposition and the risk of pre-oviposition mortality (Harding 2006). As mentioned previously, underwater basking by gerrids can also accelerate reproductive maturation. Although the benefits of thermoregulatory behaviours are likely to outweigh the disadvantages, they may also incur costs if, for example, locations that are ideal for basking may also put the individual at risk from predation.

Physiological thermoregulation (endothermy) relies on heat generated internally by metabolic

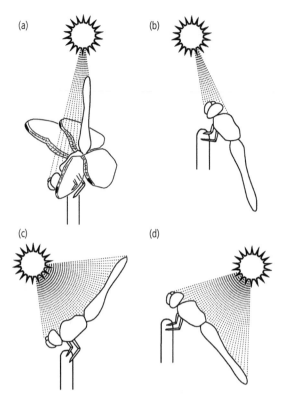

Figure 4.2 Thermoregulatory postures of some perching dragonflies (e.g. Libellulidae). (a) and (b) illustrate postures in which the body intercepts the fewest sun rays and minimizes heat uptake; (c) and (d) show postures in which the body intercepts many sun rays and maximizes heat uptake. Wings have been omitted from (b)–(d) for clarity. Typical wing positions are shown in (a) and illustrate the so-called obelisk posture.

Source: Republished with permission of the Ecological Society of America, from May, M. L. (1976). Thermoregulation and adaptation to temperature in dragonflies (Odonata: Anisoptera). *Ecological Monographs*, 46 (1), 1–32.

processes, typically through muscle movement and especially flight muscles. Endothermy is not feasible for small-bodied insects because metabolic heat is lost as quickly as it is generated. Flight is energetically demanding, but also relatively inefficient: typically, < 10% of flight energy is converted into the mechanical forces for propulsion and > 90% is converted into heat. Thus, movement of the large flight muscles (even in the absence of flight) can be an effective way of generating heat and increasing T_{th} above ambient. For example, T_{th} of the dragonfly *Libellula pulchella* (Libellulidae) during high-exertion flight can be more than 20 °C

higher than T_a, with slightly smaller temperature differences during less energetically demanding flight (Marden et al. 1996).

Most research on endothermy in aquatic insect taxa has focused on flight in adult odonates (review: Corbet 2004). The thoracic segments with flight muscles are well insulated by air sacs in the tracheal system, and this tends to minimize fluctuations in T_{th}. There are risks of over-heating during flight and individuals may require rests or cooling-off periods (behavioural thermoregulation) if T_{th} approaches upper thresholds. Some species have some capacity to transfer heat away from the thorax while flying. Dragonflies in the family Aeshnidae can increase circulation of the haemolymph to increase heat loss through the abdomen, and this allows them to fly for much longer periods than species (e.g. family Libellulidae) that must stop flying to cool off (Heinrich and Casey 1978; May 1979; Marden et al. 1996). A warm thorax and warm flight muscles are often required before an insect can start flying and pre-flight warming may be achieved behaviourally, by basking and/or endothermically through repeated contractions of the flight muscles without actually flying, also called shivering or wing whirring (May 1976, 1979). By shivering, all the energy is generated as heat and there is no movement. Again, warm-up time increases with body size and small-bodied insects may become flight-active more quickly than large ones. Only some species have the capacity to warm up via metabolic heat production, and they are often able to fly before sunrise and at T_a well below the T_{th} required for flight. These diel difference in flight activity are well established for dragonflies that theromoregulate via endothermy versus heliothermy.

4.2.2 Extreme heat

The hottest environments experienced by insects are generally terrestrial (e.g. deserts) and the adult stages of aquatic insects may experience similar extremes. This assumes, of course, that water bodies are present, at least seasonally, to allow aquatic insects to exist in such desert environments. Extremely hot aquatic environments (i.e. water temperatures in excess of 30 °C) are comparatively rare and are generally found in desert environments, where volcanism or other geothermic processes create hot-water springs, or in artificially heated waters below power plants and other sources of industrial heat effluent. In deserts, the effects of heat are often confounded with desiccation (Section 4.3). Insects in geothermal springs are less likely to desiccate if the water is fresh, but the effects of heat may be confounded by unusual water chemistry or the difficulties of gas exchange owing to low dissolved oxygen concentrations in warm water (Section 3.2). The few insects that are able to tolerate extremes of heat are predominantly Diptera (especially Chironomidae, Ephydridae, and Stratiomyidae) and Coleoptera, but also some Odonata and Hemiptera. Many more taxa may not live in the hottest water where geothermal springs emerge from the ground, but in warm water a short distance away. If high temperatures are localized and temporally variable, many insects will burrow, seek out shade or cool locations, and become quiescent until temperatures decreases (i.e. avoidance behaviour).

Physiologically, high temperatures can kill cells by denaturing proteins, altering the structure and function of membranes, enzymes, and biochemical reactions. Thus, above some threshold temperature, heat induces injury. For insects, thermal maxima generally do not exceed 53 °C and are rarely lower than $\approx$ 30 °C, except perhaps in dormant stages. The upper limits of thermotolerance can increase with acclimation or hardening, but our understanding of the biochemical processes that permit high temperature tolerance is incomplete and a topical area of research. For at least some taxa, heat-shock proteins (HSPs) are involved in this resistance, and the primary function of HSPs is to act as molecular chaperones, prevent aggregation of stress-damaged proteins, and minimize harmful interactions. HSPs may be present at all times or production can be stimulated by a range of different stressors, including extremes of heat, cold, dehydration, etc. For example, the larvae of some stratomyid flies that inhabit hot springs have extraordinarily high thermotolerance (surviving temperatures approaching 50 °C). There is no evidence of thermal acclimation in the stratomyid *Stratiomys japonica* (Stratiomyidae) and HSPs are continuously present in high concentrations for these larvae that live in very hot springs (Garbuz et al. 2008). In contrast, some diving beetles, *Nebrioporus* (Dytiscidae), living in equally hot springs do acclimate; their CT_{max} increases with acclimation time and, interestingly, with salinity (Sánchez-Fernández et al. 2010). Although there are clear benefits to HSPs, continuous expression may also incur costs, such as reduced fecundity, growth, and survival.

4.2.3 Extreme cold

The diversity of insects (aquatic and terrestrial) that occur at high latitudes and altitudes is perhaps surprisingly high given the potential stresses imposed by very low temperatures. Diversity is higher in the Arctic than the Antarctic, probably owing to the difficulties of colonizing the sea-surrounded land mass of Antarctica. There are no truly aquatic insects in Antarctica; the two endemic species of Chironomidae have terrestrial larvae (Convey and Block 1996). The physiological stresses of very low temperatures often resemble those of dehydration, because water freezes, changing from a liquid to a solid at low temperatures and becoming unavailable.

Before discussing the stresses and adaptations to extreme cold, it will be useful to review briefly the physical process of freezing. When water freezes, it goes through a phase change from a liquid to a solid. During the initial stages of cooling, individual molecules move more slowly and begin to aggregate. With continued cooling, some of the aggregated molecules are eventually oriented so as to form a rigid lattice or crystal. As soon as this nucleator crystal forms, additional molecules bind to this solid frame and freezing proceeds rapidly. Freezing can also occur without the creation of a nucleator of water molecules, if induced by foreign nucleating agents (INAs, ice nucleating agents) such as dust particles or a rough surface with small projections. The orderly arrangement of water molecules into a crystalline solid takes up more space than molecules in the liquid phase and, hence, water expands as it freezes and ice has a lower density than water. When water molecules slow down and form ice crystals, the kinetic energy of their movement is converted and released as heat (latent heat of fusion), which briefly retards the rate at which temperature drops in the surrounding medium.

Freezing of body tissues is problematic because crystal formation within cells causes irreversible damage by physically disrupting the cell contents and by reducing the water content (dehydration) required for normal cell functions. Many insects avoid freezing by moving into insulated microhabitats where temperatures remain above freezing. Other cold-hardy insects can survive freezing temperatures. Insects that are cold-hardy can be classified into one of two groups: freeze intolerant (or freeze susceptible) and freeze tolerant species. Cryoprotectants, substances that protect against freezing, are used by both groups and may include a range of chemicals such as polyhydric alcohols, glycerol, sorbitol, antifreeze proteins, and the sugar trehalose. Temperature is likely to be an important cue for cold adaptation and to stimulate the production of cryoprotectants, which are likely to be absent during warm seasons. Curiously, some taxa appear able to continue producing cryoprotectants even when frozen (Walters Jr et al. 2009). The over-wintering physiology of insects is dominated by research on terrestrial taxa (Bale 1987; Lee and Denlinger 1991); aquatic taxa are less well studied (Danks 2007, 2008), but the available evidence suggests that they will have similar physiologies.

To avoid freezing over time scales that are too long for thermoregulatory mechanisms (e.g. seasonal cycles), many insects move into over-wintering habitats or microhabitats where the risk of freezing is lower than in other areas. Many water bodies do not freeze completely, such as large lakes and groundwater-fed streams, and even small, ice- or snow-covered water bodies can have significant amounts of free water. Aquatic insects are common and many remain active throughout the winter in these places. Many will over-winter as dormant, terrestrial adults in shelters below the snow (Danks 2008) and, similarly, larvae may burrow into sediments that are unlikely to freeze (Stewart and Ricker 1997). The insulating effects of snow are illustrated well by some damselflies in the genus *Lestes* (Lestidae) that over-winter as eggs laid in the stems of emergent macrophytes, *Scirpus*. The eggs can remain viable after exposure to temperatures as low as −20 °C (Sawchyn and Gillott 1974), but winter temperatures are often much lower within the species' normal distribution range. If eggs are in *Scirpus* stems that are beneath the insulating snow, then viability remains high, whereas egg mortality is higher in exposed, uninsulated stems.

Perhaps most fascinating are the taxa with larvae that can survive frozen in the ice of ponds and streams that freeze into the substrates (e.g. anchor ice), and this includes some Odonata (Daborn 1971; Sawchyn and Gillott 1975), Trichoptera (Olsson 1981), Plecoptera (Walters Jr et al. 2009), Diptera (Danks 1971; Irons III et al. 1993), and Hemiptera (pers. obs.). Paradoxically, although encased in ice, some larvae may not actually freeze if insulating snow cover on top of the ice prevents extremely low temperatures. However, most taxa in ice may be exposed to temperatures well below 0 °C and survival is possible only if they have cryoprotectants (see below).

Freeze intolerant species cannot tolerate freezing and the overall strategy is to reduce the chance that body fluids will freeze, typically by supercooling (cooling below the freezing point without freezing) beyond the lowest environmental temperature they are likely to experience. The supercooling point (SCP, temperature below which freezing, and sometimes death, occurs) for freeze intolerant taxa is often below −20 °C, and often lower than for freeze-tolerant species (Addo-Bediako et al. 2000). How does supercooling work? Individuals typically empty the gut to remove potential ice nucleating agents (INAs), and overwinter as a non-feeding, immobile stage (e.g. pupa). Removing all INAs in an insect's body can reduce their SCP by as much as 10 °C. Before the onset of very low temperatures, individuals may move to dry locations in which to overwinter or build structures that prevent contact with any moisture, thereby reducing the likelihood of nucleation of ice on the body surface, which might subsequently extend internally. Waxy layers on the exocuticle also create a smooth surface that reduces nucleation, but this is more common in adults than larvae, which may have a thin or no waxy layer. The production of cryoprotectants, however, is the most important mechanism to prevent freezing. These substances increase solute concentrations in the body fluids and depress freezing point, and they have antifreeze properties that facilitate supercooling such that body fluids remain in a liquid state below their normal freezing point. Cryoprotectant molecules bind with water, which reduces the chance that water molecules will aggregate to form nucleating crystals and, if an ice nucleus does form, cryoprotectants significantly reduces the rate at which freezing can spread. Survival requires that the insect is able to remain supercooled for significant lengths of time.

Freeze tolerant species, in contrast, have extracellular body fluids (e.g. haemolymph in the body cavity) that can freeze without damaging the insect, and the more damaging intracellular freezing is prevented. The SCP for most freeze tolerant insects is in the range −5 to −10 °C, but once frozen, they may be able to withstand much lower temperatures. For example, larvae of the stonefly *Nemoura arctica* (Nemouridae) may be freeze tolerant and, in laboratory experiments, the majority of individuals survived being frozen in ice at −15 °C for 2.5 weeks (Walters Jr et al. 2009). The small amount of heat released by ice formation in extracellular fluids is sufficient to reduce the rate at which body tissues cool. Thus a large volume of haemolymph will reduce the likelihood of tissues freezing, as seen in some over-wintering pupae. There are two problems with this strategy: the insect must prevent freezing in extracellular fluids extending to adjacent cell surfaces (and hence into the cells), and prevent cell damage through dehydration. As extracellular fluids freeze, the osmotic pressure of the remaining fluids increases and tends to draw water out of the cells (dehydration). Cryoprotectants are used to solve both these problems. Extracellular cryoprotectants reduce the rate at which freezing spreads; intracellular cryoprotectants hold water in the cells to counteract dehydration.

Some insects are winter active and can function normally at subzero temperatures. Winter-active adults, such as some stoneflies in the family Capniidae (e.g. Bouchard Jr et al. 2009), are often brachypterous and dark-coloured, to enhance absorption of solar radiation. Both larvae and adults of the northern hemisphere chironomid *Diamesa mendotae* are active in winter, and SCP temperatures for larvae, pupae, and adults are −7.4 °C, −9.1 °C, and −21.6 °C, respectively (Bouchard Jr et al. 2006a, b). Curiously, larvae and adults of these Chironomidae have different mechanisms to prevent freezing: larvae are freeze tolerant, adults are freeze intolerant and do not feed (i.e. no INAs). Similarly, the terrestrial Antarctic chironomids also have larvae that are freeze tolerant and adults that are freeze tolerant and do not feed (review: Convey and Block 1996).

4.3 Water balance

Most insects have only a limited ability to survive water loss so there is a strong selection gradient for water conservation strategies. Water can be gained or lost via various routes, and the challenge for insects is to maintain some sort of water balance, but with a small net gain overall, as is necessary for growth. How insects minimize water loss centres on the excretory system, and in this respect there are marked differences between terrestrial and aquatic insects, and between insects that inhabit fresh versus saline waters. Maintaining water balance is inextricably linked with maintaining ionic balance of internal fluids, because a loss of water is generally associated with an increase in ion concentrations. Most cell activities can only occur within a narrow range of conditions, so the animal must somehow maintain a relatively constant level of solutes and water in the haemolymph. Desiccation-tolerant insects have special mechanisms to inhibit water loss, but most will die when water content drops below some critical threshold, which varies between species. A very few unusual species are able to survive almost completely dehydration (Section 4.5).

4.3.1 Water loss and gain

For terrestrial life stages, the major routes of water loss are by evaporation from the body surface, from gas exchange surfaces, and in excretion. Evaporation rates will vary with air temperature, humidity, and air movement relative to the insect (wind and flight), but also depend on the structure of the cuticle. Lipids within the epicuticle and the outer waxy layer restrict water loss over the body surface and the effectiveness of this water barrier will vary between species. Water can also be gained across the cuticle, if humidity is sufficiently high, but is unlikely to offset total water loss. Surfaces used for gas exchange are permeable to facilitate movement of gases and this inevitably means that they are also often permeable to water (Section 3.3). Keeping spiracles closed will reduce water loss, but also restrict gas exchange. High metabolic rates, especially when insects are active (e.g. flying), will require that spiracles are open longer and/or more frequently, with associated increases in water loss. Excretion involves production of a fluid urine to get rid of nitrogenous wastes and, although some water reabsorption occurs, there is always some water loss. Most terrestrial insects gain water through their food or by drinking. If food supply is inadequate, water loss through evaporation can dominate. Adult aquatic insects are often short-lived and eat little or nothing, but do drink and will still seek out water or food sources with a high fluid content (e.g. nectar).

Aquatic insects take up water by eating and drinking the fluids in which they live. If the water is fresh, the body fluids (haemolymph) are hypertonic to the surrounding water (i.e. the body fluids have a higher ionic or solute concentration), and water tends to move passively into the body through the cuticle. The degree of uptake will vary between species and with cuticle structure, but, if it is excessive, this excess water will have to be excreted. Some taxa, such as the adult Coleoptera and Hemiptera, have relatively impermeable cuticles, but the majority of aquatic larvae have highly permeable cuticles, or at least some areas that are highly permeable. In contrast, if the water is saline, the haemolymph may be hypotonic to the surrounding water and there is a tendency for water to be lost across body surfaces (as with terrestrial insects), as well as being lost during excretion. This loss must be offset by drinking water and somehow excreting the excess salts dissolved in that water.

4.3.2 Excretion and osmoregulation

We often think of excretion in terms of the expulsion of undigested food, toxins, and the waste products of various metabolic processes through the end of the alimentary canal (discussed in Chapter 14). Excretion is also, however, very important in maintaining water and ionic balance (i.e. osmoregulation). For most insects, the alimentary canal and associated structures are the main organs of excretion and osmoregulation, regardless of whether water and ions enter the haemolymph via the

alimentary canal (i.e. by drinking) or over the body cuticle. Many aquatic insects have tissues outside of the alimentary canal that may also be involved in maintaining water balance. The physiological mechanisms involved in water balance and osmoregulation are well known (e.g. Bradley 1985; O'Donnell 1997) and most entomology texts provide thorough descriptions for terrestrial and aquatic insects. Accordingly, our discussion of this topic will be brief.

Species can be classified as either osmoconformers or osmoregulators (analogous to the distinction between oxy-conformers and oxy-regulators discussed in Section 3.6). The osmotic concentration of the haemolymph of osmoconformers matches that of the surrounding medium and most aquatic insects probably lie in this group. Some osmoconformers can tolerate high salinities and in such situations they often have high concentrations of compatible solutes, such as the sugar trehalose, in the haemolymph and extracellular fluids. These solutes allow osmotic concentrations to rise, but do not disrupt protein and membrane structures in the way that salts do, so physiological functions are not disrupted. In contrast, osmoregulators can regulate the osmotic concentration of their haemolymph so that it remains constant, regardless of the surrounding environment. Species able to tolerate extremely high salinities and a wide salinity range, are typically osmoregulators, and they generally have specially adapted excretory systems.

The main organs of excretion and osmoregulation are the Malpighian tubules, in concert with the rectum and/or ileum of the hindgut (Figure 4.3; see Figure 14.1 for an illustration of the entire alimentary tract). Malpighian tubules are outgrowths of the gut and consist of long, thin tubules made of a single layer of epithelial cells surrounding a blind-ending lumen. They typically float free in the haemocoel and join the gut at the junction of the mid- and hindgut. The Malpighian tubules produce the primary urine, filtered from the haemolymph, which is composed of nitrogenous wastes from metabolic processes, water, sugars, salts, and amino acids. Many of these substances, including water, initially enter the haemolymph through the midgut during digestive processes (Chapter 14), or over cuticle surfaces, as discussed above. The primary

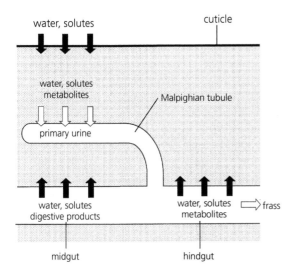

Figure 4.3 Generalized scheme of excretion showing potential movement of water and solutes into the haemolymph from the midgut and from outside the body wall; collection of primary urine in the Malpighian tubules; and reabsorption of water, solutes, and useful substances into the haemolymph from the hindgut.

urine is isosmotic with the haemolymph (i.e. no osmotic gradient between the tubule and the haemolymph), but ionically different. This primary urine then moves into the hindgut, where reabsorption of water and certain metabolites and solutes may occur, often in rectal pads that may be rich in chloride cells; these are epidermal cells on the body surface that absorb or secrete ions from/to the surrounding water, and different ions may be transported by these cells. Water loss through excretion is minimized for most terrestrial life stages, so water reabsorption in the hindgut is relatively high. For freshwater insects, ions, but not water, are reabsorbed in the rectum; for insects inhabiting saline waters, ions are secreted into the rectum.

Morphologically, chloride cells appear as a porous plate in the cuticle and, on the surface, are often difficult to distinguish from chemosensory structures (Chapter 7). Chloride cells may occur as isolated cells or small cell complexes scattered over the body of aquatic insects (e.g. some mayfly and stonefly larvae; larval and adult Hemiptera), or they may be grouped together in discrete regions. For example, caddisfly larvae have groups of chloride cells that form chloride epithelia on the dorsal surface of several abdominal segments, whereas some larval

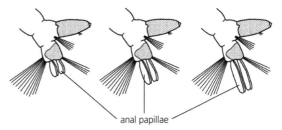

Figure 4.4 Posterior end of mosquito larvae showing anal papillae, which contain the chloride epithelium. Size of the anal papillae decreases along gradients of increasing solute concentration (e.g. water salinity).

mosquitoes and chironomids have chloride epithelia on anal papillae (Figure 4.4). In some species, the overall sizes of these papillae are not fixed, but vary inversely with the ionic concentration of the surrounding water. Chloride epithelia also occur internally in the lining of the hindgut of some aquatic species, such as some Corixidae, Odonata, and Dytiscidae.

Insects that have the necessary adaptations to survive in saline waters are often extraordinarily effective at doing so, and can inhabit marine environments and inland saline lakes, regardless of whether the dominant ions are sodium chloride (seawater and inland lakes with sediments of marine origin) or bicarbonate ions (some inland lakes). A diverse range of insect species are able to tolerate saline environments (Cheng 1976), including some Trichoptera, Hemiptera (Corixidae and Gerridae), Coleoptera (Dytiscidae), and Diptera (Ephydridae, Culicidae, Ceratopogonidae, Stratomyidae, Chironomidae). In some dytiscid beetles, high salinity tolerance is also accompanied by high thermal tolerance (Sánchez-Fernández et al. 2010). Among the Diptera, various chironomids and mosquitoes (e.g. *Aedes detritus*) can inhabit marine salt marshes with salinity levels two to three times higher that sea water. Some brine flies, *Ephydra conerea* (Ephydridae), are to osmoregulate at pH 10 and salinities three times that of seawater. These extremely salt-tolerant species are not necessarily rare oddities found in low densities and, for example, extremely high densities of brine flies occur in many inland saline lakes (Collins 1980; Herbst 1988) where they can be an important food resource for water birds (Caudell and Conover 2006).

4.4 Desiccation resistance and cryptobiosis

The loss of free water in the environment will pose challenges for many aquatic life stages. Some, however, can survive (at least short periods) by being relatively inactive or dormant in moist microclimates, such as underneath leaf litter, logs, and stones; buried in damp sediments; or in the hyporheos (Danks 2000; Boulton and Lake 2008). Dehydration risk is likely to be prolonged during dormancy (diapause, quiescence, aestivation, see Chapter 12), so such species may also be able to reduce rates of water loss and/or tolerate, physiologically, a greater loss of body water relative to species that require free water (Jones 1975; Cranston and Nolte 1996; Suemoto et al. 2004).

A very few species of aquatic insect are able to survive extreme water loss for extended periods, even after they are almost completely dehydrated, in a state called anhydrobiosis, a kind of cryptobiosis. Cryptobiosis is generally defined as the stage when the organism shows no visible signs of life and when its metabolic activity is hardly measurable (Keilin 1959; Clegg 2001; Watanabe 2006). Entering this state is usually a response to temperature extremes, but several stress gradients can be involved simultaneously, such as heat and desiccation. Individuals may be in diapause during stressful periods, but diapause is typically induced by environmental cues that are not stressful, such as photoperiod, and individuals may remain in diapause during favourable environmental conditions (Section 12.7.3). Therefore, cryptobiosis is distinctly different because it is a direct response to stressful factors.

Anhydrobiotic organisms are able to survive almost complete dehydration (95–99 per cent loss of water content) and this is common in a diverse array of generally small-bodied taxa, including plants, fungi, bacteria, worms, rotifers, crustaceans, springtails, and tardigrades (Wright et al. 1992; Clegg 2001; Watanabe 2006). The best known and perhaps most remarkable case of anhydrobiosis in aquatic insects are larvae of the chironomid *Polypedilum vanderplanki* that inhabit temporary rock pools in the deserts of West Africa (Hinton 1951). Note: many chironomid larvae appear to be drought-tolerant, but are not truly anhydrobiotic. In the case of *P. vanderplanki*, if pools dry out before the life cycle is complete, the larvae become almost completely dehydrated and remain in the pool basin where they may be exposed to extreme heat during the day and cold at night. Once re-wetted, the larvae revive rapidly, continue to feed and develop. Laboratory experiments on these dehydrated larvae indicate that, when dry, they can be exposed to temperatures ranging from −270–100 °C and still recover once rehydrated, even if they remain dry for up to 17 years (Hinton 1960a, b, 1968). How do they do it? First, dehydration rates have to be relatively slow (Kikawada et al. 2005; Nakahara et al. 2008); the tube of this tube-dwelling chironomid has high water-retention potential and this slows larval dehydration rates. Second, large amounts of the sugar trehalose accumulate in the body during desiccation (Watanabe et al. 2002). This sugar may play various protective roles during dehydration (Clegg 2001; Sakurai et al. 2008) and similar accumulations of 10–20 per cent body mass occur in other anhydrobionts (Watanabe 2006). Third, heat-shock proteins and some other molecular chaperones also appear to be involved in anhydrobiosis. Functionally, HSPs operate in much the same way as they do during high thermal stress (Section 4.2.2). Rehydration and revival from anhydrobiosis are quite rapid and are accompanied by rapid declines in trehalose concentrations.

CHAPTER 5

The biomechanics of living in and on water

5.1 Introduction

Compared with terrestrial insects, aquatic insects experience a very different physical world. The mechanics of simply moving (or not) can differ markedly between these two environments and can have profound effects on the structure and function of aquatic insects. Because human beings are terrestrial in habit and much larger-bodied than insects, many aspects of a life in water for small-bodied organisms are not intuitive. Even a general understanding may require some effort to grasp the physical principles and the challenges faced by insects moving below, within, and on top of the water column. Movement of water by currents in streams or as waves on lake edges creates problems for insects seeking to maintain their position or move about and feed, without being crushed or swept away. Moving water also transports materials that comprise the substrate, thereby changing the nature of the substrate as well, so that aspects of flow and the nature of the substrate are inextricably linked in their effects on insects.

In this chapter, we first discuss some of the physical properties of water (Section 5.2), and then consider how living in water affects the morphology and behaviour of aquatic insects, and some of the many adaptations that facilitate life in water. This discussion will be organized around the different parts of aquatic ecosystems in which insects live and the particular physical properties of water in those locations. Thus, we will first discuss organisms that live within still (or close to still) water (Section 5.3) and then those that move around on surface-water films (Section 5.4). The flowing water

of streams and rivers presents a further set of physical properties and problems that we will discuss briefly (Section 5.5) before discussing how insects overcome the attendant challenges (Section 5.6). Some of these issues regarding the biomechanics of life in water will be followed up in Chapter 8, which considers locomotion in and on water.

5.2 Some physico-chemical properties of water

Three water properties of profound importance for aquatic organisms are its viscosity, the pressures it exerts, and surface tension. A very brief discussion of these properties will be useful before any discussion of how they influence aquatic insects.

An obvious difference between water and air is that water has a much higher viscosity than air. Viscosity is a measure of the 'stickiness' of a fluid. It is formally quantified with a measure called dynamic viscosity, μ, which measures the degree to which fluids resist shearing forces. When water flows over a surface, it exerts a shearing force on the surface (directed both downstream and downward) that is proportional to the product of viscosity and how quickly velocity changes with distance from the surface. Dynamic viscosity is calculated by multiplying the force applied to the fluid (by gravity) over the distance it travels, divided by the fluid's velocity and the overall surface area of contact with the bottom. Consequently, dynamic viscosity is measured in units of kilograms per metre per second. The dynamic viscosity of fresh water at 20 °C is 1.002×10^{-3} kg m^{-1}s^{-1}, whereas sea water at the same temperature is 1.072×10^{-3} kg m^{-1}s^{-1}, and thus the addition of salts

Aquatic Entomology. First Edition. Jill Lancaster & Barbara J. Downes.
© Jill Lancaster & Barbara J. Downes 2013. Published 2013 by Oxford University Press.

increases dynamic viscosity slightly. Contrast both of these values with that of air at 20 °C, which is 180.8×10^{-3} kg m^{-1}s^{-1}; water is roughly 50 times more viscous than air. An alternative measure of resistance to flow and shearing, kinematic viscosity, is useful for describing the steepness of velocity gradients above a surface and whether turbulence develops in the flow. Kinematic viscosity, v, is the ratio of the dynamic viscosity to the density, ρ, of the fluid. Although most of us are comfortable with the notion that water is much more viscous than air, the density of air is so small that kinematic viscosity of air is actually greater than that of water. This is not at all intuitive to humans, but it is important to aquatic insects.

Before leaving the topic of viscosity, we need to consider how viscosity changes with temperature. As temperature changes, so does the density of fluids. Over temperature ranges that commonly occur in nature, density increases with cooling more quickly in air than in water. Water, however, has the interesting property that density is maximal at 4 °C and then decreases when ice forms—that is why ice floats. Again, the presence of salts changes these values, with seawater achieving maximum density at −3.5 °C, when ice crystals have already begun forming. Obviously, temperature also affects kinematic viscosity, and water is 2.5 times as kinematically viscous at 5 °C as it is at 35 °C. Thus, when insects are affected by the viscosity of water, we need to remember that these aspects change under different temperature and salinity regimes. As will become apparent later in this chapter, viscosity has several important implications for organisms living and moving around in water.

The second interesting property of water that affects organisms is water pressure. Some organisms live in deep water (e.g. at the bottom of deep lakes) where they experience considerable pressure due to the mass of the water above them. Within the range of temperatures at which most aquatic insects occur, water is a liquid and is largely incompressible, whereas air is more compressible. For insects living in the water column and carrying gas bubbles for buoyancy (and gas exchange), it is important to consider what happens when bubbles are subject to water pressure. Boyle's law states that the volume of gases varies inversely with pressure, so a tenfold increase in water pressure will mean a tenfold decrease in the volume of a gas bubble. Increased

pressure also increases solubility, so gases will dissolve into the water (unless the fluid is saturated with air or the air is in a rigid container), and Laplace's law states that the pressure is greater in small than large bubbles, because of the internal pressure due to surface tension. As discussed in Chapter 3, water pressure can influence gas exchange for insects using gas gills. It can also influence the ability of some insects to maintain their position in a water column, in other words, their buoyancy.

Finally, there is the tendency of water to form a surface layer next to air. Attraction between water molecules at the surface is strong, but water molecules are not strongly attracted to air molecules; this creates surface tension, which is apparent as a film over the surface of liquids. Surface tension can be a problem for small organisms, which may not be able to pass through surface films but, conversely, others take advantage of surface films by walking on top of them. Whether an insect is able to walk on surface films is determined by the ratio of the surface tension to the gravitational force acting on the insect (a ratio that Vogel (1994) cheekily suggests be called the 'Jesus number'), meaning that there is a predictable limit to the mass of organisms able to walk on water.

5.3 Living in still (or close to still) water

For insects living in still water, the key physical properties to consider are water pressure, density, and buoyancy. Although viscosity also has a plethora of effects on individuals, these are better presented in the context of the effects of flowing water (Section 5.5).

Insects living in standing water have three choices: they can live on the bottom, in the water column, or on top of the water surface. Insects that live predominantly at the bottom of the water column, even in deep lakes, do not have a direct problem with water pressure, although there are indirect effects in the way water pressure influences gas gills. Although water is heavy, because of its density, and exerts considerable pressure on surfaces at depth, insects are not generally compressible. The density of their tissues, coupled with a chitinous exoskeleton, means they are slightly denser than water (Matthews and Seymour 2008) and hence do not get compressed, even at depth. Water pressure, however, does present problems for insects when they climb up into the water column, either to live there or to reach the surface, as when metamorphosing from aquatic larvae or pupae into terrestrial adults. Their density means they tend to have slightly negative buoyancy, and when motionless they will sink. For insects that do not have a solution to this buoyancy problem, the only way to hold position in the water column is to swim actively, which can be energetically expensive.

One way of achieving neutral buoyancy is to have a density equivalent to water, because it is the density difference that creates problems with buoyancy. Neutral buoyancy could be achieved by incorporating into the insect body a small amount of lipid with a density of approximately 900 kg m^{-3}. Given that lipids have roughly the same compressibility as water, buoyancy would then not depend on depth (Vogel 2006). How lipids can influence buoyancy is not straightforward, however, as discussed for copepods (Campbell and Dower 2003), and, as mentioned in Chapter 3, many insects carry gas bubbles and therefore have positive buoyancy. Gases provide more buoyancy than any liquid, but they come with the problem that, unlike liquids, they are compressible and volume decreases with pressure. Consequently, as an insect descends through the water column, its gas bubble shrinks in volume, providing less buoyancy, and even greater sinking, the reverse of what is required to maintain position. Likewise, as an insect rises, gas expands and buoyancy increases. When gas bubbles are used for respiration, bubble volume decreases as oxygen is used up (Chapter 3), buoyancy decreases accordingly and, again, insects are likely to sink. Large gas bubbles can create too much buoyancy, such that the insects must cling on to some object to maintain position. This difficulty with controlling buoyancy is suggested as one of the main reasons why there are few insects that live predominantly and comfortably in the mid-water column (Matthews and Seymour 2008).

One insect group with impressive abilities to maintain neutral buoyancy in mid-water are the larvae of phantom midges, such as *Chaoborus* and *Corethra* (Chaoboridae). These larvae control buoyancy through tracheal sacs, which are often arranged in two pairs at either end of the body, and they have seemingly perfect control over their position in the water column. The sacs are semi-rigid and contain some gas. The larvae are able to change sac volumes, and therefore buoyancy, in response to changes in pressure or movement of the surrounding liquid (Teraguchi 1975b, a). The mechanism is not by directly secreting gas into the sacs or by moving fluid in and out (as in the ballast tanks of ships). Instead, the larvae appear to have the capacity to expand and contract the walls of the sacs by physical force. When the walls are expanded, gases dissolved in the surrounding fluids pass through the membranes and accumulate within the sacs, thus increasing buoyancy. The movement of gases into the sacs appears to be a secondary effect of wall expansion; the actual mechanism is unclear (Teraguchi 1975b). These midges are ambush predators and are virtually transparent when in the water column, a characteristic that presumably allows them to 'hide' in low-light situations. In many lakes, phantom midges remain far below the water surface during the day and rise to the upper water layers at night. The tracheal sacs of *Chaoborus* have chromatophores (pigment cells) on their dorsal surfaces, and these chromatophores change in both size and location on the sacs depending on background illumination. In light, the chromatophores appear as black dots or strings of beads, but in darkness the chromatophores flatten out and move apart (using amoeboid-like movement), so that the entire surface of the sac is covered in dark pigment. This may also help camouflage the animal during the day when the tracheal sacs might otherwise be reflective and make them apparent to visual predators (Weber and Grossmann 1988).

5.4 Standing on the surface of water

A variety of insects live on the surface film of water in lentic systems (lakes, ponds) and the quieter backwaters of streams and rivers, as well as in marine environments such as estuaries and mangrove swamps. Insects living on the water surface include many families of Hemiptera: the water striders (Gerridae), water measurers (Hydrometridae), some of the velvet water bugs (Hebridae), and some species of Veliidae and Macroveliidae. Perhaps the masters of this life style are gerrids in the genus *Halobates*, which are exclusively marine and include at least one species that spends its entire life cycle out on the open ocean, distant from the land (Andersen and Cheng 2004). All of these insects are superbly adapted to a life that takes advantage of the properties of surface films. The science of surfaces and how they behave when interacting with fluids is complicated and our treatment is thus rather superficial. Interested readers are referred to the erudite (and meaty) review papers (Bush et al. 2008; Hu and Bush 2010) for a detailed introduction to this area and a paper on locomotion by animals living on surface films (Bush and Hu 2006), which may be more accessible to biologists. Our treatment of the topic that follows is based strongly on those works. Essential criteria in determining whether an insect can stand on the water surface are body size (especially weight), and the surface areas and water-repellent properties of the body parts in contact with the water.

As long as animals are not too heavy, they can be supported by the tension in the surface film. The surface film distorts as an animal stands on it, analogous to the way a trampoline is deformed when a child stands on it. The distortion generates curvature forces in an upward direction (Figure 5.1) that support the insect's weight. This basic attribute of surface films explains how most of the water-walking insects are able to stand and move around (Bush and Hu 2006). Interestingly, as well as a maximum size, there is also a minimum size for being able to walk on water. Water-walking arthropods cannot be less than approximately 0.5 mm in length because they are unable to generate sufficient force to manipulate the surface film, i.e. they can stand on water, but moving is much more difficult. The physics of water-walking is sufficiently well understood that it is possible to build mechanical robots able to stand and walk on water (Figure 5.1c) (Hu et al. 2003).

Also pivotal to the capacity of arthropods to walk on water are the hydrofuge or water-repellent properties of their leg surfaces. Without water repellency, insects would penetrate the surface film and sink. They would also have considerable trouble moving, because they would have to extract their legs from the surface tension at every step and this would require forces in the order of 10–100 times their body weight. Consequently, water repellency is critical. The degree to which an object can be wetted

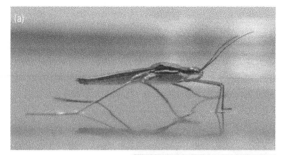
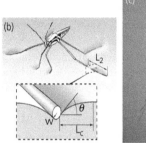
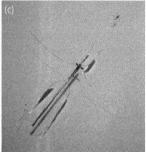

Figure 5.1 Water strider standing on the water surface. (a) An adult gerrid (*Gerris remigis*) at rest, showing that the last segment of each leg is flat to the water's surface. (b) Drawing of a water strider resting on a surface film. Its weight is supported by a combination of buoyancy and curvature forces, whose magnitudes are given by the weights of the fluid volumes displaced inside and outside its line of contact with the water. For a long, thin water strider leg (radius of W), weight is supported principally by the curvature force per unit length and is related to the width of the leg in relation to the capillary length of water (L_c—a characteristic length beyond which force from gravity makes a body too large to be supported by surface tension), which is 2.7 mm, and the angle (θ) of distortion of the water film. (c) An adult gerrid (top right) facing a mechanical copy with hydrophobic legs constructed from stainless steel wire and a body of lightweight aluminium, 'Robostrider'.

Source: From Hu et al. (2003). Reproduced with permission from Nature Publishing Group.

is related to the chemistry and roughness of its surface. Water-repellent surfaces have what is called a high contact angle, which means that water in contact with the surface is forced into rounded droplets, which can be easily shed, rather than spreading to saturate the surface (consider how raindrops look and behave on a waxy leaf or other highly water-repellent surface).

Typically, water-walking arthropods have a waxy coating over the epicuticle and dense coverings of hairs (macrotrichia), which are distributed over much of the body and make the surface rough. Both the waxy coating and the macrotrichia contribute to the hydrophobic or water-repellent nature of the surface, but the hairs are particularly important. There may also be a layer of finer hairs, microtrichia, between the macrotrichia. In water striders (the most well-studied group), the density of macrotrichia varies between the front, mid, and hind legs. This arrangement is probably related to different functions of each leg during locomotion and density is typically highest at the tips of legs where water pressure (created from thrust as the insect moves) is greatest. The density of hairs also varies between species, with the fastest runners having the greatest hair density, but differences also depend on the weights of species. The outermost layer of macrotrichia resists fluid impregnation and traps a layer of air. Microtrichia, if present, may also trap air and may function as a gas gill if they cover a spiracle (Chapter 3). Macrotrichia are also able to shed water droplets, which is a second important component of water repellency. These hairs project at an angle from the body, such that they meet the water surface at a tangent, and they are also curved inward toward the leg at the tips, which reduces their capacity to

penetrate the water's surface (Figure 5.2a, b). Both of these characteristics confer water repellency. Moreover, the macrotrichia have grooves down the length of each hair (Figure 5.2c) and these too increase the hydrophobic nature of the surface and hence water repellency. Such nanogrooves have also been seen on *Microvelia*, suggesting these might be common to water-walking arthropods, not just gerrids. Even more amazing, nanogrooves can either maximize thrust or reduce drag on legs in contact with the surface film depending on the angle at which the legs are deployed. As the insect glides over the water surface, the grooves on the hairs will reduce drag when the hairs are aligned with the direction of movement, but greatly increase contact forces when at an acute angle, thus allowing the insect to generate thrust.

Organisms like gerrids are at very little risk of sinking into the water column and the most serious risk of getting wet comes from raindrops. When bombarded with raindrops, gerrids have been observed to bounce off the surface film to avoid drowning (Gao and Jiang 2004). Some body parts of these insects are, however,

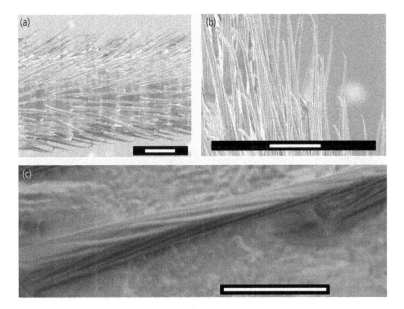

Figure 5.2 Scanning electron micrographs of the hair layer of a water strider leg, *Gerris* sp. (a) Leg thickness is comparable to that of a human hair; it resembles a hairy brush, with hairs angled at 30 ° away from the leg surface. Hairs are typically 30 μm long, 1–3 μm thick at the base, and tapered; their density is 12 000–16 000 hairs mm⁻². White scale bar = 20 μm. (b) A closer view of the hairs showing that the tips are bent inwards, towards the leg. White scale bar = 10 μm. (c) Each hair is patterned with grooves of characteristic width 400 nm that run its length. White scale bar = 5 μm.

Source: From Bush et al. (2008). Reproduced with permission from Elsevier.

designed to get wet, such as the smooth-surfaced claws at the ends of the legs, or ungues. These retractable, hydrophilic claws are able to penetrate the surface film, allowing the insect to grasp the surface of the water and anchor itself in strong winds. Ungues are also important to insects that do not walk or run on the surface film in the same way as gerrids, but use a movement method called meniscus-rising (Section 8.6.3). Another arthropod using a hydrophilic body part and the properties of surface films are the springtails, which have a ventral tube that is wetting (not water repellent), whereas the rest of the animal is non-wetting (water repellent). Springtails use this tube to leap off the water's surface (Section 8.6.1).

5.5 The physics of flowing water

Water that flows presents numerous challenges. Consequently, most stream-dwelling insects live at, or close to, substrate surfaces where they can cling to objects and also avoid some of the strongest and most turbulent flows. The forces of flow operating at small spatial scales and among the topographically rough bottoms of streams and rivers are not at all intuitive, so we present some basic fluid mechanics. By necessity, the treatment is brief and the reader is referred to the excellent writings of Steven Vogel (Vogel 1994, 2006) and alternative texts (e.g. Gordon et al. 2004) for more detail and a plethora of interesting information and examples. Here, we will briefly introduce several important and interrelated concepts, including Reynolds numbers, drag, shear stress, streamlining, laminar and turbulent flows, and boundary layers.

5.5.1 Reynolds numbers, drag, and streamlining

Water that flows over a surface exerts a force against the bottom that is directed both downwards and in the direction of travel (this is called shear stress: Section 5.2). This force creates drag on any object that projects into the flow, but before we can discuss drag, we need to introduce a dimensionless ratio called the Reynolds number (Re). The Reynolds number is the simplest way to convey what happens when a fluid encounters a solid object. One formula for Re is:

$$Re = \frac{\rho l U}{\mu}$$

where ρ is the density of water, l is a 'characteristic length' (roughly speaking, the greatest length, in the direction of flow, of an object immersed in the fluid), U is the velocity of the water, and μ is the dynamic viscosity (Section 5.2). Recall that the ratio of the density of a fluid to its dynamic viscosity is called the kinematic viscosity, and it becomes evident that Re simplifies to:

$$Re = \frac{l U}{v}$$

That is, the velocity of fluid multiplied by the characteristic length of a body immersed in it is compared to the kinematic viscosity of the fluid. This ratio has no units and it is especially interesting because the properties of flow change at different values of Re. In fresh water, values of Re between roughly 10^3 and 10^4 (these values are not fixed, because they depend on a few other variables) are accompanied by a change from laminar flow where streamlines are in parallel—particles in the water will flow more or less in parallel and in smooth paths—to turbulent flow, where particles move in many different directions in an irregular manner (Vogel 1994). This shift in flow type has many effects upon organisms living in it, but the fascinating aspect about Re is that this transition point is not just a property of the fluid and how fast it is travelling, but also reflects the 'characteristic length' of an object. For an organism aligned with the direction of the flow, the 'characteristic length' will typically be the body length. Obviously, this means that the way flows are experienced by an organism is strongly affected by its size. Thus a tiny organism in slow flows experiences life at low Reynolds numbers, in which case it experiences flow as laminar, whereas a larger organism exposed to greater speeds of flow (or travelling at speed) has a high Re and experiences flow as turbulent. The interesting aspect of this is that both these organisms can be living in the same general area and relatively close to one another.

Another perspective is to consider that Re is also the ratio of inertial forces (the forces that tend to keep fluids moving) to viscous forces (these are those that slow fluids down). From this perspective, it is apparent that a twentyfold reduction in the size of an organism is equivalent to a twentyfold increase in viscosity. In other words, small organisms moving slowly experience water in a way that for humans would be like swimming in an extremely viscous fluid, such as honey. This is why flows at small scales and at slow velocities are not at all intuitive or within the typical experience of human beings. Calculations such as these also allow us to contrast the experiences of organisms in water with those that live in air. Thus, for an object of a given size, achieving a particular value for Re requires a velocity 15 times higher in air than in water, because of the differences in viscosity and density. Consequently, in air, speeds of 30 m s^{-1} are considered to be very fast, whereas in water, speeds of 2 m s^{-1} are very fast.

Having established that forces exerted on an organism are related to Re, we can return to the concept of drag. This is still a complex topic because the relation between Re and drag is highly changeable and depends on a suite of variables (shape of the organism and orientation to flow, for example). However, there are two forms of drag that warrant discussion in order to understand how organisms reduce the effects of drag. The first is skin friction, which is caused by the shear stress of water acting against a surface and is related to the area of that surface. The drag is related to the so-called 'non-slip' condition of fluids, whereby the layer of fluid molecules directly next to the surface do not move (even at high velocities), and so the fluid exerts a shearing force on the surface. The second kind of drag is pressure drag and it is directly related to an important aspect called separation of flow. Separation of flow, which occurs at a separation point, is where fluid moving across a surface loses its momentum because of viscosity and moves downstream, typically in a turbulent manner, without following the surface of the body any further. Pressure drag is caused by the dynamic pressure on the front of an object not being counterbalanced by an equal, and opposite, pressure on the rear. It is present at all Re, but tends to be most important at high Re, whereas skin friction is most significant at low Re. Understanding the two forms of drag is pivotal to understanding how organisms reduce drag. At high Re, a round body will experience high pressure drag because the flow separates at the rear, which causes a sharp pressure drop. Alternatively, having a rounded front followed by a long tapering tail (a fusiform shape) causes gradual deceleration and no separation of flows. This is called streamlining and is most effective at high Re where it significantly reduces pressure drag. At low Re, however, a fusiform object has a relatively high surface area and hence experiences greater skin friction. For organisms living at uniformly low Re, streamlining confers no advantage. Another related topic is the roughness of the surface, which again has different effects at different Re. At low Re, the roughness of the surface may have no particular effect, whereas at high Re it can increase the drag on streamlined objects. At intermediate values of Re, roughness can result in turbulence at the surface that delays separation, and this can reduce drag.

5.5.2 Boundary layers and velocity gradients

The final essential phenomena regarding the physics of flowing water are boundary layers, or how flowing water behaves at, and close to, the surface of objects. For aquatic insects that live closely associated with the surface of substrate particles, how water flows close to these surfaces may influence many aspects of their life. As water flows over surfaces, friction induced by the surface reduces current speed close to the surface and we see a steady decrease in velocity towards the bed (Figure 5.3). A boundary layer is the layer of water affected by the surface, i.e. the fluid 'feels' the resistance of the underlying surface. If the water is deep enough, flow patterns (e.g. turbulent eddies) in the water far above the surface (i.e. above the outer layer in Figure 5.3) may be unaffected by the surface, and the free-stream velocity of this uppermost layer is characteristically faster than the movement of water within the boundary layer.

It is important to recognize that boundary layers occur at multiple scales. At an extremely large scale, the Earth itself has a boundary layer: air movements close to the planet's surface are affected by friction with the surface. Only at altitudes in excess of 1000 m above the ground are air movements unaffected by the Earth's surface topography. In an aquatic context, boundary layers are often viewed from two scales. In a stream or river, the movement of water over the channel bed creates a boundary layer in which velocity increases away from the stream bed, sometimes called the 'benthic' boundary layer. In most real streams, this benthic boundary layer extends through the full depth of the flow to the water surface and there is no true free-stream velocity. Research questions about how the physical environment affects populations or communities of aquatic invertebrates over large areas of stream often focus on benthic boundary layers and these relatively large-scale patterns of flow. Within a stream, however, many smaller 'local' boundary layers exist on the surfaces of submerged objects such as individual rocks, logs, and even insects themselves. These local boundary layers create the microenvironments often occupied by individual aquatic insects and to which they may be adapted. Flow patterns at these smaller scales are most pertinent to our discussion of flow adaptations (Section 5.6), so the following discussion will focus on the velocity gradients of local boundary layers on the surfaces of individual objects. We refer readers interested in the benthic boundary layer of streams and rivers to other sources (for a useful review, see Carling 1992).

Within a boundary layer, several other flow layers can be identified. For most of the lower part of the boundary layer, velocity changes logarithmically with height above the surface (the logarithmic layer of Figure 5.3); the change is more gradual further away from the surface (the outer layer), until it reaches free-stream velocity, which is unaffected by the surface. Very close to the surface, in the viscous sublayer (also called the laminar sublayer), velocity gradients may approximate a linear relationship with height above the bed. Viscous forces dominate in this sublayer and flow remains laminar very close to the bed because the no-slip condition means that velocity is very, very low. The thickness of the viscous sublayer depends on the shear velocity and kinematic velocity (Section 5.2) and, thus, as velocity increases the sublayer becomes thinner. In real streams, the viscous sublayer of local boundary layers (e.g. on submerged objects) is often very thin and, importantly, much thinner than the bodies of most insects. Accordingly, for an insect sitting on a stream rock, most or all of its body may be within the local boundary layer, but above the viscous sublayer. Hence, from an insect's perspective, flow

within the viscous sublayer may be relatively unimportant, but the thickness of the sublayer relative to the roughness of the surface is very important because this influences flow in the rest of the boundary layer.

Over very smooth surfaces, such as a glass sheet, flow throughout the local boundary layer may be laminar or near laminar (Section 5.5.1). Over rough surfaces, such as grained or pitted rocks, flow is turbulent through most of the local boundary layer (i.e. the logarithmic and outer layers of Figure 5.3). Turbulent flow can be smooth-turbulent over finely grained surfaces (roughness elements do not protrude above the viscous sublayer), or rough-turbulent over coarser grains (roughness elements protrude above the viscous sublayer). Important things to note are that surface roughness and Reynolds number are important in determining whether flow is laminar, smooth-turbulent, or rough-turbulent, and that hydraulically smooth flow is not necessarily the same as laminar flow. In smooth-turbulent flow, viscosity has a strong influence on flow and turbulent stress is relatively small, but not absent. It is also worth noting that the 'model' velocity gradient of laminar flows depicted in Figure 5.3, and many other texts, may be encountered only rarely in real streams (e.g.

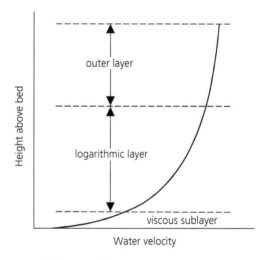

Figure 5.3 Illustration of a boundary layer over a smooth surface. Flow in the logarithmic and outer layers may be laminar or turbulent depending on hydraulic conditions (e.g. Reynolds number and velocity), and the thickness of the viscous sublayer relative to surface roughness elements. Flow in the viscous sublayer is invariably slow and laminar, although this sublayer may be very thin.

Hart et al. 1996), but it is generally true that velocity increases with distance above the surface.

5.6 Adaptations to living in water

The morphology and behaviour of many aquatic insects are clearly related to the constraints of living in water, especially those inhabiting flowing water and living on or close to the substrate surface, and affected by local boundary layers (Section 5.5.2). There are two points to make before considering some examples of adaptations. First, there is often a strong association between the nature of the substrate and the flow patterns above it—as with boundary layers, this surface roughness can be examined at different scales, such as the roughness of individual rocks or collections of rocks over larger areas of stream bed. The important point is that adaptations to flows will often reflect the properties of the substrate as well as the overlying flow, although we will not deal with substrates in much detail. Second, relatively little recent research has focused on adaptations to flow, and the classic text by Hynes (1970) is still one of the best sources of examples. Unfortunately, there is much speculation regarding how (or whether) particular morphologies confer benefits with respect to flow, and such speculation can be quite wrong when examined in the context of fluid dynamics (Statzner 2008), so caution is required.

Although insects are terrestrial in their evolutionary origins, aquatic insects have adapted to life in water in diverse ways. Our discussion will be structured primarily around the concept of drag and the need to reduce drag. As will be clear from the previous sections, any discussion about drag must be predicated on knowing the Reynolds numbers at which organisms live, and this is a function of both organism size and the typical velocities of water in which they live. Consequently, the following sections will consider adaptations to drag at high versus low *Re*, and then examples of how some insects use flowing water to feed.

5.6.1 Dealing with drag at high *Re*

A dorso-ventrally flattened shape is one characteristic common to some species from a range of insect groups living in fast-flowing water (Figure 5.4).

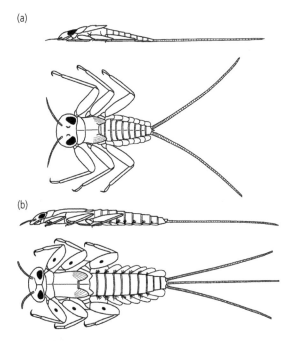

Figure 5.4 Lateral and dorsal views of two, highly dorso-ventrally flattened mayflies. (a) The European mayfly *Rhithrogena* (Heptageniidae), and (b) the North American mayfly *Epeorus* (Heptageniidae). Both species are typically found on the tops of stones exposed to currents.

Source: Adapted from Hynes (1970). Reproduced courtesy of Liverpool University Press.

Some of the earliest work on prospective adaptations to flow was carried out by Steinmann (e.g. Steinmann 1907), who suggested that flatness is an adaptation to reduce drag. This claim has been repeated multiple times with some authors also (incorrectly) calling this flatness streamlining (Dodds and Hisaw 1924; Hora 1930). It is true that being flat allows an individual to live close to the substrate surface and in a low velocity microenvironment, which helps reduce drag. However, flat shapes are also likely to create lift (discussed in Section 9.3), which means stream flows can also whip individuals off surfaces. A flattened body shape alone may not be sufficient to prevent dislodgement. For example, the mayfly *Epeorus* is relatively flat, but it experiences significant lift in fast flows and clings to the substrate with tarsal claws, whereas another flattened mayfly, *Ecdyonurus*, tips its large head shield into the current to generate downward forces that counter lift (Weissenberger et al. 1991). Being flat also increases the surface area of the body in contact with the surface, which can aid attachment. Additionally, flatness is not only an adaptation to fast flows because some very flat species live predominantly on the bottom of stones (e.g. the mayfly *Ecdyonurus*) where being flat may yield other advantages, including relatively free movement through cracks and crevices (Hynes 1970). Thus, flatness is not necessarily an adaptation for reducing drag and this is a salient example of the dangers of jumping to conclusions about adaptations to flow.

Some insects living in fast flows are streamlined and have a characteristic fusiform shape that reduces pressure drag by delaying flow separation (Section 5.4). These species typically face into the current to get maximum benefit of this streamlining. Larval baetid mayflies are among the best examples of streamlined body shapes (Figure 5.5). Baetid mayflies even have legs that are fusiform in cross-section, being rounder at the anterior and thin at the posterior edges, and at least one species (*Baetis bicaudatus*) can withstand velocities of 3 m s⁻¹ (Dodds and Hisaw 1924). Many other aquatic insects have fusiform shapes, including some caddisflies that construct conical cases, pupal simuliids and blephaericerids (Figure 5.5). Water pennies (psephe-

nid beetles) are relatively flattened, but are also streamlined (Smith and Dartnell 1980).

In only a few studies has drag on stream insects been measured in a semi-realistic setting. Statzner and Holm (1989) measured flow velocities close to the body surfaces of stream organisms, including three species of cased caddisflies (Figure 5.6). Of these three species, *Micrasema longulum* (Brachycentridae), with a relatively smooth conical case, was the only one not to experience flow separation at the rear of the case (Figures 5.6e, f). This species experienced less drag than the other two, indicating that it can withstand high flows without being swept away. In contrast, both *Anabolia nervosa* (Limnephilidae) and *Silo nigricornis* (Goeridae) experienced flow separation at the rear of their cases, indicating more significant pressure drag (Figures 5.6a, c), and similar results have been found for other cased caddis (Waringer 1993). *Silo* adds pairs of small stones to the sides of its case with the largest at the front, whereas *Anabolia* builds a case from vegetation and incorporates long sticks into the sides (Figure 5.6). It is tempting to speculate that the stones provide ballast, whereas sticks help the animal orient head forwards into the current (e.g. Hynes 1970), but neither piece of speculation is supported by empirical evidence (Statzner and Holm 1989). The stones added to the case of *Silo* actually increase lift, because they increase overall surface area, and so their presence is unrelated to providing ballast (compare Figures 5.6c, d). Cases of *Anabolia* with sticks do have slightly lower water velocities over the top surface compared to cases from which sticks had been removed (compare Figures 5.6a, b). Theoretically, removing sticks from cases would reduce lift, but because the sticks add to the total surface area, total lift may not be lower, and thus the sticks do not necessarily help *Anabolia* to orient in an upstream direction. These examples further illustrate why speculating about adaptations to flow is problematic, and empirical studies are required.

So why do some caddisflies incorporate relatively large stones or sticks into their cases? There are likely to be multiple answers to this question. For at least some species, details of case construction may be related primarily to reducing predation risk (Otto and Johansson 1995; Otto 2000) rather than offset-

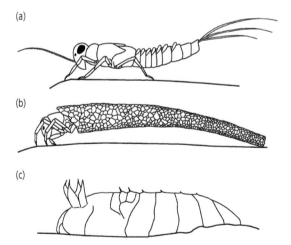

Figure 5.5 Profiles of some insects that live atop stones in fast-flowing streams and that have a fusiform shape: (a) a mayfly larva, *Baetis* (Baetidae), (b) a cased caddisfly, *Neothremma* (Uenoidea), and (c) pupa of a net-winged midge, *Blepharicera* (Blephariceridae).

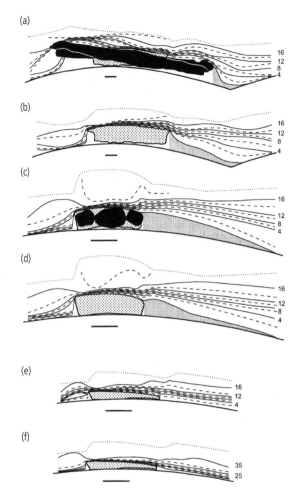

Figure 5.6 Isovels (lines connecting points of equal flow velocity) around cased caddisflies, demonstrating differences in flow separation caused by case shape. Numbers on each isovel are velocities in cm s⁻¹. Flow direction is from left to right. (a) *Anabolia nervosa* with complete case and (b) a case with the side sticks removed; (c) *Silo nigricornis* complete case and (d) a case with lateral gravel removed; (e) *Micrasema longulum* at medium flows, and (f) fast flows. The stippled areas of the water column in (a)–(d) indicate the area of flow separation, which begins at the upper posterior tip of the cases of *Anabolia* and *Silo* and causes significant pressure drag. In contrast, flow does not separate along the smooth, conical case of *Micrasema*. Dotted lines indicate the upper limit of the velocity profile measured. Scale bars below each case are 3 cm.

Source: Springer-Verlag (Statzner, B. and Holm, T.F. (1989). 'Morphological adaptation of shape to flow: microcurrents around lotic macroinvertebrates with known Reynolds numbers at quasi-natural flow conditions'. *Oecologia*, 78, 145–157, Figure 7). With kind permission from Springer Science and Business Media.

ting drag. However, cased caddisfly larvae have often been found in the gut contents of predatory insects and some fish (e.g. J. Lancaster, unpublished observations; Meissner and Muotka 2006), which suggests that cases do not necessarily provide complete protection against predation.

For organisms living at high Reynolds numbers, a variety of structures may help them cling to surfaces. Perhaps the champions of holding on are larvae of the net-winged midges, family Blephariceridae (Figure 5.7), which are typically found on smooth surfaces and in fast flows. These larvae have no true legs, but on the ventral surface of each segment (except the last) they have a sucker, comprising a tube with a soft edge and a fringe of small hooks. In the centre of the tube is a fleshy structure that can be pushed down by muscles and that acts like a piston, driving water out of the cavity through a small anterior notch and creating negative pressure when the piston is then raised. These suckers provide extraordinary clinging power. For example, *Hapalothrix lugubri* occurs in velocities of 2 m s⁻¹ (Frutiger 1998).

Friction pads, peripheral hairs, and gill lamellae are some other structures that may help stream insects to remain attached to surfaces. Friction pads are areas of the ventral surface with backwardly directed hairs or spines. The idea is that these spines create substantial friction with the substrate, increasing the capacity

Figure 5.7 Larvae of Blephariceridae: (a) side and (b) ventral views showing suckers along the centre of the body.

Source: (a) From (Zwick 1977). Reproduced with permission from CSIRO Publishing.

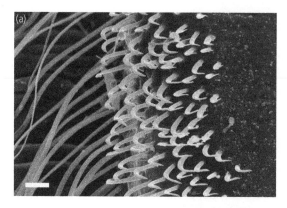
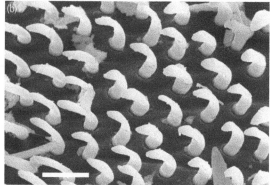
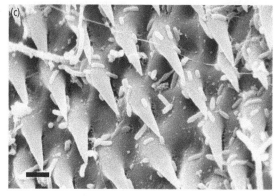

Figure 5.8 Friction pads on a mayfly larva, *Epeorus assimilis* (Heptageniidae). Setose pads on the ventral side of the gill lamellae: (a) Short setae in the centre, with longer setae on the left, which lie on the lateral and distal ends of the pads. Scale bar = 10 μm. (b) Close up of the short setae on the thickened outside rim of the first gill lamellae. Scale bar = 10 μm. (c) Friction pad and the abdominal sternites with areas of spiky acanthae. Scale bar = 2 μm. In each case, the posterior direction is the bottom of the frame.

Source: From Ditsche-Kuru and Koop (2009). Reproduced with permission from Taylor & Francis.

of the animal to stay in place (Figure 5.8). Some insects have fringes of hairs or setae arranged around their periphery, which allow the animal to make close contact with the substratum and reduce their chance of being lifted off (Figure 5.9). Another strategy is to use the gill lamellae to form close contact with the substrate. For several species of mayflies, the gill lamellae were thought to be arranged against the surface in such a way as to act like a sucker, for example, species of *Iron* (Heptageniidae) and *Drunella* (Ephemerellidae) (Dodds and Hisaw 1924). Claims that gills can act like suckers are often repeated, but recent studies have demonstrated these assertions are incorrect (Ditsche-Kuru and Koop 2009; Ditsche-Kuru et al. 2010). The gill lamellae do not provide close surface contact nor is the uninterrupted seal required to create negative pressure. Indeed, mayflies were able to

maintain position even with lamellae removed. Instead of a sucker-like mechanism, there are friction pads on some gills (Figure 5.8) and the anterior gill lamellae are tilted downward, into the current such that water flowing over the gill may help to push the animal down onto the substrate.

Suction, but not suction cups, is used to aid attachment by some larval water pennies, *Scleorocyphon* spp. (Psephenidae) (Smith and Dartnell 1980). At low *Re*, the shape of the larva actually increases the thickness of the local boundary layer immediately downstream of its body, which reduces the potential for flow separation and drag. When the boundary layer is thinner than the larva and drag forces become potentially problematic (e.g. at high *Re*), a small amount of water passes through slots between the lateral laminae and underneath the animal

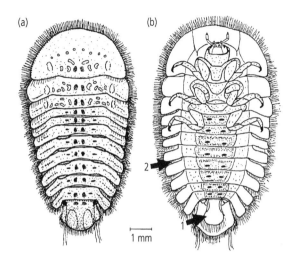

Figure 5.9 Illustration of the (a) dorsal and (b) ventral surfaces of the water penny *Sclerocyphon*. Water passing over the dorsal surface moves through the slots between lateral lamellae (arrow marked 2) and passes under the animal. The last ventral tergite, to which an operculum over the retractable anal gills (arrow marked 1) is attached, can be lifted up to allow movement of water past the rear of the animal, thus regulating the volume of water underneath and aiding respiration and excretion also.

Source: From Smith and Dartnell (1980). Reproduced with permission from Taylor & Francis.

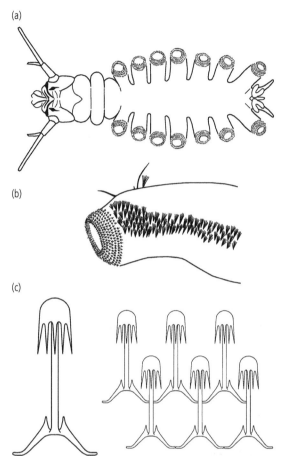

Figure 5.10 Illustrations of a hypothetical larva of the mountain midge *Deuterophlebia* (Deuterophlebiidae). (a) Ventral views showing ventrally directed prolegs on the ends of the abdominal segments, each with multiple rings of attachment hooks. (b) Lateral view of a single proleg with rings of hooks at the tip and bristles along the side. (c) Close-up, frontal view of a single hook from a proleg, showing the comb-like plate with five teeth (left), and (right) a series of hooks showing how they are arranged in rows.

(Figure 5.9). It was thought that this creates suction and delays flow separation along the dorsal body surface, hence reducing pressure drag.

Many insects have tarsal claws or hooks on the ends of the abdominal prolegs with which the insect can clasp projections on rough substrates and hold on. The above study of *Sclerocypon* (Smith and Dartnell 1980) showed that these water pennies can grip a roughened substrate with sufficient force to resist drag forces at typical *Re*. Several types of Diptera have circlets of hooks on their prolegs that aid attachment, but being able to detach and move is also important. For example, the larvae of mountain midges, *Deuterophlebia* (Deuterophlebiidae), have paired lateral lobes on seven abdominal segments, each of which is curved downwards and forms a proleg ending in 8–14 circlets of toothed, hooked spinules (Figure 5.10). The number of circlets varies between species and instars, and with wear and tear. The proleg terminates in a chitinous pad that can be drawn in and out of the proleg tip, along with the hooks (Pulikovsky 1924). Thus, these larvae are able to cling to the substrate, often in very fast flow, yet also move about effectively. They move in a zig-zag fashion, alternately releasing the anterior and posterior ends of the body and rotating the free end through an arc before reattachment (Kennedy 1958).

Finally, some aquatic insects use silk or other sticky secretions to maintain position. Blackfly larvae, for example, spin a pad of sticky silk that adheres to the substrate. Like the mountain midges, *Deuterophlebia*, blackfly larvae also have circlets of hooks on their prolegs, one on the proleg just beneath the head and one on a very short proleg at the end of the abdomen. These hooks can be engaged into the silk allowing the larvae to stand upright and face into the current to filter-feed, and hooks can be removed from the silk when the larvae want to move. Most caddisfly larvae attach their cases to the substrate with silk during moulting and most use silk to anchor pupal cases during metamorphosis, even species that are otherwise free-living. Some caddisflies have their cases almost permanently attached to the substrate with silk, such as some Hydroptilidae and Brachycentridae, for example *Brachycentrus americanus* (Gallepp 1974). Indeed, a variety of insects construct retreats or cases that are attached to the substrate with silk: several genera of Chironomidae and multiple families of caddisflies (Hydropsychidae, Polycentropididae, Psychomyiidae).

One final issue about drag at high *Re* concerns organisms that may occasionally find it advantageous to increase drag. Many insects swim through the water column at relatively high *Re* and for these animals the capacity to move through water relies on creating thrust, which in turn depends on using their limbs and bodies to create drag for that purpose. This will be discussed in Chapter 8.

5.6.2 Flowing and still water: Dealing with drag at low *Re*

Life is a low *Re* experience for many aquatic insects, including those that inhabit very slow flows, such as in backwaters of rivers, or the still water of lakes and ponds. They occupy a very strange world indeed, wholly unlike much within human experience. For them, the effects of viscous forces dominate. Significant drag can be created by objects that are many body lengths away. Although pressure drag is not completely absent, the main source of drag is skin friction, and the most direct way to reduce drag is to reduce surface area. Consequently the bodies of these animals are more likely to have hemispherical shapes than the flattened or fusiform shapes typical of animals living at high *Re*, which have greater relative surface areas and would result in high skin drag at low *Re*. There are few documented examples that illustrate these low *Re* effects on body shape for aquatic insects, but adult dytiscid

beetles (with different species ranging in body length from 2–35 mm) are quite rounded at small body size, and become less rounded and more angular with increases in size, which would be expected if skin friction is less significant for larger beetles (Vogel 1994).

The most interesting implications of low *Re* for aquatic insects are the effects on the structure and function of appendages, such as legs, mouthparts, and gill plates, that must be moved through the water to create movement of the body, or movement of water over the body (e.g. during filter-feeding and gas exchange). Organisms living at low *Re* often have very hairy appendages that may be involved in either creating forward movement, i.e. designed to create thrust, or for feeding, in which case they act more like sieves or filters. Whether a hairy appendage is a paddle or a sieve depends critically on three things: the width of the gap between the hairs compared to the diameter of the hairs, and the Reynolds number. At *Re* of 10^{-3}–10^{-5}, almost no water moves between hairs, and a hairy limb will act like a paddle, regardless of hair spacing. As *Re* increases from 10^{-2}–1, a shift occurs. Water starts to move more between the hairs, but particularly if the ratio of hair gap to diameter reaches 10 or more, at which point the limb acts more like a sieve and can be used to filter particles out of the water column. At *Re* greater than 1, hair spacing has almost no effect unless the hairs are very close together. Other outcomes depend on whether the limbs are deployed near the body surface or not. At *Re* of 10^{-2} and lower, the leakiness of an appendage can be increased by moving it near the body, whereas at *Re* of between 10^{-2} and 1, leakiness is increased by speeding up the appendage. Examples from insects are rare (except perhaps in the context of filter-feeding, Section 5.6.3) so to illustrate the implications we give an example from the crustaceans (Koehl 2000). Calanoid copepods use the second maxillae to capture food. The mechanism of action appears similar for many species, but some operate their maxillae at very low *Re* where the maxillae are paddles that move parcels of water containing food towards the mouth. In other species, the maxillae operate at much higher *Re* where they act like sieves (on which food is trapped) and consequently they are more like filter feeders (see also Section 13.10.1).

(a)

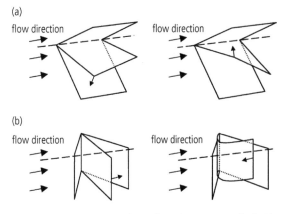

(b)

Figure 5.11 An illustration of the different kinematics involved with (a) flapping and (b) rowing appendages. Left: power stroke; right: recovery stroke.

Source: Adapted from Sensenig et al. (2009).

How an appendage is moved—whether it is rowed or flapped—also depends critically on *Re*. In flapping, the flat surface of the appendage is oriented parallel to the flow and moved vertically, whereas in rowing the appendage is moved back and forth in the direction of flow or movement (Figure 5.11). At relatively high *Re*, viscous forces are trivial and flapping is an effective way to create current and movement, because circulatory lift forces dominate to create forward thrust. At low *Re*, viscous forces dominate and so circulatory lift forces do not occur, making flapping an ineffective means for creating a current. Instead, rowing is more effective at producing thrust at low *Re*. Rowing means using the limb to create drag through the first half of the stroke (hence producing thrust), while minimizing drag to avoid creating backwards thrust in the second half of the stroke, by changing the way a limb is deployed (of course, rowing can also work at high *Re*, as will be discussed in detail in Chapter 8). Between *Re* of approximately 1–20, a transition from rowing to flapping is required and this transition has been demonstrated in the mayfly *Centroptilum triangulifer* (Baetidae). Larvae of *C. triangulifer* have seven pairs of plate-like gills that are moved to facilitate gas exchange (Chapter 3). Small individuals (approximately 1 mm long) use rowing movements of the gills, which create a current that runs along the body. Larger larvae (2 mm long) flap the gills, creating a current that runs dorsally and across the body. The

transition occurs over a very narrow change in *Re* from 3–8, corresponding to the increase in body size with growth (Sensenig et al. 2009, 2010).

What happens as insects grow and move from low *Re*, as neonates, to higher *Re* as more mature larvae? A simple expectation would be for body shape and size to change with growth to reflect the changing effects of *Re* (Koehl 2000). Despite this prediction, many stream invertebrates appear to maintain constant body shape with increasing size (Statzner 1988), with some exceptions, such as some caddisflies that change case construction with larval size. The studies on *Centroptilum* (Sensenig et al. 2009, 2010) illustrate that the issues are complex and require detailed examination, but this topic has attracted little research.

5.6.3 Using flow to feed

No discussion of the mechanics of life in water would be complete without discussing insects that feed by removing particles from flowing water. This diverse group of filter feeders includes those that filter with body parts (e.g. blackfly larvae, some caddisflies and mayflies), and those that construct filtering devices (e.g. some chironomids, hydropsychid caddisflies). This section will consider how these structures operate; the implications for feeding and diet are discussed further in Section 13.9. Filtering systems are well understood in terms of fluid mechanics (Rubenstein and Koehl 1977; Cheer and Koehl 1987). There are three elements to the efficiency of any filtering system: the particles (size, shape, mass), the fluid medium (velocity, turbulence, viscosity, temperature, as discussed in Section 5.2), and the filter itself (fibre thickness, pore space size, orientation, surface properties). Maximum efficiency of most filters occurs in laminar or near-laminar flow, and turbulent flows can remove particles that might otherwise have been captured. Although insects cannot influence the particles that are present in suspension, they may be able to influence which particles are captured and the efficiency of the filtering device by altering their location in the stream (and hence velocity and turbulence), or by altering the filter itself.

Perhaps the most well-studied filter-feeding group are the blackflies, which filter particles using a pair of cephalic fans. The fans are constructed from several primary rays armed with microtrichia, which the blackfly holds up into the current (Figure 13.11). A detailed study of *Simulium vittatum* (Chance and Craig 1986) shows how complex the flows are around individual larvae during filter-feeding. The body projects at an angle from the sub-strate (rather than being upright) so that much of the body lies in the local boundary layer (Section 5.5.2). The larva twists its body so that one fan is projected near the top of the boundary layer and one is closer to the substrate surface. The fan at the top of the boundary layer filters particles directly out of the free-stream current, but the fan at the bottom filters particles driven up from the substrate by a vortex that is created by the shape and position of the larva, and that forms immediately downstream of it. The fans thus operate at lower Re than suggested by the location of blackflies, which are invariably found in fast flows. Nevertheless black-flies cannot operate in stream flows that are too slow because flow around the fans themselves become too viscous, and drag becomes a significant problem if flows are too fast (Braimah 1987b, a). Particles are captured with a mix of different mechanisms but some of these are quite unclear. For example, Koehl (1995) suggested that *S. vittatum* fans are 'leakier' than theory would predict given

the Re at which they operate. Alternatively, Braimah (1987c) found that flow passes between the rays of the fans of *S. bivittatum* but not the microtrichia, which act 'like a wall'. Nevertheless, blackflies can filter particles as small as 0.091 μm, i.e. colloidal size (Wotton 1976), which is smaller than the spaces between microtrichia, and Braimah (1987a) suggested that adhesion of particles via mucosub-stances might play a role, but the evidence for this is poor (Section 13.9).

Another well-studied group of filter-feeders are those that construct silk nets to intercept particles from the water column, and particularly the hydropsychid caddisflies. Flow velocity can influence the mesh size with larger mesh sizes for faster flows, which affects the numbers and sizes of particles caught (Loudon and Alstad 1990, 1992; Tachet et al. 1992). Again Re can be an issue. Although hydropsychid caddisflies that feed in this way are found in fast flowing locations, they may experience low Re (e.g. 10) near the substrate surface where they construct their nets. At low Re, water can slide around the net rather than pass through it. Because water becomes less viscous with increasing temperature, nets become 'leakier' with a rise in temperature (Loudon and Alstad 1990), which may have implications for distribution patterns along river networks.

PART 3

Sensory Systems, Movement, and Dispersal

The chapters of Part 2 considered how physical and chemical gradients impinge directly on aquatic insects, affecting aspects such as survivorship and growth. However, that information can only explain the distribution of insects to a limited degree, because animals make behavioural choices in response to cues and can move away from unsuitable locations. In Part 3, we consider how insects sense their environment and discuss their capacity to move within and between habitats in response to sensory cues. Questions about such behaviour underpin much ecological research. Insects, like other animals, make choices that reduce their likelihood of being killed by physical or chemical extremes, or by predators, while ideally improving their chances of locating the right kind of food, a mate, a place to lay eggs, a place to pupate, and so forth. Such information is fundamental to many basic, ecological investigations into predator–prey interactions, interspecific competition, mating success, and many others. Understanding how insects respond to cues in their environment is also a pivotal aspect of many applied research questions. For example, understanding the attraction of mosquitoes to different types of water bodies or chemicals in the air (such as CO_2) can mean knowing how to reduce risks of human infection with *Plasmodium*, the parasite that causes malaria. Another aspect of Part 3 is that movement includes dispersal that results in insects moving from one population to another (i.e. migration). As a result, our presentation of how, where, and when insects move begins also to address the information needed

for studies of population dynamics (discussed further in Part 4).

Understanding how insects can sense their environment begins with knowing what sensory organs they possess, how these sensory organs are deployed, and the major differences between species. The major sense organs revolve around photoreception (light), mechanoreception (touch and sound or vibrations) and chemoreception (taste, smell), but each of these types encompasses an enormous range of structures and consequent abilities. The environment can also influence how effectively the different sensory systems operate. At its most basic, sensory information may trigger insects simply to move in a particular direction (toward or away from a cue), but they can also enable complex responses that accomplish activities such as food acquisition, mating, or predator avoidance. In dragonflies, for example, which possess two kinds of photosensors (ocelli and compound eyes), photoreception is involved in aiding level flight, locating mates of the right species (often signalled through colour and movement), and in detecting and chasing down prey. Sensory organs are also often involved in communication where signals must be sent as well as received. Thus, chemoreception may involve the detection of specific chemicals released by prospective mates for attracting the opposite sex (pheromones), and mechanoreception the detection of specific vibrations made through the substrate for the same purpose.

Moving through water presents some interesting challenges to insects, especially when water is flowing. Life for relatively small organisms in a relatively dense medium like water is not intuitively simple to understand and, consequently, their abilities to move in particular directions at will is also not intuitive. A basic grasp of the physical properties of water and fluid dynamics is needed to understand why the majority of insects have difficulty maintaining position in a water column, why some are flat and others spherical when they live in the same habitats, why some are competent swimmers and others poor, and how some insects can run on the surface film of water without any danger of getting wet. We review also some less well-known means by which aquatic insects move around, in, and on water: jet propulsion, undulations, meniscus-rising, and Marangoni propulsion, which all illustrate interesting solutions to moving in and on a fairly dense medium. Of course, many aquatic insects have terrestrial adults, and of these most are winged and disperse by flying. The basic principles of how wings generate lift and thrust are needed to grasp why some species,

such as dragonflies, are superb flyers whereas others are relatively poor. Additionally, the aquatic insects contain some unique solutions not seen among their terrestrial counterparts. Some adult insects use their wings to sail across the water using the wind. Others skim, where they flap their wings but maintain contact with the water surface with two, four, or all six legs. All of this variation results in vastly different, species-specific abilities to move around the environment and locate food or mates successfully. It also creates differences that allow some species, many of them dragonflies, to migrate over hundreds to thousands of kilometres, sometimes over open ocean.

To understand then why aquatic insects are found in particular environments and not others means appreciating how insects sense their environment, the cues to which they respond, and whether and how they are able to act upon those cues through movement—the topics addressed in the next four chapters.

CHAPTER 6

Sensory systems—photoreception

6.1 Introduction

Most insects have some ability to detect light (photoreception) and some taxa have sophisticated visual systems. There are three different kinds of receptor organs used in photoreception, with one or two in operation on any individual insect. The most conspicuous visual organs of insects is a pair of compound eyes with multiple facets, which are found on virtually all adults and the larvae of hemimetabolous taxa (Figure 6.1). Insects with compound eyes may also have light-sensitive ocelli (singular ocellus), which vary in number from zero to three (Figure 6.1). In contrast, the larvae of holometabolous insects lack compound eyes and ocelli, but often have up to 12 photosensitive stemmata (singular stemma). Elaborate eyes are required to form images (i.e. pattern detection) and to detect movement. The camera eyes of humans and most vertebrates are composed of a single optical unit with many nerve cells, and they are ideal for image formation. The compound eyes of insects, with their many optical units, are less well-suited for image formation, but are capable of pattern detection. They are, however, extraordinarily good at detecting motion and can be superior to the eyes of most other animals in this respect. Ocelli probably detect the presence or absence of light, and its quality or

intensity, but they cannot form images. The stemmata may have the potential to focus an image, but the stemmata are few and relatively small, so the images are likely to be small and poorly resolved (i.e. out of focus). A few exceptional insects have no external photoreceptors and are unable to detect light, including some subterranean or groundwater taxa (stygofauna) and some endoparasites. Some apparently 'blind' species may be able to detect light through the body surface, even though there are no obvious visual structures or optical systems with focusing mechanisms. Photoreception, however, is only one of three primary sensory systems; mechano- and chemo-reception (Chapter 7) may provide sufficient sensory information. One blind stonefly larva, for example, has a profusion of mechanoreceptors on the head where the eyes would normally occur, and these may provide adequate sensory information (Rebora et al. 2010).

At the most basic level, vision requires that light is focused onto cells containing light-sensitive pigments (photoreceptors). Light energy is converted into electrical energy, which is then relayed via the optic nerve(s) to a complex central nervous system capable of processing the information (e.g. optic lobe of the brain). The general plan for light-sensing organs in most insects consists of an optical, light-gathering structure (a dioptric apparatus or lens) above a sensory structure that contains photosensitive pigments that transduce light to electrical energy. The compound eyes, ocelli, and stemmata of insects respond to light wavelengths from about 350 nm (UV) into the red range at about 700 nm. For comparison, the typical human eye responds to wavelengths of 390–750 nm and, in plants, light in the 600–700 nm range promotes photosynthesis, via chlorophyll pigments. All invertebrate and

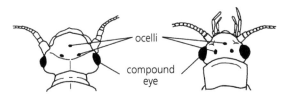

Figure 6.1 Head of a stonefly larva (left) and adult (right), showing location of compound eyes and ocelli.

Aquatic Entomology. First Edition. Jill Lancaster & Barbara J. Downes.
© Jill Lancaster & Barbara J. Downes 2013. Published 2013 by Oxford University Press.

vertebrate photoreceptors use the same basic visual pigment, rhodopsin, although typically there are several other photopigments present. Even though the morphological structures of visual systems evolved independently and many times in animal lineages, some form of rhodopsin was probably present in very ancient animals and has been conserved through evolutionary time, as has the basic structure of chlorophyll in plants.

There is an enormous body of literature on vision in insects and it is beyond the scope of this book to review it all. Instead, this chapter provides a relatively brief overview of the structure and function of insect photoreceptors, highlighting some problems (and solutions) peculiar to aquatic insect taxa and providing examples of research on aquatic insects (where they exist). It is notable, however, that there is relatively little written on the visual capabilities of insects that live underwater. This is unfortunate because the properties of light in water (Section 6.2), may limit the potential for even the most sophisticated insect eyes to operate underwater. Compound eyes have probably received the most attention, so we start with a discussion of the basic structure and the different types of compound eye, and the basic optical units, the ommatidia, which have much in common with the ocelli and stemmata (Section 6.3). How compound eyes function, including image formation, resolution, sensitivity to light levels and wavelength, will be discussed in Section 6.4, with specific reference to aquatic insects. The following sections will consider the other eye types, ocelli (Section 6.5) and stemmata (Section 6.6), which are much less complex than compound eyes, but still serve important kinds of photosensory information to aquatic insects. Finally, Section 6.7 provides a brief discussion of bioluminescence (the ability of insects to produce light), which a few aquatic insects use in communication.

6.2 Properties of light in water

Most limnologists will be very familiar with the idea that sunlight is a critical factor in structuring aquatic ecosystems. Light is essential for photosynthesis and primary production (e.g. growth of algae, macrophytes, etc.), but light penetration decreases with depth, so primary production is depth-limited. This is important to aquatic insects, because many feed upon photosynthetic organisms (Chapter 13), but light attenuation also has profound consequences for their ability to see underwater and, consequently, the structure and function of photosensory systems.

Some sunlight that reaches the water surface will be reflected back upwards and the rest will penetrate the air–water interface. In water, light can be scattered or its radiant energy absorbed by diverse molecules and particles; the amount of scatter and absorption will depend on the wavelength of light and the characteristics of the water. Scatter occurs when light is intercepted by bubbles or particles in the water (e.g. fine sediment, algae, zooplankton) and deflected or reflected in many directions. Distilled water has the lowest scatter coefficient, so this is a convenient baseline to examine wavelength-specific absorbance (Figure 6.2). In the visible spectrum, absorbance decreases with wavelength, so that blue light penetrates much farther than red light, but absorbance increases again in the ultraviolet spectrum, so blue light also penetrates farther than UV light. Relatively few water bodies are crystal clear or as pure as distilled water, and light attenuation increases with water colour (e.g. from dissolved organic matter), but in a wavelength-specific manner. Perhaps more important, however, is the effect of scatter or light attenuation, which will also vary with wavelength. For very turbid freshwater systems, such as eutrophic ponds and glacial rivers, light may penetrate only a few centimetres, and the rest of the water column will be very dark. Remember also that the amount of light reaching the water surface (i.e. before attenuation through absorption and scatter) will vary: highest irradiance will be on the brightest sunny days when the sun is directly overhead; reflection will increase (and irradiance decrease) as the sun approaches the horizon; in the shade (e.g. dense forest canopy) or on overcast days, incident light will be only 10–20 per cent of the maximum; on very overcast days, at dawn and dusk, incident light could be less than 1 per cent of the maximum.

From this very brief discussion of the properties of light in water, it should be immediately obvious that most insects that live underwater spend most of their lives in the dark or in poorly lit environments.

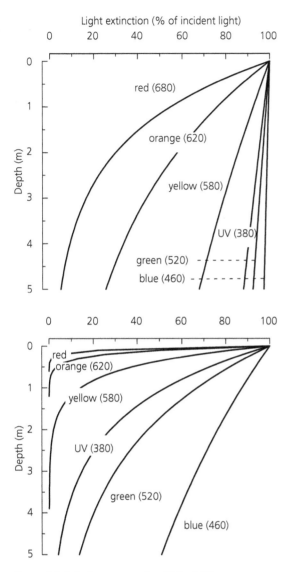

Figure 6.2 Extinction or attenuation of light of different wavelengths with increasing depth of water. Numbers in brackets indicate wavelengths in nm. Top panel: distilled water. Bottom panel: moderately coloured and/or slightly turbid water.

Therefore, we might predict that aquatic insects that rely on photosensory information are likely to live in shallow water and to have adaptations that facilitate photoreception in low light. In the terrestrial environment (e.g. adult aquatic insects) and on the water surface (e.g. semi-aquatic insects), the potential for acute vision and a reliance on photosensory information is very much higher.

6.3 Structure of compound eyes

Each facet of a compound eye is a single optical unit called an ommatidium, and many ommatidia that are closely packed together form an array of hexagonal facets that comprise a compound eye. The number of ommatidia ranges from hundreds to tens of thousands in some adult dragonflies (Pritchard 1966; Sherk 1978a). In hemimetabolous taxa, the number of ommatidia per eye generally increases as larvae develop, reaching a maximum number in adult eyes (Sherk 1978b; Cloarec 1984; Gupta et al. 2000). Each ommatidium is covered by a transparent and colourless cuticle that often forms a corneal lens. A second lens, the crystalline cone, often lies below the cornea. There are three main kinds of compound eye, with ommatidia that have distinctly different structures and optical mechanisms. Most diurnal or day-active insects have apposition eyes. The other two eye types, optical superposition and neural superposition eyes, are better adapted for photoreception at low light levels and are characteristic of nocturnal, diurnal, and crepuscular insects. They differ, however, in the mechanism that allows operation at low light levels.

In apposition eyes, a single ommatidium typically consists of a dioptric apparatus (a corneal lens and sometimes a cone) and eight retinula cells, which contain light-sensitive pigment. This pigment is contained in microvilli arranged parallel to one another (rhabdomeres) (Figure 6.3a, b). The retinula cells are arranged so that the rhabdomeres are adjacent to one another and collectively form a single fused rhabdom. It is common for retinula cells to be twisted along their length and around the central rhabdom, so the orientation of the microvilli of each rhabdomere changes through the depth of the eye. Retinula cells are sometimes layered or tiered within the rhabdom, as in some dytiscid beetles (Horridge et al. 1970; see also Section 6.4.4). The rhabdom extends up to the cone, which means that the sensory cells are close to the lens. Shielding cells that surround the bundle of retinula cells contain other pigments that are not light sensitive. Pigment in these shielding cells is uniformly dispersed and prevents light entering adjacent ommatidia. Thus, light can strike the rhabdom only by entering directly above (Figure 6.3c). Optically, the rhabdom behaves

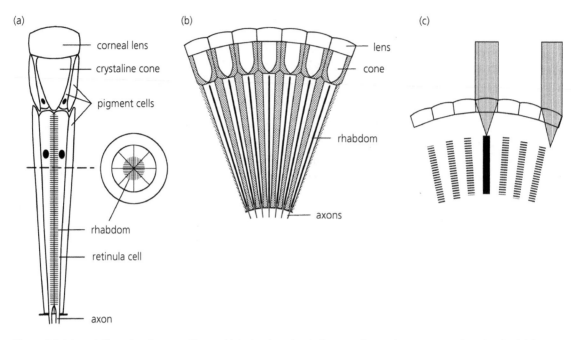

Figure 6.3 Schematic illustration of an apposition eye. (a) Section through a single ommatidium, with a cross-section through eight rhabdomeres in a closed or fused rhabdom. (b) Section through multiple ommatidia of the eye showing rhabdoms extending from the basal lamina to the cones. (c) Light capture: Stippled beams show from an image at a distant point. The left light beam with an axis aligned with the rhabdom of an ommatidium contributes to the neural signal, the right light beam does not. Illuminated rhabdom is shown in black.

as a light guide: light entering the rhabdom and aligned close to its long axis is totally reflected internally and contained within the rhabdom; light entering at more oblique angles may be only partially reflected or lost. The image is focused on the distal or outer tip of the rhabdom, so any spatial information from the image is lost. All the retinula cells within one rhabdom and ommatidia have the same field of view, and the rhabdoms of abutting ommatidia have slightly different fields of view. A nerve axon extends basally from each retinula cell, passes out through the lamina on the back of the eye and into the optic lobe of the brain (cerebral ganglion). The axons from each ommatidium are generally kept together in a cartridge and connect with the same area of the optic lobe.

Neural superposition eyes are largely a variation on the plan of apposition eyes, but differ in the arrangement of the rhabdom and axons. First, the rhabdomeres of the eight retinula cells are not fused into a single rhabdom, but remain separate in an open rhabdom, typically with six outer rhabdomeres

surrounding a two-tiered central rhabdom from two retinula cells (Figure 6.4a). The image is focused just below the dioptric apparatus on the distal tips of the retinula cells, but each retinula cell within an ommatidium has a slightly different, separate visual field that is shared by retinula cells in adjacent ommatidia (Figure 6.4b). Second, the axons emerging

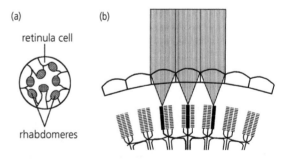

Figure 6.4 Schematic illustration of a neural superposition eye. (a) Cross-section through an open rhabdom showing retinula cells with separate rhabdomeres. (b) Light capture: Light beams with axes aligned with any of the multiple rhabdomeres within an ommatidium contribute to the neural signal. Illuminated rhabdomeres in black.

from a single ommatidium are not kept together in the same cartridge. Instead, each cartridge contains axons from retinula cells with the same field of view, even though they arise from different ommatidia. Thus, the photo signal received by the optic ganglion is the same as it would be in an apposition eye, but the signal is seven times stronger, hence allowing the insect to see in low-light conditions.

In contrast to apposition compound eyes, the ommatidia of optical superposition eyes have a shorter, fused rhabdom that typically extends only two-thirds of the distance from the basal lamina and the nerve axons, to the cone (Figure 6.5a). The remaining space or clear zone may be largely empty or filled with thin strands of the retinula cells, forming a crystalline tract or thread, which may act as a light guide. Crystalline threads serve an important function in maximizing how much of the light passing through the dioptric apparatus reaches the rhabdoms. Another distinguishing feature of optical superposition eyes is that pigment in shield

cells can migrate vertically in response to light intensity (Figure 6.5b). In bright light, the superposition eye may function much like an apposition eye, with pigment dispersed or perhaps concentrated in the lower part of the pigment cells and shielding the rhabdom against light from adjacent ommatidia, i.e. light beams must be directly aligned with the rhabdom. At low light, however, pigment gathers in the distal part of the cell and this allows light from adjacent ommatidia to strike the rhabdom (Figure 6.5c). The result is increased sensitivity at low light, in an analogous manner to neural superposition eyes. At or near the base of some optical superposition eyes is a thick mat of air-filled tracheoles called the tapetum. The shiny surface of the tapetum reflects light back into the upper parts of the eye and may increase light sensitivity. Light reflected from the tapetum also produces the so-called 'eye shine' seen in some insects when lights are directed at them, in the same way that the tapetum of a cat's eye reflects light.

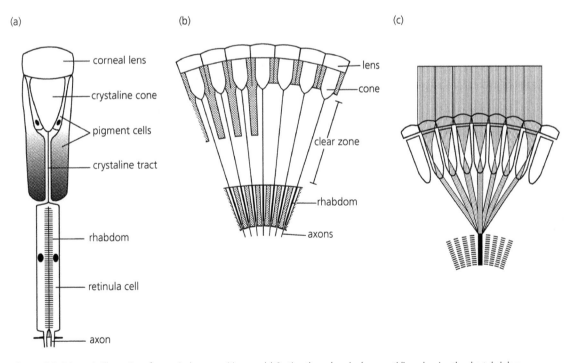

Figure 6.5 Schematic illustration of an optical superposition eye. (a) Section through a single ommatidium showing the short rhabdom. (b) Section through multiple ommatidia in high and low light (left and right within the diagram, respectively) showing position of migrated pigment in the pigment cells. (c) Light capture at low light intensity, showing light beams from several adjacent ommatidia all striking a single rhabdom. Illuminated rhabdom in black.

6.4 Functioning of the compound eye

In discussing how the compound eye functions, we are interested in how the information captured by the many independent facets can combine to form an image, how much fine detail can be incorporated in the image (resolution), how well the eyes function in very low or very bright light (sensitivity), and the extent to which they can adapt to fluctuations in light intensity, and whether they can discriminate between different wave lengths or kinds of light (colour vision, UV sensitivity, polarized light). Some of these functions involve specialization within some or all ommatidia, others relate to the configuration or spatial arrangement of the many facets.

6.4.1 Image formation

The first step of image formation is to focus light from the image on the photosensitive structures within the eye. This requires that light is bent or refracted into the eye, and this depends on the optical properties of the dioptric apparatus (i.e. the corneal lens and cone). In terrestrial insects, the simplest way to focus light is via a curved cornea, so that the large difference in the refractive index between air and the much denser cornea bends the light rays by diffraction (i.e. a convex lens) (Figure 6.6a). Unlike air, the refractive index of water is more similar to that of the cornea, so a simple curved corneal lens is unable to focus an image underwater. This presents a potentially significant problem for semi-aquatic insects that spend time in both air and water, because images refracted by a curved cornea would be in focus when the insect is in air, but out of focus when underwater. To solve this problem, the ommatidia of many aquatic and semi-aquatic insects have a corneal lens that is flat on the external surface, and a cornea (and sometimes the cone) composed of multiple layers that differ in their refractive index (Figure 6.6b) (e.g. Meyer-Rochow 1978; Schwind 1980). Thus, most refraction occurs as light beams are bent within the dioptric apparatus, regardless of whether the insect is in air or water.

Although light is focused and each ommatidium produces a small image of the object in its field of view, the image pattern is not preserved within an ommatidium. Instead, each individual ommatid-

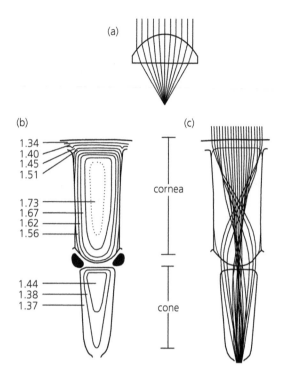

Figure 6.6 (a) Refraction of light rays from directly above by a convex lens, analogous to the corneal lens of a terrestrial insect. (b) Schematic illustration of the dioptric apparatus in an ommatidium of the adult beetle *Cybister fimbriolatus crotchi* (Dytiscidae) with a flat cornea and showing the refractive index of different layers. The refractive index of water is approximately 1.34. (c) Approximate defraction of light rays from directly above by the dioptric apparatus in (b).

Source: Meyer-Rochow, V.B. (1978). The dioptric system of the eye of *Cybister* (Dytiscidae: Coleoptera). *Proceedings of the Royal Society, London Series B*, 183 (1071), 159–178, by permission of the Royal Society.

ium or rhabdom 'sees' roughly a single spot of light with an intensity roughly equal to the average of the image 'seen' by that ommatidium. Light intensity varies between ommatidia, however, depending on the amount of light in each particular field of view. Thus whereas each individual ommatidium or rhabdom 'sees' roughly a single spot of light, the rhabdoms collectively receive a mosaic of light spots that differ in intensity and make up an entire image. This mechanism works in much the same way as digital images are made up of many pixels, each of which is uniformly coloured (Figure 6.7).

Compound eyes clearly have the potential to create an image, but it is less obvious whether insects can recognize particular images (pattern recogni-

(a)   (b)

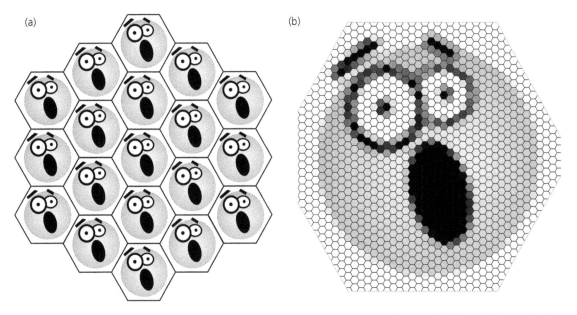

Figure 6.7 What does an insect see with a compound eye? (a) An incorrect representation, commonly portrayed in films made by entomologically naive directors, assumes that each ommatidia produces its own copy of the image. (b) A more probable representation in which each ommatidium detects a single, uniform colour or light intensity; collectively, the information from all the ommatidia forms the image. This is analogous to a digital image made of many pixels, each of which is a uniform colour.

tion) and whether particular patterns elicit certain behaviours. Some insects are capable of pattern recognition but, unfortunately, aquatic insects are rarely the subjects of such behavioural research. The one exception is the adult Odonata, which do use visual cues to recognize potential mates, rivals, and competitors (Corbet 2004). Logically, this suggests that many dragonflies and damselflies are capable of pattern recognition, but does not mean that other aquatic groups use similar strategies. Many locate mates using chemical and mechanical forms of sensory information (Chapters 7, 10); whether they can also recognize images is largely unexplored.

Detecting motion may be a more important function of compound eyes than image formation or pattern recognition. The pattern or image created by a compound eye will change as the insect moves or as objects in its field of view move. Detecting a change in pattern, rather than the pattern itself, may be important to many insects, especially to avoid predation. Most insects can detect movement visually, and some are extraordinarily good at it, as will be familiar to anyone who has tried to catch a fly or a dragonfly.

6.4.2 Resolution

Optically, resolution refers to the degree of fineness, or the amount of fine detail, with which the eye forms an image. With high resolution comes the potential to form complex images. The key determinants of resolution in compound eyes are the number of ommatidia, the interommatidial angle, the facet diameter or diameter of the lens that light passes through to reach the rhabdom, and the diameter of the rhabdom. Resolution or fineness of the image increases with a decrease in the angle between the visual axes of adjacent ommatidia (i.e. the interommatidial angle). Large-diameter lenses have higher resolution than small ones, and light loss through diffraction can be a problem in ommatidia with small apertures. It is possible to maximize lens aperture yet maintain a small interommatidial angle by increasing the curvature of the compound eye. For hemimetabolous insects, such as Odonata and Hemiptera, resolution is likely to increase with each larval instar, in accordance with the increases in the number of ommatidia, interommatidial angles, and facet diameter that occur during development

(Sherk 1978b; Cloarec 1984). In many insects, inter-ommatidial angle, lens diameter, and eye curvature can vary in different parts of the compound eye. The area of the eye with the highest resolution is called the fovea or acute zone, and may be pronounced in predatory species that must detect and fix prey locations, such as most odonates, and in life stages or sexes where detecting mates is important, as in many male mayflies and flies that form mating swarms (see also Section 6.4.6).

6.4.3 Light sensitivity and adaptation

Like most organisms, vision or the ability of an insect to respond to light cues is better in high- than low-light environments, but this is not a simple linear relationship (Figure 6.8) and depends on whether the eye is dark- or light-adapted. Sensory cells do not respond if light intensity is too low, and there is an upper threshold where cell output is at a maximum (i.e. saturated). As described in Section 6.3, basic eye structure will determine to some extent whether an insect's eyes are more receptive at low or high light intensity, but there are many variations to the patterns outlined above. For example, some mosquitoes have appositional eyes (best suited for vision in daylight), but are active at night. In contrast to the elongate rhabdoms of model apposition eyes, such species often have short, conical rhabdoms, which can intercept wider cones of light than cylindrical rhabdoms and thereby increase light sensitivity (Land et al. 1999).

Many insect eyes have various devices that can respond to ambient light conditions and regulate the amount of light reaching the receptors, thereby allowing the insect to see over a wide range of intensities. There are two main mechanisms for light adaptation: regulating the amount of light that reaches photoreceptors and changing the sensitivity of the receptors. As discussed already, movement of pigment within the shield cells of optical superposition eyes is one mode of regulating how much light reaches the sensors. Comparable movement of pigment and movement of the rhabdom itself also occurs in some apposition eyes. Some adult craneflies (Tipulidae), for example, have apposition eyes and an open rhabdom that is fused for a short length at the distal end. In the dark-adapted condition, the rhabdom is close to the cone, but in brighter light, the entire rhabdom migrates proximally and is connected by a narrow cone cell tract (Figure 6.9) (Williams 1980; Ro and Nilsson 1994). Two primary pigment cells surround the tract and act like an iris altering the size of a pupil, i.e. they move in and out laterally, in concert with pigment movement in the secondary pigment cells, thereby regulating the

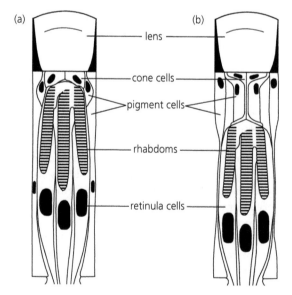

Figure 6.9 A single ommatidium from the compound eye of an adult cranefly, *Ptilogyna spectabilis* (Tipulidae) in which the entire random migrates between dark- and light-adapted states. (a) Dark-adapted, with rhabdoms raised close to the cone, (b) light-adapted with rhabdoms lowered.

Source: Springer-Verlag (Williams, D. S. (1980). Organization of the compound eye of a tipulid fly during the day and night. *Zoomorphologie*, 95, 85–104, Figure 1). With kind permission from Springer Science and Business Media.

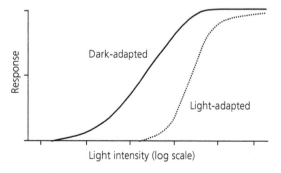

Figure 6.8 Non-linear relationship between the responsiveness of dark- and light-adapted visual sensors to light intensity.

amount of light reaching the photosensors. Similar pupil adjustments and proximal-distal migration of the rhabdom have been recorded in other insects, including aquatic Hemiptera, *Lethocerus* spp. (Belostomatidae) and *Notonecta glauca* (Notonectidae); Megaloptera, *Archichauliodes guttiferus* (Corydalidae); and Coleoptera, *Dineutes assimilis* (Gyrinidae) and *Dytiscus marginalis* (Dytiscidae) (Walcott 1969, 1971; Walcott and Horridge 1971; Ro 1995). This cell movement may be slow, for example, 30–40 minutes in *Lethocerus* (Walcott 1971) and 40–50 minutes in *Notonecta* (Ro 1995), and is thus unlikely to operate as insects swim through patches of shade and sun. Instead it is more likely to 'tune' the eye to average light intensity, which may vary with cloud cover, water colour, and turbidity (Section 6.2), or to be a diel response or circadian rhythm that accommodates foraging activities during the day and aerial dispersal by night.

6.4.4 Colour vision

Colour vision, or the ability to discriminate between light of different wavelengths, has been documented in several insect orders and has been present in insects since the Devonian period, 360–415 million years ago (Briscoe and Chittka 2001). In order for colour vision to occur, retinula cells must have different photopigments that vary in the wavelength at which peak absorption or light sensitivity occurs. All insects, including those with monochrome vision, have a photopigment with maximum absorption in the green range (≈ 530 nm). Many insects have two other pigments with maximum sensitivity in the blue and ultraviolet regions of the spectrum. At the extreme end, some Odonata have five pigments, with peak sensitivity at 330, 410, 460, 525, and 630 nm (Meinertzhagen et al. 1983; Yang and Osorio 1991), and this potential for sophisticated colour vision is consistent with the important role visual cues and colour can play in odonate behaviour. The wavelengths of light absorbed by visual pigments will depend initially on which wavelengths reach the receptors. For insects living underwater, wavelength-specific light attenuation (Figure 6.2) will determine which wavelengths reach the eye at all. It is perhaps fortuitous that the most widespread photopigment

has maximum sensitivity in the green spectrum, which also has relatively good penetration through water.

Some corneal lenses in some terrestrial insects, including adult aquatic insects, are coloured and filter the light. It is because of these filters that insect eyes are often coloured red, yellow, blue, green, or even multiple colours, if different-coloured filters are arranged in rows. Such patterned eyes are quite striking in some adult Diptera, including the Tabanidae. The result of having different-coloured filters is that the rhabdoms of different ommatidia will be differentially activated. This arrangement can provide a mechanism for wavelength discrimination between ommatidia, even if only one photopigment is available. Screening pigments in pigment cells of the ommatidium also can selectively absorb particular wavelengths and prevent them reaching the rhabdoms.

Rhabdom structure (fused, open, or tiered) influences wavelength discrimination, because light absorbed by one rhabdomere is not available to be absorbed by another. Accordingly, retinula cells with different photopigments can occur in the same ommatidium. In fused rhabdoms, rhabdomeres that carry spectrally different photopigments act as lateral filters for one another. In tiered rhabdoms the most distal retinula cells can act as filters for the proximal cells, thereby restricting the range of wavelengths reaching the proximal receptors.

It is often claimed or assumed that the larvae of aquatic insects, with compound eyes, cannot detect red light and therefore behavioural observations can be carried out under red light without concern that the light itself might alter behaviours. However, photoreceptors in the red spectrum do occur in some insects (Briscoe and Chittka 2001) and some aquatic larvae do respond to red light (Heise 1992; Turnbull and Barmuta 2006), even though others may not (Barmuta et al. 2001). The absence of behavioural responses to red light, however, does not necessarily mean that the insect is unable to perceive red light. There is relatively little research on the spectral sensitivities of the aquatic life stages of insects (but see Maksimovic et al. 2011), and it would be imprudent to generalize in the absence of evidence.

6.4.5 Polarization sensitivity

In addition to spectral differentiation within the rhabdom (responding to different wavelengths of light), there may also be polarization resolution, i.e. responding to polarized light (Horváth and Varjú 2004). Before discussing the significance of polarized light to aquatic insects, it is useful to consider, briefly, how light becomes polarized. The various properties of light are sometimes best understood if light is characterized as a wave or as a stream of particles (photons). All light waves vibrate at right angles to the ray direction. Vibrations may be equally distributed throughout the 360° range (unpolarized light) or a high proportion of the vibrations may be in a single plane (polarized light), or all the vibrations may be in one plane (plane-polarized light). A common mechanism of plane-polarization occurs when light is reflected off smooth surfaces, such as water, and vibrations are parallel to the surface. Polaroid filters (as in sunglasses) selectively absorb polarized light and hence reduce the glare it creates. Fine particles in the atmosphere can scatter out blue light and also preferentially scatter light in a plane at right angles to the direction of the light rays. The result is a pattern of polarization in the sky that varies with the position of the sun, even if the sun is obscured by cloud. Consequently, the polarization pattern of the sky can be used to determine the position of the sun.

The potential for insects to detect polarized light exists because the photopigment molecules are oriented along the microvilli of the rhabdomere, microvilli can be arranged in the same plane, and maximum absorption occurs when light is polarized in the same plane. If the rhabdom is twisted, as in many ommatidia, then the microvilli are not oriented in the same plane and polarized light cannot be detected. As a result, only some ommatidia within a compound eye are able to detect polarized light and these typically have straight rhabdoms. Photopigments that detect polarized light are also usually receptors of UV light. This makes sense because the shortest wavelengths are most strongly scattered and polarized. Polarization-sensitive ommatidia typically make up a small proportion of the eye surface.

Insects can use polarization patterns in the sky (also called the sky's pattern of e-vector orientation) as an aid to navigation, as first demonstrated for honey bees (review: Frisch 1967). Whether aquatic insects use such polarization patterns in the sky is unclear, even though some have ommatidia on the eye's dorsal surface that can detect polarized light, and some are capable of long-distance migrations (Chapter 9). Sun-related e-vector patterns extend underwater, even though it is commonly and erroneously reported that e-vector orientation in natural waters is largely horizontal (Waterman 2006). It is theoretically possible, therefore, that polarization patterns could also be a celestial compass underwater, at least in clear water lakes up to a depth of ≈ 50 m.

Light polarized by water surfaces is used by many adult aquatic insects to locate water bodies during aerial dispersal, as described first for the backswimmer *Notonecta glauca* (Schwind 1983). These taxa typically have polarization-sensitive ommatidia in the ventral portion of the compound eye. This is critical ecologically as a mechanism to locate potential egg-laying sites (Chapter 11), especially for taxa that have long-lived adults and/or adults that may disperse far from the water body in which they spent the larval life. Light can be plane-polarized by surfaces other than water and this can account for the paradoxical and often maladaptive landing of aquatic insects on plastic sheeting, cars, road surfaces, windows, and oil slicks (Horváth and Zeil 1996; Kriska et al. 1998; Kriska et al. 2006). Sensitivity to light polarization may also be an asset in prey detection underwater, as has been demonstrated for some marine invertebrates (Shashar et al. 1998; Shashar et al. 2000). The potential for some aquatic stages of insects to detect polarized light has been demonstrated for some dytiscid larvae (Stecher et al. 2010), but the behavioural significance of this phenomenon is unclear.

6.4.6 Divided and specialized compound eyes

Superficially, compound eyes may appear to comprise an array of identical ommatidia. This is rarely the case, however, and the arrangement, structure, and function of ommatidia typically varies in different regions of the eye. Such divisions can be relatively subtle, as in the fovea of predators and fast fliers (Section 6.4.2) and the limited distribution of polarization-sensitive ommatidia (Section 6.4.5), or very pronounced, as in the divided eyes of whirligig beetles (Gyrinidae) (Figure 6.10). Dorsal and ventral eyes of adult gyrinids are both composed of optical superposition ommatidia, but they have different photopigments and differ in spectral sensitivity, and differ in the degree of light adaptation (Bennett 1967; Burghause 1976; Horridge et al. 1983).

Fovea, or areas of especially high visual acuity, are located in three main areas of the eye depending on their function. When visual acuity is required during forward locomotion or flight, fovea are often aligned with the direction of movement and are particularly good at detecting the 'backward movement' of objects in the landscape relative to the insects' direction of travel. For animals living on flat surfaces or near the water surface, fovea often form 'visual streaks' and are aligned with the horizon (Wehner 1987), as seen in water striders (Gerridae) and backswimmers (Notonectidae) (Schwind 1980; Dahmen 1991). Acute zones may also be associated with the need to locate prey or mates, and sometimes occur on divided eyes (below). Adult odonates that hunt prey whilst in flight have acute zones with a variety of configurations, and often have two different acute zones: one forward-facing and concerned with flight and one directed dorsally to detect prey. In contrast, perching dragonflies that detect prey from a stationary position have a highly developed, dorsally directed acute zone (Labhart and Nilsson 1995). Larval odonates are also predatory and, similarly, the fovea or zones of visual acuity in their compound eyes correspond to modes of prey capture, with the greatest structural specialization found in taxa that are dependent on vision for prey capture, whereas the least specialization occurs in taxa or instars using tactile cues (Sherk 1977).

Sexual dimorphism in eye morphology of adults is also common, especially for species that swarm, such as some blackflies (Simuliidae) and most mayflies (Figure 6.11) (Arvy and Brittain 1984). Being hemimetabolous insects, mayfly larvae and adults have compound eyes; the number and size of ommatidia increase during larval development, but sexual dimorphism is not usually apparent until the subimago moult (Gupta et al. 2000). Male mayflies typically have specialized turbinate eyes in which all the ommatidia are directed dorsally, whereas both males and females have lateral eyes that see laterally, ventrally, forward, and also dorsally. Although the morphology of male dorsal eyes varies between and even within species (Müller-Liebenau 1973; Lord and Meier 1977), there are some general patterns in form and function (Horridge et al. 1982). In gross anatomy, the dorsal mayfly eye is separate from the lateral eye and the ommatidia differ in structure, with superposition ommatidia dorsally

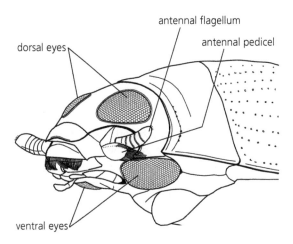

Figure 6.10 Head of a gyrinid beetle (oblique view), showing the four compound eyes. The antennal pedicel floats on the water surface (Chapter 7), so the ventral eyes are in the water and the dorsal eyes are above the surface.

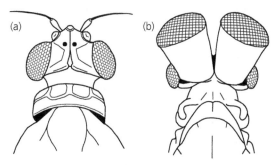

Figure 6.11 Dorsal views of the head of a female and male mayfly, *Cloeon* (Baetidae). (a) Female head with a single pair of compound eyes. (b) Turbinate compound eyes of a male, consisting of one pair of eyes with small facets on the side of the head, and a second pair with large facets on the top.

Source: Adapted from Verrier (1956).

and apposition ommatidia laterally. The facets of the lateral eye are hexagonal, whereas those of the dorsal eye are large, square, and with an obvious convex surface, and have small interommatidial angles. Finally, the dorsal eye is sensitive only to UV light. This particular suite of characteristics means that the specialized dorsal part of the male eye has greater lens-resolving power and spatial resolution (Section 6.4.2), and is well-suited for males to spot and pursue females against the bright background of the sky.

6.5 Ocelli

Typically, there are three ocelli forming a triangle on the dorsal surface of the head (Figure 6.1), although one or all ocelli have been lost in some taxa. Although located on the dorsal surface of the head, the ocelli are not directed upwards, but point in different directions: the medial ocellus is directed forward and the lateral ocelli are directed to the sides. Each ocellus typically has a single large corneal lens, many retinula cells packed close together, but without regular arrangement, and the rhabdomeres of several retinula cells combine to form rhabdoms (Figure 6.12). Importantly, any image passing through the corneal lens would be focused some distance behind the receptor cells, suggesting that ocelli cannot function in image formation. Instead, the lens functions as a condenser lens, which ensures that the retinula layer is evenly illuminated, so each rhabdom receives and sends the 'same' signal to the CNS, and signal strength increases with the number of retinula cells. Pigment cells are sometimes prominent in ocelli. A tapetum may be present behind the receptor cells, but may consist of urate crystals in a layer of cells rather than a tracheole mat as in compound eyes. Morphological and functional differences of ocelli between species are inevitable, but in these morphologically simple photoreceptors, there are few modifiable units (lens, retinula cells, pigments), and the range of potential variation is limited accordingly. Among aquatic insect species, variations in the morphology of ocelli typically centre on the shape of the corneal or cuticular lens and its refractive properties, whether the retinal layer lies immediately below the corneal lens or a short distance away (Figure 6.12), and whether the rhabdoms within the retinal layer are arranged into discrete units or in a random meshwork (Hallberg and Hagberg 1986).

Although the ocelli cannot form images and their photosensory capacity is limited to dark/light perception, for many insects they nevertheless have important functions as a steering agent, assisting with spatial orientation during flight and walking, phototaxis, circadian activities, etc. Many insects have no ocelli, so clearly there are alternative ways of solving these particular problems. Highly mobile and fast-flying insects (such as many Odonata) often have all three ocelli that together provide fairly sophisticated spatial information. Basically, the ocelli allow the insect to detect the horizon by looking exactly at the horizon: the medial ocellus points in the animal's forward direction, the two lateral ones at 90° to the left and right of the forward direction. By detecting rapid changes in overall light intensity in these different directions, the ocelli can inform a flying insect about rolls and tilts of the body, i.e. orienting the body in the horizontal plane. In some dragonflies, however, the ocelli are more complex with various anatomical and optical strategies to ensure high sensitivity, wide fields of views, and high spatial resolving power (Stange 1981; Stange et al. 2002; Berry et al. 2007). What role the ocelli play in aquatic life stages is largely unexplored. The need for spatial orientation still exists in water, but this function may also be associated with other sensory organs, such as mechanoreceptors (Chapter 7).

6.6 Stemmata

The only visual organs of larval holometabolous insects are the stemmata, and they are absent in some apparently 'blind' species. Stemmata are sometimes called 'lateral ocelli', but this term should be avoided because they are often structurally and functionally different organs. Stemmata occur on the lateral sides of the larval head and vary in number among taxa, from one to seven on each side. On each side of the head, stemmata are generally clustered and may be very close together or fused into what appears to be a single compound eye, although they are morphologically and functionally different from true compound eyes. The number of stemmata usually remains constant between instars. The basic structure of a stemma has much in common

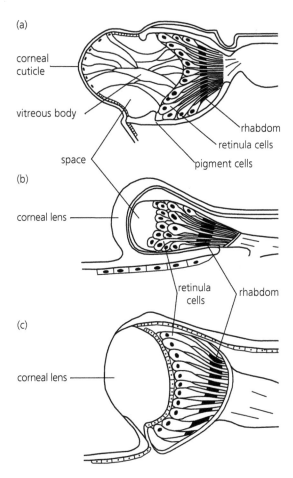

(a)

corneal
cuticle

vitreous body

space

rhabdom
retinula cells
pigment cells

(b)

corneal lens

retinula
cells

rhabdom

(c)

corneal lens

Figure 6.12 For three species, a schematic representation of the adult ocelli that vary in complexity: (a) the mayfly *Caenis robusta* (Caenidae), (b) the caddisfly *Limnephilus flavicornis* (Limnephilidae), and (c) the caddisfly *Agrypnia varia* (Limnephilidae). In *Caenis*, the corneal surface is convex but a true corneal lens is lacking, *Limnephilus* has a moderately developed corneal lens, whereas *Agrypnia* has a well-developed, biconvex lens of considerable thickness. The retinal layer lies immediately below the corneal lens in *Agrypnia*, but is separated from the cornea by a vitreous body and/or spaces in *Caenis* and *Limnephilus*.

Source: Adapted from Hallberg and Hagberg (1986).

with ommatidia and ocelli: it consists of a translucent cornea or dioptric apparatus above photosensitive retinula cells, in which the microvilli containing the light-sensitive pigment form rhabdoms, and neural signals are transmitted via axons to the central nervous system. Dioptric structures on the stemmata are often reduced or simplified for taxa inhabiting murky or turbulent water, as in some larval Simuliidae (Nyhof and McIver 1987), but a translucent cuticle is required for light penetration.

There are two main kinds of stemmata that differ in having either one or many rhabdoms (Figure 6.13), and both forms occur within the aquatic insects. In

the first model, the dioptric structure of each stemma consists of a relatively unmodified corneal lens (translucent cuticle), often with a crystalline cone underneath. Light is sensed by seven retinula cells, with microvilli typically aligned along the optical axis to form a tiered rhabdom, generally with three distal and four proximal photoreceptors (Figure 6.13a). The photoreceptor axons join and enter the optic lobe as a single optic nerve. Larval Trichoptera, Mecoptera, and most Diptera typically have this model of one rhabdom per stemma. Stemmata can be fused in some species and give the appearance of more than one rhabdom per stemma, as in the

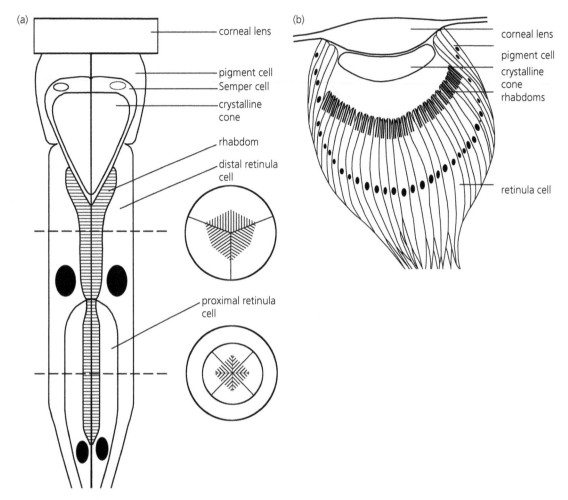

Figure 6.13 Schematic structure of stemmata in larval aquatic insects. (a) One of seven, morphologically similar stemmata from the eye spot of a larval caddisfly (*Rhyacophila*) showing cross-sections through the distal and proximal retinula cells. (b) One of six, morphologically different stemmata from the eye spot of a larval dytiscid beetle (*Dytiscus marginalis*). The multiple rhabdoms are shown as solid black bars.

Source: Adapted (a) from Paulus and Schmidt (1978) and (b) from Günther (1912).

caddisfly *Sericostoma* (Sericostomatidae) where stemmata I and II are fused (Paulus and Schmidt 1978), and the phantom midge *Chaoborus crystallinus* (Chaoboridae) where three stemmata appear to be fused under a single dioptric structure to form a single, complex primary stemma with a fused but branching rhabdom (Melzer and Paulus 1991). In the second model, common in the larvae of some aedephagous beetles (e.g. Dytiscidae), there are more rhabdoms and retinula cells per stemma, but the general structure is otherwise broadly similar (Figure 6.13b). Within these two basic models, however, there is con-

siderable variation in structure among taxa, and the different stemmata on a single insect often vary in form (Paulus and Schmidt 1978; Gilbert 1994).

The visual function of stemmata is variable among taxa and, although stemmata are often described as 'simple' eyes, there is potential for quite sophisticated visual perception. Within the holometabolous orders, including terrestrial taxa, there is evidence that larvae with stemmata can orient to a light stimulus, perceive simple patterns, detect movement of objects, orient to the plane of polarized light and discriminate colour. In terms of spatial vision, it is generally stated that stemmata provide only coarse, mosaic vision because the number of photoreceptors underlying the lens is insufficient to form a fine mosaic. This is true for some taxa (e.g. most caddisfly larvae) and such stemmata achieve broad, but incomplete, coverage of the environment at best. Where many photoreceptors underlie a lens, as in some Dytiscidae (Figure 6.13b), there is potential for higher spatial acuity and pattern vision. The non-overlapping visual fields of adjacent stemmata generally leads to poor spatial sensitivity, but side-to-side movements of the head (scanning), however, can markedly improve spatial perception (see below).

The general morphological structure of stemmata has been described for a range of aquatic insects, and there is a general presumption that they provide limited visual function, but detailed investigations are few. This is unfortunate because some detailed investigations suggest that stemmata may provide surprisingly high visual perception for some taxa. For example, larvae of the sunburst diving beetle, *Thermonectus marmoratus* (Dytiscidae) are visually oriented predators and their visual system consists of a cluster of six elaborate stemmata and one eye patch on each side of the head. However, these six stemmata are morphologically and functionally different from one another, although they conform to the general model in Figure 6.13b (Buschbeck et al. 2007; Mandapaka et al. 2007), and collectively they are capable of spectral discrimination and potentially detecting polarized light (Maksimovic et al. 2009; Stecher et al. 2010; Maksimovic et al. 2011). The physical organization of these eyes results in a visual field extending laterally in the horizontal body plane, but with only a narrow vertical field. Nevertheless, it is an effective visual system, as one might predict for a visually oriented, mobile predator that must locate, track, and attack mobile prey. The two most dorsal and anterior eyes on each side of the head are tubular and directed approximately anterio-dorsally; the remaining four are more spherical and point in different directions in the horizontal plane. There are multiple rhabdoms and retinula cells per stemma but, unusually, they are arranged into distinctly different groups (or retinas) with different three-dimensional orientations, which suggest that specific visual tasks are distributed among the eyes. The two pairs of anterior stemmata are the principal eyes for prey capture and the peripheral eyes might function in the early detection of potential prey. Only a narrow image can be resolved by the anterior eyes at any particular time and the visual field cannot be enlarged by moving the eyes, as occurs in jumping spiders that can move their tubular eyes (Land 1969). Instead, scanning via vertical up-and-down movements of the head increases the effective vertical visual field of the tubular eyes by about one order of magnitude. Scanning behaviour routinely precedes attack strike on prey and undoubtedly increases prey capture success. This scanning behaviour may result in loss of visual resolution due to motion blur, especially at low light levels, but is clearly advantageous for prey location in many circumstances. However, *T. marmoratus* is also a scavenger able to locate food resources using chemical cues (Velasco and Millan 1998) and, therefore, light level may alter foraging mode, but not preclude foraging.

Curiously, stemmata are retained in the adult stage of some holometabolous insects, including some Trichoptera, potentially providing three light sensory systems simultaneously (compound eyes, ocelli, and stemmata). Although much reduced in comparison to the larval stemmata, they appear to be functioning photoreceptors (Hagberg 1986). What behavioural role they play, if any, is unclear.

6.7 Bioluminescence

As well as being able to perceive visible light or various wavelengths of electromagnetic radiation, some insects can also generate light, a process called bioluminescence. There are several distinctly different biochemical systems for bioluminescence in insects, and they can generate white, yellow, green, or red light. Chemically, light is produced by the oxidation of luciferin by the enzyme luciferase. Bioluminescence occurs in a few Diptera and some beetles, commonly called fireflies, glow worms, or lightning bugs. As a means of communication, light is more likely to be important in terrestrial environments than underwater, because light energy is rapidly absorbed by water and therefore will not travel far. However, some beetles in the family Lampyridae have aquatic larvae (e.g. species in the genus

Luciola), with both adults and larvae capable of light generation in paired lanterns on the ventral side of the distal abdominal segments (Ho et al. 2010). Larval luminescence occurs frequently and spontaneously during locomotion and is often induced by disturbance. The pulses of light emitted by these beetles are conspicuous signals that could play multiple roles. Luminescence has an obvious 'attractive' function in mate recognition, and species- and sex-specific signalling is well-known in lampyrid fireflies (Lewis and Cratsley 2008). Light may also attract potential prey, a lure used by some cave-dwelling Diptera, and may also be employed by aquatic lampyrid larvae that forage at night (Ho et al. 2010). Alternatively, light may serve as a warning to would-be predators of distastefulness and, indeed, some aquatic fireflies have glands that produce and release compounds that function as a broad-spectrum defence against various predators (Fu et al. 2007).

CHAPTER 7

Sensory systems—mechano- and chemoreception

7.1 Introduction

In addition to vision, the other major kinds of sensory information used by insects are mechanical and chemical signals. Mechanical signals include touch and sound; chemical signals include smell (olfaction) and taste (gustation). In the insects, temperature and humidity are also considered part of chemoreception because of the structural similarities in sensory receptors. Mechanical and chemical signals can provide a lot of information about the abiotic and biotic environment in which insects live. In fact, mechanoreception may be the most important source of sensory information for aquatic insects, especially given the difficulties of vision underwater (Chapter 6). Of course, insects must have the appropriate sensory structures to detect signals, discriminate between different signals, and a central nervous system that can process multiple kinds of information simultaneously. As well as being able to detect sensory information, many insects have some capacity to send signals and potentially communicate with other organisms. Mechano- and chemoreception are functionally important to the autecology of insects inhabiting both running and still water environments, as will be become apparent in the sections to follow.

The organs that insects use for sensing mechanical and chemical signals (generically referred to as sensilla, singular sensillum) are often small and inconspicuous, but they are enormously diverse in form (Zacharuk and Shields 1991), occur internally and externally on virtually all parts of the body, and can occur in extremely high densities. For example, the beetle *Graphoderus occidentalis* (Dytiscidae) has nine different kinds of sensilla on the terminal antennal segment alone (Jensen and Zacharuk 1991). This diversity and abundance of sensilla suggests that mechanical and chemical cues are extremely important, and insects must have sophisticated neurological systems that can simultaneously process a broad array of sensory input. With the development and widespread use of electron microscopy came a virtual explosion in the number of morphological descriptions of sensilla, but the behavioural and electrophysiological experiments required to determine their function are much more laborious, so the function of many sensilla awaits investigation.

There is no obvious 'best' way to organize a discussion of these sensory systems. The various morphological kinds of mechanoreceptor do not map neatly onto different kinds of sensory stimuli, and detecting particular kinds of sensory information can involve several kinds of sensor. Similarly, a single sensillum may function as both a chemo- and mechanoreceptor. To complicate matters further, particular structures or combinations of structures may correspond to different life stages (adult versus larva) and taxonomic groupings, although in some situations sensilla may be useful characters in taxonomic and phylogenetic studies (Wells 1984; Gaino and Rebora 1999a, b). Unlike photosensors, which are located on the head, mechano- and chemoreceptors can occur anywhere on the body, often with multiple kinds of sensilla grouped close together, as for example on the antennae, mouthparts, and genitalia (Zacharuk and Shields 1991; Gaino and Rebora 1999b). In this chapter we will first consider mechanoreception, describing the gross morphology of the major kinds of sensilla and how they work, and then provide

Aquatic Entomology. First Edition. Jill Lancaster & Barbara J. Downes.
© Jill Lancaster & Barbara J. Downes 2013. Published 2013 by Oxford University Press.

some examples of how aquatic insects use sensors to perceive their environment (Section 7.2). Many aquatic insects can also send mechanical signals to communicate with other insects. The diverse ways in which they generate these signals and the functional roles they play in communication will be discussed in Section 7.3. Similarly, the discussion of chemoreception will first consider the morphology of sensilla and how they work (Section 7.4), and then how insects use chemical cues to perceive their environment and to communicate (Section 7.5). In all these sensory systems, mechanical or chemical energy is transformed or transduced into electrical energy and ultimately into a nerve pulse that is processed by the central nervous system. The underlying neurophysiological processes are unlikely to differ between terrestrial and aquatic insects, and interested readers are referred to other entomological texts.

7.2 Mechanoreception

Mechanoreception is the perception of any distortion of the body, including touch, movement and acceleration of the body, movement of air or water over the body, movements of various body parts (proprioception), gravitational sense, sensing vibrations of the substrate or vibrations transmitted through water or air (including hearing). The distinction between mechanoreception and sound reception (hearing) is often vague and will be discussed in Section 7.2.2. Briefly, sounds are waves of pressure produced when particles are made to vibrate and cause displacement of adjacent particles, regardless of whether vibrations occur in air, water, or solids. Thus, sound waves have much in common with various forms of air and water movement (e.g. waves on the water surface), but differences in the physical properties of air and water also mean that sound waves behave differently in the two media. Some insects have specialized auditory sensors, but insects lacking such structures can clearly hear because they often respond in a characteristic manner to particular sounds.

Morphologically, there are three main types of mechanoreceptors in insects: internal stretch receptors, cuticular structures, and chordotonal structures

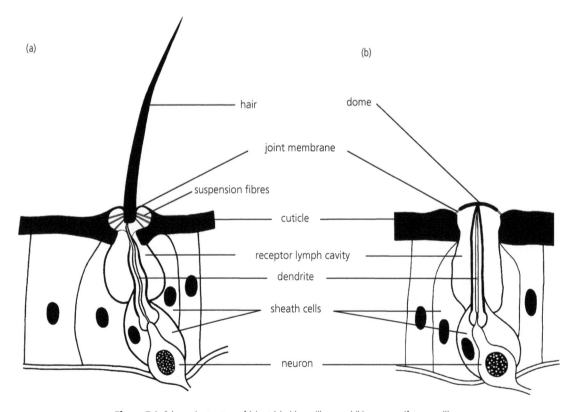

(a)

(b)

hair

dome

joint membrane

suspension fibres

cuticle

receptor lymph cavity

dendrite

sheath cells

neuron

Figure 7.1 Schematic structure of (a) a trichoid sensillum, and (b) a campaniform sensillum.

(also called subcuticular organs). With the possible exception of internal stretch receptors, these are typically protrusions or extensions of the cuticle. Stretch receptors are internal and usually associated with muscles. Functionally, they are probably most important in providing information on rhythmic events, such as movements associated with locomotion (walking, swimming, flying), gas exchange (beating gills, abdominal ventilation), and waves of peristalsis along the gut. Importantly, stretch receptors also provide information on gut fullness. The form and function of stretch receptors is largely independent of the medium in which an insect lives (i.e. air versus water) and will not be discussed further.

7.2.1 Cuticular mechanoreceptors

There are two main forms of cuticular mechanoreceptors: hair-like trichoid sensilla that are stimulated when the hair is bent, and dome-like campaniform sensilla that are stimulated by a compression of the dome (Figure 7.1). These two basic forms, however, can be modified in various ways to detect an astonishingly broad range of sensory information.

Hairs of varying lengths cover the bodies of insects to varying degrees and many of the long hairs are mechanoreceptors. Mechanoreceptor hairs may be scattered over the body, clustered in groups of small hairs known as hair plates, and the highest densities are on body parts that frequently come into contact with the substrate or other objects in the environment. These body parts include tarsal segments of the legs, antennae, mouthparts, and genitalia (Figure 7.2). Each trichoid sensillum consists of a hair-like structure with the hair shaft forming a lever that is flexibly suspended in a socket within the exoskeleton (Figure 7.1a). The hair shaft is easily deflected by forces in the external medium. One or more sensory neurons surrounded by sheath cells and attached to the hair base pick up the mechanical stimulus and transfer the information to the central nervous system. Although the concept of hairs as levers is simple, it is a complex series of events that transform environmental input into a mechanical stimulus that finally elicits a nervous signal (Humphrey and Barth 2008). To keep the hair in a normal position yet still provide flexibility, the

Figure 7.2 Scanning electron micrograph of trichoid sensilla (hairs), or possibly chaetica (bristles), on the antennae of the mayfly *Habrophlebia eldae* (Leptophlebiidae). These sensilla are common along the distal border of each flagellar segment.

Source: From Gaino and Rebora (1999a). Reproduced with permission from John Wiley & Sons.

socket usually has an external cuticular joint membrane that is continuous with the body cuticle and with the hair. Immediately below that is a ring of suspension fibres and the socket septum. These suspension devices return the hair to its normal position once the mechanical force displacing the hair has ceased. Many hairs are restricted to moving only in particular directions, which provides directional sensitivity. Although trichoid sensilla are generically referred to as 'hairs', they can take diverse shapes, including hairs, setae, and bristles. The hairs of some sensilla have pores and this generally denotes a chemosensory function (Section 7.4). Mechanosensory hairs also differ in their sensitivity and the signals to which they respond; long delicate hairs respond to very small forces, such as changes in air pressure and airborne sounds, whereas short, thick spines are only stimulated by much greater forces.

Campaniform sensilla are structurally similar to trichoid sensilla except that instead of a hair, they have a domed area of thin cuticle (Figure 7.1b). Laterally, the dome is connected to the main body cuticle by a flexible joint membrane that forms a ring around the dome and is made of thin cuticle. Attached to the underside of the dome is the sensory neuron. Thus, the dome can be pushed inward, thereby stimulating the neuron, and will spring back once the mechanical force ceases. Campaniform sensilla typically occur on areas of the body that are subject to stress, and are common close to the joints of appendages, on the mouthparts, on basal segments of the antennae, and on the ovipositor, as described for some Odonata (Gorb 1994; Matushkina and Gorb 2002).

Cuticular mechanoreceptors differ in types of response or neural input depending upon their function. Hair sensilla with a tactile function respond only when the hair is initially bent and there is no further response if the hair remains bent. This is called a phasic response. Hairs with a proprioceptive function and all campaniform sensilla have a phasic response followed by a tonic response of continued neural input (provided that the hair remains bent or the campaniform dome is depressed), usually described as a phasic/tonic response.

Sensing movement of air or water over the insect body, whether the insect is moving through the medium or the medium is moving over the insect, is the primary function of many trichoid sensilla. The length of sensory hairs relative to boundary layer thickness is related to their function, i.e. hairs must extend beyond the boundary layer in order to detect movement of the insect relative to the surrounding medium (for an explanation of boundary layers, see Section 5.5.2). Hairs that detect movement are often long relative to the insect's boundary layer, and the angular deflection of the hair (i.e. its sensitivity) is a function of hair length, movement speed, and direction. However, because the density and viscosity of air and water are quite different (Section 5.2), flow-sensing hairs are likely to function optimally in only one medium and their morphology will vary accordingly (Devarakonda et al. 1996). For example, water has a smaller boundary layer thickness (0.22 times that of air), which means that hairs of the same length will be more sensitive in water than in air. Drag is 43 times higher in water than air, however, so hydrodynamic sensors are typically shorter than aerody-

namic sensors (drag forces increase with hair length). Most flow-sensing hairs in water and air are filiform (straight shafts); branched hairs are subject to greater drag and occur most commonly in some terrestrial arthropods.

Flow sensors often occur in groups and may form characteristic patterns. Flying insects, for example, often have hair plates (or hair fields) on the head. Arrays of hairs may provide better sensory information than single hairs in some situations, such as directional information, and hairs within the array may differ in length and sensitivity. When multiple hair fields are present on the head, fields may differ in their directional sensitivity. Hairs that are close together, however, can potentially interact with one another, so the shape and the spatial arrangement of hairs within an array is important. Most research to date has focused on single hairs. Work on mechanosensory hair arrays has focused on terrestrial insects and air flows (Humphrey and Barth 2008), but hair arrays are also likely to function in detecting water movements.

In aquatic life stages, trichoid sensilla are especially effective at detecting water movement or pressure waves; they occur all over the body, and are often abundant on legs, antennae, and cerci. Being able to detect pressure waves or particular flow patterns (also called hydrodynamic cues) and also to locate the origins of such waves has many potential adaptive functions. For stream-dwelling insects, for example, the forces of flowing water (Section 5.5) affect almost every aspect of their aquatic life so there may be many advantages to being able to detect water movements and, importantly, changes in water movement. There is ample evidence, for example, that some benthic insects can detect changes in water velocity near the substrate surface and respond by altering their movement (e.g. Lancaster 1999; Lancaster et al. 2006; Lancaster et al. 2009). How they detect those changes in velocity has not been investigated.

For insects that live in tubes or burrows, trichoid mechanoreceptors may provide sensory information regarding the dimensions of the retreat. Larvae of the burrowing mayfly *Ephemera danica* (Ephemeridae) have long, trichoid sensilla on the antennae, which probably function in determining the dimensions of the space in the burrow and around the larva's head (Rebora and Gaino 2008). Case-building caddisfly larvae add new material to the anterior end of the case as they grow, and trim material from the posterior end to keep the case a manageable length. They rely on the anal hooks and sensory hairs on the terminal abdominal segments to provide information about case length—when hooks and hairs are ablated, caddisflies build much longer cases (Merrill 1965; Hansell 1973). It seems likely that these hairs are mechanosensory, but we are unaware of any study that has examined their structure or carried out electrophysiological experiments, although mechanically manipulating these hairs does elicit characteristic behavioural responses (Sakhuja et al. 1983).

In aquatic insects, trichoid sensilla can also function as pressure sensors to detect water depth and provide orientation with respect to gravity. For example, the stream-dwelling hemipteran *Aphelocheirus aestivalis* (Aphelicheiridae) uses a plastron for gas exchange, which functions most efficiently in highly oxygenated, shallow water, i.e. at low pressure (Section 3.4.3). On the ventral surface of the abdomen and within the plastron are a pair of depressions containing hydrofuge hairs that are longer then the plastron hairs, and some of which are trichoid sensilla (Thorpe and Crisp 1947a). If the insect moves into deep water, the volume of air in the plastron decreases due to the increased water pressure, the hairs are bent, and the sensilla stimulated. The insect responds by swimming upwards, which reduces pressure and the hairs return to their normal position. Another hemipteran, *Nepa* (Nepidae), also has three pairs of pressure sensors on the ventral surface of the abdomen (Figure 7.3a), which provide information on whether the animal is oriented horizontally, head up, or head down. Each of these sensors consists of over 100 trichoid sensilla with tips that are expanded into thin plates, forming mushroom-shaped structures (Figure 7.3b) (Thorpe and Crisp 1947a). Adjacent sensilla overlap, enclosing an air space underneath, which is connected to the tracheal system via a spiracle. These receptors do not respond to changes in pressure per se, but posterior and anterior sensilla respond differently to one another if the animal is not horizontal. If the head is tilted upward, air in

the tracheal system tends to rise in an anterior direction, the hairs in the anterior end are pushed out and those at the posterior end collapse, and the reverse happens when the head is tilted downward (Figure 7.3c). In this way the insect can determine the orientation of its body.

Proprioceptive functions can be carried out by both trichoid and campaniform sensilla (position and movements of the body). At the joints between leg segments or at the neck (joint between head and thorax), there are often groups or rows of small sensory hairs. The hairs are arranged so that movement of one segment relative to another causes the hairs to bend. Campaniform sensilla usually occur in groups and are often located at joints or other parts of the body subject to distortion. Sensilla on the legs, for example, are often oriented so that they are stimulated when the foot is on the ground and the leg bears the insect's weight.

7.2.2 Chordotonal organs

Chordotonal organs (also called scolopophorous organs) exist as strands of tissue stretched between two points (Figure 7.4). A change in the relative position of the two attachment points will alter the length of the strand and initiate a nerve signal. Being subcuticular, there is often no visible external evidence of chordotonal organs. Each organ consists of one or more functional units called chordotonal sensilla (also called scolopidia, scolopidium singular). These are anatomically more complex than trichoid and campaniform sensilla, and similarly varied in form and function (Field and Matheson 1998). Briefly, a single sensory neuron is ensheathed within two or three other cells, including a distinctive scolopale cell and cap. Any stress or pull on the cap results in a stimulus of the neuron. Simple chordotonal organs, each with one or a few scolopidia, occur widely over the body of adult and larval insects in most orders. Soft-bodied dipteran larvae such as the Chironomidae, for example, have many such organs arranged segmentally along the body.

Multiple, simple scolopidia may be organized into more complex organs that are often associated with joints (e.g. on legs, at the neck, or wing bases). Three kinds of complex chordotonal organs are worthy of further discussion: subgenual organs and

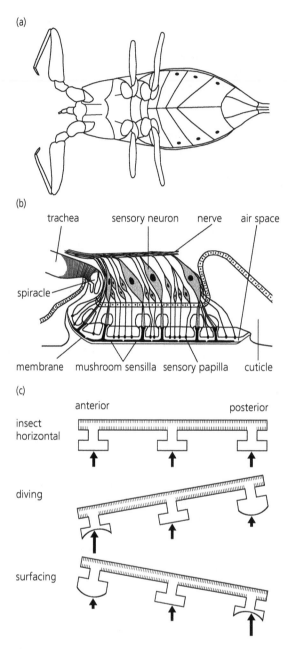

Figure 7.3 Pressure sensor of the water scorpion *Nepa* (Nepidae). (a) Schematic, ventral view of water scorpion showing location of paired, closed spiracles on three abdominal segments, and their associated pressure sensors. (b) Illustration of receptor system associated with one spiracle of the ventral surface of the abdomen (open water lies below the sensor). Overlapping ends of the mushroom sensilla form a continuous flexible sheet (or membrane) enclosing an airspace that is continuous with air in the tracheal system. (c) Illustration of how the pressure receptors of adjacent spiracles are joined through the tracheal system and the manner in which pressure differences arise when the body is tilted and that cause differential compression of the mushroom sensilla. When the head is lower than the tail (e.g. when diving), the mushroom layer of the most anterior spiracle is pushed in by water pressure and the mushroom layer of the most posterior spiracle is pushed out by air in the trachea; vice versa when the head is higher than the tail (e.g. when swimming toward the surface).

Source: (b) and (c) Reproduced with permission from the *Journal of Experimental Biology*, (Thorpe and Crisp 1947a).

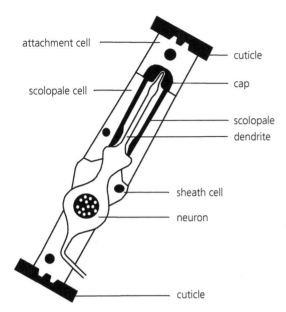

Figure 7.4 Schematic structure of a scolopidium or chordotonal sensillum.

metatarsal organs, Johnston's organ, and tympanal organs. These structures are especially important in detecting vibrations and sounds waves, and can be important in receiving information about the environment as well as signals transmitted by other organisms (Section 7.3).

In order to understand how chordotonal organs function, it is important to grasp some of the physical properties of vibrations and sound waves. Vibrations are transmitted by a variety of wave forms and their transmission properties (e.g. transmission velocity and attenuation rate) depend upon the medium, i.e. air, water, or solid. All vibrations involve the movement of particles within a medium. In transverse waves, particle movement is perpendicular to the direction of wave motion; in longitudinal waves, particle movement is parallel to the wave direction. Vibrations of hard substrates, taut strings of webs, and the water surface are predominantly transverse waves, as are the familiar waves often seen on the surface of lakes and the ocean. Transmission of transverse surface waves is relatively slow and most effective over short distances, because attenuation rates are high and wave patterns are readily disrupted by objects in the landscape. In contrast, sound waves are longitudinal

compression waves, created whenever a mechanical disturbance occurs in a compressible medium, such as air or water. The characteristics of sound differ in air and water because these media have different physical properties. Sound will travel roughly 4.3 times faster in water than in air, and a sound of a given intensity will usually travel much farther in water. The ways in which sounds are sensed and created are related to two physical features associated with the longitudinal waves. As a sound wave passes a given point in the medium, fluid pressure oscillates up and down (the pressure aspect) and fluid molecules are displaced away from and then toward the propagating source (the particle velocity aspect). The particle velocity aspect is more significant than the pressure aspect close to the propagation source (the near field), but their relative importance changes with distance and the pressure aspect is more important far from the source (the far field). Because the wavelength of a given sound frequency is roughly five times longer in water than in air, the extent of the near field is also greater underwater. Thus, the structures of the acoustic organs of insects will influence whether they are most effective in the near field or the far field, in air or in water.

Aquatic insects that live at the air–water interface, such as some Hemiptera (e.g. Gerridae, Notonectidae), often have well-developed sensors that can detect transverse waves associated with vibrations of the water surface. Their vibration sensors are typically subgenual organs located on the proximal part of the tibia (i.e. close to the femur), but not associated with a joint, or other vibration sensors located on the tarsi and tibia, such as metatarsal organs. Notonectids often sit immediately below the water surface, touching the underside of the surface with the first and second pairs of legs, and tip of the abdomen (Figure 7.5a). The main vibration receptors are scolopidia on the tarsi of the pro- and mesothoracic legs (Figure 7.5b), mechanoreceptive hairs (trichoid sensilla) located along the tip of the abdomen, and perhaps also the subgenual organ on the tibia (Wiese 1972; Murphey and Mendenhall 1973; Lang 1980b). Gerrids sit on top of the water surface, but similarly detect water-surface vibrations via chordotonal organs in the tarsi and mechanoreceptor hairs on the ventral side of the tarsi (Murphey

(a)

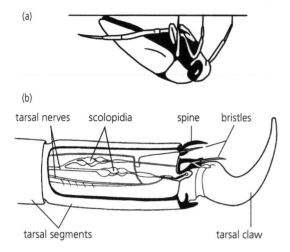

(b)

tarsal nerves   scolopidia   spine   bristles

tarsal segments   tarsal claw

Figure 7.5 (a) Typical position of a backswimmer at rest below water surface. (b) Vibration sensors (scolopidia or chordotonal organs) in the tarsi of *Notonecta* (Notonectidae). The scolopidia detect movement of the tarsal claw relative to the bristles on the ventral side of the claw.

Source: (b) Adapted from Wiese (1972).

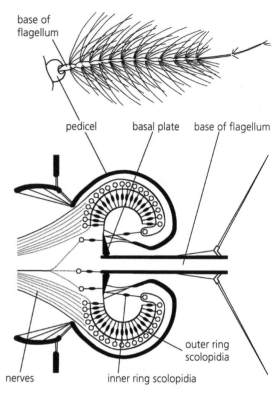

base of flagellum

pedicel   basal plate   base of flagellum

nerves   inner ring scolopidia   outer ring scolopidia

Figure 7.6 Johnston's organ housed within the enlarged pedicel of a male mosquito antenna. Upper: external view of whole antenna; lower: schematic cross-section of the pedicel. Mechanosensory neurons receive sound-induced vibrations as the antennal flagellum oscillates up and down. Oscillations of the flagellum are transmitted to the scolopidia through processes connected to the basal plate. The scolopidia are arranged in two rings around the axis of the antenna, with several single scolopidia joining the flagellum and the scape.

Source: (b) Adapted from Autrum (1963).

1971; Goodwyn et al. 2009). Being able to detect surface waves, however, is only the first step. Insects must also have the capability to discriminate between vibration frequencies created by the physical environment and by other organisms. Functionally, discrimination is often based on characteristics of the wave itself, such as duration and frequency, and the nature of the signal generated by the receptor neurons, such as the frequency and amplitude of the signal transmitted to the central nervous system.

One of the most fascinating, complex chordotonal organs is Johnston's organ, which occurs on the antennae of most adult insects and in a reduced form in many larvae (Figure 7.6). Although located on the antennal pedicel (second segment of the antenna), the scolopidia are also attached to the flagellum (antennal segments three and above) and detect movements of the pedicel relative to the flagellum. Note: the flagellum lacks muscles so movement can only occur via external forces. Movement of these segments could arise in many different ways and, consequently, Johnston's organ can serve a variety of functions, even in a single insect. In many flying insects, the position of the antennae changes with flight speed, and thus Johnston's organ senses changes in air speed. In contrast, many backswim-

mers use Johnston's organ to orient the body relative to the direction of gravity when under water. When correctly oriented on its back (Figure 7.5a), the antennae are deflected away from the head and into an air bubble that lies between the head and the antennae. If the insect's body is oriented differently, the antennae are drawn toward the head and the Johnston's organ registers the position. In adult whirligig beetles (Gyrinidae), one function of Johnston's organ is to detect transverse waves or ripples on the water surface. The antennal flagellum of gyrinids is club-shaped and is typically raised above the water; the pedicel has a fringe of hairs that rest on the water surface and move up and down with any surface waves (Figure 7.7). The pedicel and flagellum move

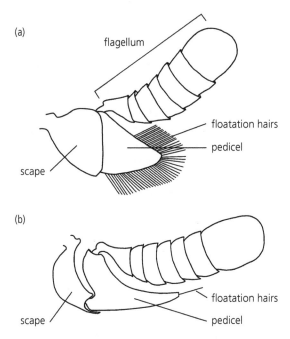

(a)

flagellum

floatation hairs

pedicel

scape

(b)

floatation hairs

scape

pedicel

Figure 7.7 Schematic illustration of the antenna of a gyrinid beetle, showing the clubbed flagellum (main part of the antenna), the enlarged pedicel with flotation hairs, and the scape, which is attached to the head capsule. (a) Dorsal view, and (b) side view (cross-section). Typically, the flagellum is raised above the water, whereas the pedicel and flotation hairs rest on the water surface, moving up and down with any surface waves. The Johnston's organ at the base of the pedicel detects movement of the pedicel relative to the flagellum.

freely relative to one another, and Johnston's organ lies at the joint between these two segments and detects movements of the one relative to the other, i.e. they form a vibration system. The amplitude of any neural signals created by Johnston's organ increases proportionally to the relative motion between flagellum and pedicel. These beetles can detect ripples on the water surface with an amplitude of only a few μm and are also able to detect waves that differ in frequency (Rudolph 1967). Ripples will vary in wavelength and frequency depending upon what causes them, and beetles may use this information to identify the source and alter their behaviour accordingly (Tucker 1969; Bendele 1986).

Tympanal organs, the third major kind of complex chordotonal organ, are specialized for sound reception in the far field (i.e. they respond to the pressure aspect of sound waves), and are most likely to occur in adult insects. They have evolved repeatedly in at least eight insect orders and, unsurprisingly, taxa

differ with respect to where tympanal organs are located on the body (e.g. legs, thorax, abdomen) and in their ultrastructure (Yager 1999). Nevertheless, all tympanal organs have three basic features: an area of thin cuticle (the tympanum or tympanic membrane) backed by a tracheal air sac so that it can vibrate (this acts as a drum), and one or more scolopidia attached to the inside of the tympanum that are stimulated when the membrane vibrates. The amplitude of the vibrations varies with sound intensity and structure of the membrane, and species differ in the sound frequencies to which they are most sensitive. Tympanal organs primarily act as pressure receivers or pressure-gradient (pressure-difference) receivers. In pressure receivers, sound impinges only on one side of the tympanum. In pressure-gradient receivers, sound reaches both sides of the membrane and the nature of the vibrations depends on differences in sound pressure between the two sides, resulting from differences in vibration phase and the potential for the membrane itself to absorb some frequencies. Tympanal organs are typically paired (on left and right sides of the body), but because insects are small relative to the speed of sound, they are unable to determine the direction from which sound comes based on differences in arrival time at the two organs. Instead, directional sensitivity arises from properties of the tympanal organs as pressure-difference receivers and the diffraction of high-frequency sounds. The physical principles of acoustic reception are complex and we refer readers to other detailed accounts (e.g. Ewing 1989; Greenfield 2002).

Well-developed tympani are typically found in the taxa that can also generate sounds for communication in the air and/or under water. Among the aquatic insects, this includes the adults of many Hemiptera and some Coleoptera. Most of the aquatic Hemiptera that have been examined appear to have two or three paired tympanal organs on the thorax (Anderson 1980). Adult stages of these insects are semi-aquatic and the tympani may operate as far field pressure receivers when in air. Underwater, however, the tympani of many Corixidae may also operate as particle-velocity receivers. An external sclerotized club-shaped process on the edge of the membrane (Figure 7.8) may be stimulated to vibrate by displacement of the surrounding fluid molecules. The air bubble that surrounds the tympanal organ

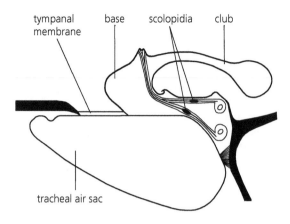

tympanal membrane   base   scolopidia   club

tracheal air sac

Figure 7.8 Schematic longitudinal section through the tympanal organ on the mesothorax of a female *Corixa punctata*. Sound waves travelling through water cause the tympanal membrane to vibrate against the tracheal air sac, thereby stretching the scolopidia.

Source: Springer-Verlag (Prager and Streng 1982) Prager, J. (1976). Das mesothorakale Tympanalorgan von Corixa punctata III. (Heteroptera, Corixidae). *Journal of Comparative Physiology A*, 110, 33–50. Figure 2. With kind permission from Springer Science and Business Media.

(i.e. the insect's compressible gas gill, Section 3.4), amplifies the pressure changes within the bubble and converts the sound pressure wave into greatly amplified particle displacements (Prager and Streng 1982; Prager and Theiss 1982) which, ultimately, may cause membrane vibrations.

7.3 Communication via mechanical signals

Mechanoreceptors are clearly important for insects to receive information about their environment, but being able to generate mechanical signals may also be important for communicating with other organisms. The previous section emphasized how mechanosensory sensilla detect environmental signals. In this section we focus on how mechanical signals are used in inter-individual communication, including how aquatic insects create mechanical signals and their behavioural roles. We consider communication to occur when both signaller and receiver expect to benefit in some way from the exchange. Some signals may be adventitious or unintentional and result from the insect's normal activities, such as the signals created by prey and detected by predators. Others may be intentional, such as mating signals. Communication signals are often characterized by

being discrete pulses of vibrations or sounds that are repeated in specific temporal patterns. The primary means used by insects to generate mechanical signals (vibrations and sounds) include percussion, stridulation, tymbals, and air expulsion. The frequency of acoustic or vibratory wave motion represents the quality of a mechanical signal and receptor organs are often 'tuned' to receive certain frequencies, such as those produced by conspecifics. For completeness, we consider first a somewhat different form of mechanical signalling via hydrodynamic cues—different in that the signal is created by water moving over an insect's body.

7.3.1 Hydrodynamic cues

Aquatic insects can detect the movement of water relative to their bodies (e.g. using trichoid sensilla, Section 7.2.1), and there is evidence that they can also use hydrodynamic cues to detect potential predators and prey. Fundamentally, this form of communication requires that the signaller creates a characteristic flow pattern as it moves through the water (or water moves over its body) and that signal receivers are able to identify this particular flow pattern. From a fluid mechanics perspective, this is straightforward as certain shaped objects are known to produce characteristic flow patterns (e.g. vortices) in moving water, or as they move through still water (Chapter 8). Some predatory aquatic insects do indeed detect prey via the flow patterns that prey create, including some larval phantom midges (Chaoboridae) (Giguère and Dill 1979), perlodid stoneflies (Peckarsky and Wilcox 1989), dytiscid beetles (Formanowicz 1987), and aeschnid and libellulid dragonflies (Kanou and Shimazawa 1983; Rebora et al. 2004). In the same manner, some prey may be able to use hydromechanic cues to detect and avoid potential predators (Peckarsky and Penton 1989; Culp et al. 1991). Exactly what the signals are (e.g. particular vortices) and which sensilla are involved in receiving these signals is less clear, but evidence suggests that trichoid sensilla are involved for some taxa (Gaino and Rebora 2001). More complex and less well understood is the notion that the insect receiving signals, whether predator or prey, must be able to discriminate between wave patterns generated by various sources, to detect and recognize the signal of interest over any background noise. This discrimination may also require multiple kinds of sensory information, as in the larvae of *Dytiscus verticalis* (Dytiscidae) that use both hydromechanical and chemical cues to elicit the strike and capture phases of an attack on prey (Formanowicz 1987).

7.3.2 Substrate vibrations and percussion

Communications via vibrational signals commonly involve the signaller manipulating the water surface or more solid surfaces to create transverse waves. The receiver, depending on the taxon, could use a range of different sensilla to detect the vibrations, including trichoid sensilla, metatarsal organs (e.g. many Hemiptera), and Johnston's organ (e.g. Gyrinidae) (Section 7.2). The most common behavioural roles for such communication include detection of predators and prey, courtship (mate detection), and territorial behaviours. The attenuation of surface waves is considerable, and high-frequency waves attenuate more rapidly than low-frequency waves, so this form of communication at the water surface is most effective over short distances (typically < 1 m). The duration, amplitude, and frequency of the vibration signals differ depending on the source and, provided that the receiver can discriminate between these wave characteristics, the insect can identify the source. For example, the surface waves generated by insects struggling on the water surface are typically of high frequency, long duration, low amplitude, and temporally irregular; stochastic signals of abiotic origin (e.g. a pebble falling into the water) are generally of low frequency, short duration, and high amplitude; communication waves generated by insects are often low frequency, long duration, and very regular.

It is common for terrestrial insects to land accidentally on water surfaces and be preyed upon by aquatic insects living at the air–water interface. Whilst struggling to free themselves from the surface film, terrestrial insects create surface waves that may attract predators. Having multiple sensors (e.g. on different legs) allows the predatory insect to detect phase differences in the signals received by the different sensors, and thus identify in what direction the prey lies relative to the predator (Wiese 1974; Lang 1980a).

Surface waves may be created intentionally and allow insects to communicate with one another and, as mentioned above, the frequency of waves generated is typically different from waves generated by the abiotic environment or by potential prey—an obvious advantage to ensure that one is not mistakenly consumed by a potential suitor. During courtship, males typically send signals, females receive them and, in a few taxa, females may return signals (see also Chapter 10). Surface-wave communication is well developed in many Gerridae, and is used for courtship, mate-guarding, and associated territorial behaviour (Wilcox 1972; Hayashi 1985; Wilcox and Di Stefano 1991). Both males and females of some species, such as *Rhagadotarsus anomalus* (Gerridae), can send signals when on the water surface or attached to an object (Wilcox 1972). Males produce relatively large amplitude waves by pushing or pumping the midlegs against the water surface (calling signals), and low amplitude waves with the forelegs (courtship signals). Females also produce courtship signals with the forelegs. In contrast, courting males of the giant water bug *Abedus indentatus* (Belostomatidae) generate low-frequency surface waves in what is called a 'pumping' display (Kraus 1989). The male hangs onto an object at the water surface, such as a macrophyte, with its metathoracic legs, and the body is moved up and down by alternately extending and contracting the mesothoracic legs.

Surface-wave communication is also well documented for adult whirligig beetles (Gyrinidae) to detect prey, locate mates, detect conspecifics in rafts, and echolocate to avoid colliding with one another or with other obstacles (Rudolph 1967; Heinrich and Vogt 1980; Kolmes 1983, 1985; Bendele 1986). The surface-wave signals are created by the beetles swimming in particular ways, such as altering speed, direction, turn frequency; and this is somewhat similar to the way in which hydromechanic signals are generated. Signal detection, as discussed in Section 7.2.2, is via the antennal vibration system and Johnston's organ. Adult gyrinid beetles occur in aggregations (also called rafts) and communication with members of the raft is via surface waves and ripples. For example, one individual 'fleeing' in response to a visual stimulus, such as a perceived threat of some sort, will generate surface waves that prompt other individuals in the aggregation to flee also (Vulinec and Miller 1989).

Signalling via percussive substrate-borne vibrations (i.e. vibrations produced by the impact of the body against the substrate, or by the impact of two body parts) occurs in some adult Trichoptera (Ivanov and Rupprecht 1992; Wiggins 2004) and Megaloptera (Rupprecht 1975), and is well documented in adult Plecoptera (Stewart and Sandberg 2006). Some dobsonflies (*Sialis*, Sialidae) create substrate vibrations by tapping the wings and abdomen on the substrate, or by abdominal vibrations that are passed through the legs and into the substrate (Rupprecht 1975). Drumming stoneflies (Arctoperlaria of the northern hemisphere) tap or rub the tip of the abdomen on the substrate, thereby creating low-frequency vibrations with a frequency, or a pattern of frequencies, that is both species- and sex-specific. Males of some Plecoptera may have vesicles, lobes, knobs, or hammers on the ventral surface of the abdomen that are used for drumming, but these are not essential and species without any such structures also drum (Stewart and Maketon 1991). These drumming signals can be detected over many metres and against the various background noises found beside streams (Szczytko and Stewart 1979). As with *Sialis*, the abdominal movements of some stoneflies do not result in percussive impacts, but tremulating (trembling) the body may be transmitted to the substratum via the legs, resulting in substrate vibrations. Generating surface vibrations via tremulation is common in many terrestrial insects and spiders (Ewing 1989; Greenfield 2002), but examples of such behaviour in aquatic taxa are more scarce. Substrate vibrations can also be created through stridulation, and this will be discussed in the next section.

7.3.3 Stridulation

Stridulation refers to the production of vibrations using a file–plectrum system, in which a sharp object (plectrum) is drawn across a series of pegs (file, strigil, or pars stridens). In insects, files and plectrums occur on different body parts that can be moved against one another. The plectrum is often a rough surface made up of one or more parallel, cuticular ridges. Stridulation can cause air- or water-

borne sounds in the far field (this is how grasshoppers and crickets create their distinctive sounds), and may also give rise to substrate-borne vibrations if vibrations of the insect body are transmitted to the substrate, perhaps via the legs. File–plectrum systems vary enormously among taxa (Aiken 1985; Polhemus 1994), and some species may have more than one stridulation system.

Many aquatic Hemiptera stridulate underwater and in the air (Polhemus 1994). In both male and female Corixinae (perhaps the most-studied aquatic insect group for sound production), the main mode of stridulation is by rubbing a patch of pegs on the fore femora (the pars stridens) against sharp ridges (the plectrum) on the maxillary plate of the head (Jansson 1972). When underwater, the head is the sound-radiating structure and vibrations of the head are amplified by the air bubble, which is also the gas gill for these air breathers (Theiss 1982). Because the air bubble is compressible (i.e. bubble size changes with depth and dive duration), the sounds produced can also vary accordingly. If both sexes stridulate, males and females may have different 'songs' and this is related, at least in part, to sex-specific differences in the structure of the peg pattern on the fore femora (Jansson 1973; Aiken 1982). Some species may have a second stridulation system, such as males of *Palmacorixa nana*, where a series of pegs on the femora of the middle leg is drawn against the edges of the hemielytra (Aiken 1982). Other species may have a single, but quite different, system such as various species of *Micronecta* that have a strigil and plectrum on different abdominal segments (King 1976). Males of *Micronecta* can chorus and synchronize their calls (King 1999), generating an extremely loud (for their size) courtship song (Sueur et al. 2011), which may attract potential mates from far away.

Stridulation has been documented or proposed in adults of several groups of aquatic Coleoptera, including the families Hydrobiidae, Dytiscidae, Haliplidae, Amphizoidae, Hydrophilidae, and Hydraenidae, but there is still some confusion about the stridulation mechanisms (Aiken 1985). Stridulation in the Hydrophilidae is perhaps best documented, and sound is produced when a grooved area on the dorsal surface of the first abdominal segment is rubbed against a field of fine teeth on the underside of the elytra. Inevitably, details of the file–plectrum processes vary among species within the family (Maillard and Sellier 1970).

With the exception of some Hydropsychidae (Trichoptera) (Johnstone 1964; Jansson and Vuoristo 1979), stridulation is rare in larval aquatic insects. When disturbed, hydropsychid larvae stridulate by rubbing striations on each side of the head (pars stridens) against special tubercules (plectrum) on the fore femora (Figure 7.9). Although the sound signals produced have been described as 'messy', species-specific sound patterns are apparent. These are net-spinning larvae and only individuals in possession of a net appear able to stridulate. This makes sense for two reasons. First, the function or behavioural role of stridulation lies in defending nets from would-be intruders. Hydropsychid larvae often compete for net-spinning sites and the defender of a net can be displaced by a superior competitor. Second, stridulation signals are transmitted as vibrations through the silken thread of the net, not as sound waves through the water. Feather-like and plume-like hairs on the forelegs appear to function as vibration sensors, which would only operate if the intruding larva is touching the net (Jansson and Vuoristo 1979). Thus, communication via stridulation in hydropsychid larvae can only operate over very short distances and there is no need for a mechanism to amplify signals, which is the function of the air bubble in stridulating adult corixids. This example is an interesting reminder that a particular mechanism of signal generation (stridulation) can produce quite different kinds of signals: acoustic or substrate vibrations.

7.3.4 Sound via vibration and tymbal mechanisms

All sounds must be produced by the vibration of some structure. We have already considered stridulation and percussive mechanisms. The remaining mechanisms are those in which body parts are vibrated or somehow manipulated by direct muscle action to produce sound, and often without an amplification system or interaction with the substrate. Sounds produced by vibrations are typically in the high-frequency range and are perceived as variations in particle velocity (i.e. near field signals).

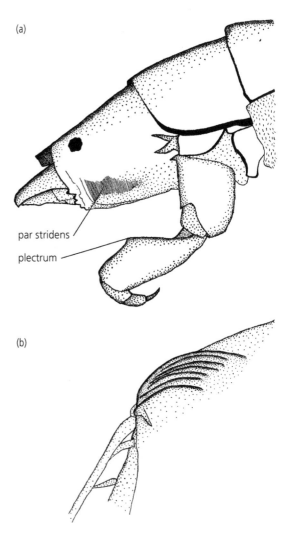

(a)

par stridens

plectrum

(b)

Figure 7.9 Stridulation system in larval hydropsychid caddisflies. (a) Ventrolateral view of the head and prothorax of the larval caddisfly *Hydropsyche pellucidula*, showing the stridulation structures (par stridens and plectrum). (b) Close-up, lateral view of the plectrum of *H. pellucidula*.
Source: From Johnstone (1964). Reproduced with permission from John Wiley & Sons.

Sound produced by the vibration of wings in air is common, and many people will be familiar with the characteristic whine of mosquitoes or the buzz of horseflies. Such sounds are often an incidental consequence of flight rather than an insect's 'intention' to send out communication signals, but they can play a role in communication, especially among conspecifics. One of the best-known examples is indeed that of the flight tone of mosquito wings and the role it plays in mating (Downes 1969; Clements 1999; Gopfert et al. 1999). These sounds are generated by the females and detected by the male antennae via Johnston's organ. Male wings do create sounds, but the flight tones are typically higher frequency than those of females and generally lie outside the females' sensitivity range. Furthermore, the antennae of female mosquitoes are morphologically different from those of the males, and not well suited for sound detection (Figure 10.5). The antennae of these males are plumose, and the many fine hairs vibrate when exposed to sound waves and cause the entire flagellum to move, thus they are near-field, particle-velocity detectors that only operate at short range (1s–10s cm). The amplitude of these vibrations varies with the sound frequency and the behavioural response of the insect varies accordingly. For example, a male may attempt to mate if sound frequencies correspond to the wing beat of a conspecific female, but other frequencies may evoke different behaviours such as cleaning movements, flight, or remaining motionless. The precise nature of the sound generated by females varies with wing beat frequency and wing size (thus sounds may be species-specific). The sound may also vary with insect age, and the sounds produced by newly emerged, reproductively immature females are often below the auditory range of males.

A tymbal is an area of thin cuticle, often backed by an air sac and surrounded by a rigid frame—a bit like a drum (tymbal is in fact an archaic name for a kettledrum). Vibrations, and sounds, are produced when the tymbal buckles (like metal click toys), usually as a result of the movement of muscles attached to it. Typically, these are airborne sounds in the far field and, hence, perceived by suitable receptors such as tympanal organs. We are unaware of tymbals being used by any aquatic insect, but include this mechanism for completeness. Tymbals are common in some Lepidoptera and Hemiptera, with the classic example being the cicadas (Young and Bennett-Clark 1995; Bennett-Clark 1997). Normally only male cicadas produce sounds and one or more sound pulses can be created by a single contraction of the powerful tymbal muscles attached to a pair of tymbals on the first abdominal segment.

7.3.5 Air expulsion

A relatively common mechanism for generating mechanical signals is rapid air expulsion through a small opening. Hissing cockroaches (*Gromphadorhina*), for example, make a hissing sound by expelling air from the tracheal system through spiracles on the fourth abdominal segment, whilst other spiracles are closed. These sounds are produced when the cockroach is disturbed but also during courtship. In the aquatic insects, sounds of this sort have been reported for some beetle larvae (Dytiscidae, Hydrophilidae), usually produced when the larva is disturbed (Aiken 1985). Although these larvae live in the aquatic medium, they are air breathers, completely dependent upon atmospheric air, and periodically replace air in the tracheal system. Reports suggest that these sounds are generated when the larva is out of the water; rapid expulsion of air under water would presumably produce a stream of bubbles, which may be no less surprising to any animal responsible for disturbing the beetle.

7.4 Chemoreception

In chemically mediated interactions, chemicals are released by the transmitting entity, transported through a fluid medium (short or long distances), received by an organism (chemoreception), and this stimulates a physiological or behavioural response. Chemoreception (sensory stimulation by chemicals) involves olfaction (smell) and gustation (taste). The differences between the two are not clear cut, but are commonly distinguished by the distance between the signal source and the receiving insect. Olfaction implies the ability to detect chemicals in a gaseous state that are typically in low concentration and potentially travel far from the source; gustation implies the ability to detect chemicals on objects or substances that come into contact with the insect. Olfactory and contact chemoreceptors are often morphologically different and, importantly, the central nervous system processes the two kinds of information quite separately. In humans, taste is associated with sensory structures in the mouth; olfaction with the nose. For insects, 'contact chemoreception' may be a more appropriate description of this process than 'taste'. Detection of chemicals in an aqueous state (i.e. dissolved in water) should, strictly speaking, be classified as contact chemoreception, but olfactory and gustatory sensilla can occur together on some aquatic insects and operate in both air and water. There are many similarities in the way that chemicals behave and are detected in air and water, so it is efficient to consider many aspects of receptor function together.

Insects can have chemoreceptors on any part of the body. Olfactory sensors are structurally diverse and typically abundant on the antennae of both larvae and adults, whereas contact chemosensors are often abundant on antennae, mouthparts and tarsi of adults and larvae, and on the genitalia and ovipositors of some adults. Thermo- and hygroreceptors (temperature and humidity) have some structural similarities with chemoreceptors, so we will discuss them here.

7.4.1 Olfactory and contact chemoreceptors

The general structure of chemoreceptors is similar to cuticular mechanoreceptors, but they typically have pores that allow the entry of chemicals. Olfactory sensilla have numerous small pores (Figure 7.10a), which are often referred to generically as multiporous sensilla. In contrast, contact chemoreceptors typically have a single pore at or close to the tip and are often called uniporous sensilla (Figure 7.10b) (Mitchell et al. 1999). It is common to assume that sensilla have a chemosensory function if pores are present, even though direct experimental evidence may be lacking. Although virtually all chemoreceptors have pores, the gross external form of sensilla is very variable (Zacharuk and Shields 1991). Ten general forms have been identified (Schneider 1964), at least four of which are relatively common (Figure 7.11). Sensilla chaetica or trichoidea are relatively long bristles or hairs and look similar to trichoid mechanoreceptors, except for the presence of pores and the absence of a specialized socket that allows movement. Basiconic pegs (sensilla basiconica) are short finger-like projections, also lacking a socket for movement. Plate or placoid sensilla (sensilla placoidea) are flat plates level with the cuticle surface. Coeloconic sensilla or peg pits

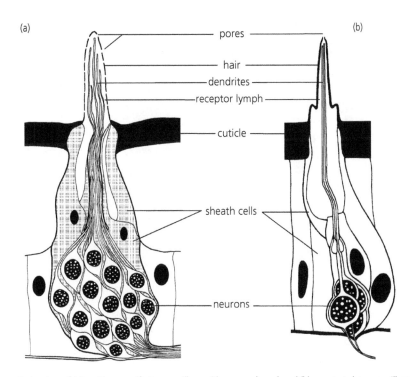

Figure 7.10 Schematic structure of (a) multiporous olfactory sensillum, with pores enlarged, and (b) a contact chemo sensillum with a single pore.

(sensilla coeloconica) are depressions or pits in the cuticle, often with a narrow opening, and with one or more short pegs in the depression. Complex, compound chemosensory organs may be formed from the fusion of several sensilla, as seen for example in some stonefly larvae (Kapoor 1987). Uniporous, flat-tipped or spatulate sensilla occur on the antennae of some mayfly larvae (Gaino and Rebora 1999a) and are perhaps best classified as a kind of basiconic peg (Figure 7.11a), but they almost certainly detect chemicals dissolved in the aquatic medium. Multi- and uniporous sensilla often occur together, as seen on the antennae and palpi of some water beetles (Figure 7.11c), and both are likely to have a chemosensory function underwater and in

air (Behrend 1971; Jensen and Zacharuk 1991, 1992; Baker 2001).

There are two main steps in the function of chemoreceptors: first, the capture of a chemical molecule and transport to the membrane of the nerve cell (perireceptor events), and second, the conversion of chemical to electrical energy (transduction). Stimulating molecules diffuse through the pores and the likelihood that a molecule enters a pore will increase with the strength or concentration of the chemical signal. Molecule capture will also depend on many other factors such as molecule size, pore size, and the fluid flow field in the immediate vicinity of the sensillum (e.g. the boundary layer and viscous sub-layer, Section 5.3.2). Molecules must also move

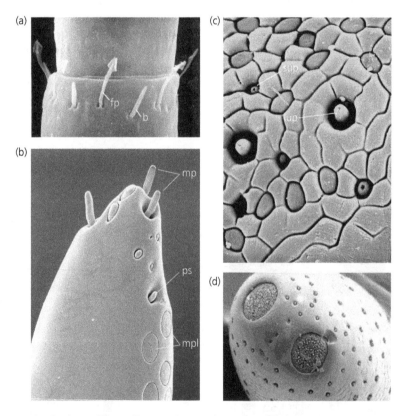

Figure 7.11 Electron micrographs of some different olfactory and contact chemoreceptors. (a) Basiconic pegs: Uniporous, sensillum basiconica, b, and arrow-shaped, flat-tipped basiconic sensilla, fp, on the seventh antennal flagellum of the mayfly larvae *Cinygmula mimus* (Heptageniidae). Although not obvious at this magnification, these sensilla typically have a single pore at the tip (see also Gaino and Rebora 1999a, b). (b) Placoid sensilla: Multiporous plate, mpl, on antennal apex of beetle *Peltodytes duodecimpunctatus* (Haliplidae). Also a mulitporous peg, mp, and a uniporous peg in a shallow depression, ps,. (c) Coeloconic sensilla: large and small uniporous peg sensilla (sup and lup, respectively) recessed in pits on the antennae of the beetle *Amphizoa lecontei* (Amphizoidea). (d) Distinct sensory fields on the tips of the labial palps of the adult beetle *Amphizoa lecontei* (Amphizoidea).

Source: (a) Kindly provided by B. L. Peckarsky, unpublished; (b)–(d) From Baker (2001). Reproduced with permission from John Wiley & Sons.

from a gaseous or aqueous state to become dissolved in the lymph fluid within the receptor, and there may be odour- or pheromone-binding proteins in the receptor that facilitate this process. In order that the sensillum remains sensitive to chemical stimuli, molecules must be removed once a nerve signal has been triggered. Molecules of the chemical signal may be degraded by enzymes within the lymph, or deactivated by odour-binding proteins once the neuron membrane has been activated.

Neurons in chemosensilla are typically specialized and respond only to certain chemicals, to a single compound, or even a specific isomer. Multiple chemosensilla often occur in sensory fields or arrays, with different sensilla tuned so that they respond to different chemicals. Concentration of the chemical signal influences transduction, such that the neuron will not respond if concentration is below a lower threshold, and there will be a maximum threshold above which neurons cannot respond any more strongly. Threshold concentrations vary enormously and pheromone or species-specific chemicals may have lower thresholds than chemicals that occur widely in the environment. Sensitivity of the insect as a whole will depend on the number and arrangement of sensilla over the body surface. Threshold levels differ between neurons: high concentrations of common compounds (e.g. nutrients such as sugars) may be required to elicit nerve responses, whereas minimum threshold concentrations may be lower for toxins or chemicals with specific ecological relevance. Most contact chemoreceptors have a phasic or short-lived response to stimuli and neuronal input can cease very quickly, even though continuous sensory input may be required to maintain some behaviours, such as feeding. This is perhaps why many insects make drumming movements, repeatedly touching the substrate with the antennae, legs, or palps to maintain the flow of sensory information.

7.4.2 Fluid mechanics of olfaction

Chemical signals in the environment must reach the surface of a sensillum before perireceptor events can occur. This is likely to be a straightforward event for contact chemoreceptors, where the sensillum is in direct contact with the origin or producer of the chemical cue, and the chemical itself. This process is more complex, however, when the chemical is volatile and molecules are dispersed in the air, or dissolved in water, because the fluid flow field in the immediate vicinity of the sensillum will strongly influence capture rate (Koehl 1996). Fortunately, both air and water behave as fluids and, in principle, are governed by the same physical laws (some of these phenomena are discussed in Chapter 5). Factors that might influence small-scale fluid mechanics and the likelihood of a chemical molecule reaching the sensillum include the shape of the sensillum and how far it projects into the boundary layer (e.g. hair-like sensilla versus peg pits), the shape of the body part on which the sensilla are located (e.g. antennal morphology), density and viscosity of the medium (e.g. air versus water), magnitude of drag forces, and thickness of the boundary layer around the animal. The chemosensory structures of many aquatic insects tend to be short (in contrast to the long hairs and plumose antennae of terrestrial insects) and this may minimize drag forces. Particular behaviours, such as antennal vibrating and flicking or sniffing (Schneider 1964; Koehl 2006), can reduce boundary layer thickness and increase access of sensilla to chemicals. Fluid flows will also vary depending on whether sensilla are distant from one another or clustered into chemoreceptor arrays. Sensilla in arrays may interact with one another and influence the penetration of odour-carrying water into the array—a problem analogous to the potential interactions between multiple mechanosensory hairs within a hair field (Section 7.2.1), and a complex problem in fluid mechanics (Koehl 1996, 2006).

Much further away from the insect, fluid conditions and flow patterns determine the dispersal of odour molecules in the environment (their spatial and temporal distribution), the likelihood that an insect will encounter a chemical cue, and its ability to orient and respond to the signal. Chemical concentrations generally decrease as molecules disperse and spread away from their sources. Typically, only a small fraction of this dispersal is by molecular diffusion, and most diffusion occurs within the boundary layer at the release point (in this context, also called the diffusive sub-layer). Beyond the boundary layer, fluid movement (e.g. wind and water currents) has a major effect on the dispersion of chemical signals and the shape of the odour plume. Fluid flow is rarely stable (i.e. laminar flow) and the predominantly turbulent fluid motion means that odour plumes have a patchy structure and the strength of the chemical signal fluctuates markedly within the plume (Vickers 2000; Webster and Weissburg 2009).

If insects can detect the overall chemical gradient of an odour plume, they may respond by moving toward, or away from, the odour source. How this actually works depends on the insect's mode of movement and its mechanisms for chemoreception. Using chemical information only, there are two primary mechanisms for orienting to a chemical gradient (chemotaxis). In tropotaxis, the insect uses two or more sensilla to detect the chemical signal in two spatially separate locations, simultaneously, and responds to relative differences in signal input. In klinotaxis, the chemical signal is sampled at one location, but at two or more different times. Chemical information also may be integrated with fluid flow information (from visual or mechanosensory structures) to locate the source of the chemical cue. Many adult Diptera and Lepidoptera, for example, rely very heavily on visual information when flying upwind in an odour plume. Organisms travelling along an odour plume follow very similar movement paths as they navigate toward the odour source, even though the behavioural mechanisms producing those paths may be quite different. Typically, they travel against the current direction (upwind or upstream) and zig-zag back and forth across the plume (Vickers 2000).

7.4.3 Thermo-hygroreceptors

Receptor cells that respond to temperature and humidity often form a sensory triad (two hygroreceptors and one thermoreceptor) in a single sensillum; they are located on the antennae and typically take the form of an aporous peg, without specialized or articulated sockets (Figure 7.12). Thermohygroreceptors are usually not very numerous; they can take on a wealth of different shapes (structures similar to olfactory and contact chemosensors) and, consequently, it is difficult to define them according to external appearances (Altner and Loftus 1985).

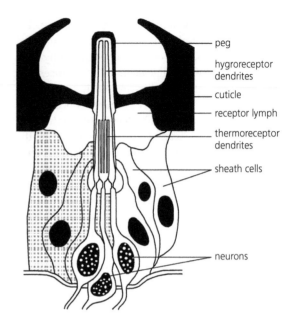

Figure 7.12 Schematic structure of a thermohygroreceptor in the form of a coeloconic sensillum with a poreless peg and two different kinds of sensory cells (two hygroreceptors with long dendrites and one thermoreceptor with a short, branched dendrite).
Source: Adapted from Altner and Loftus (1985).

Although they lack pores, the cuticle is thin over all or most of the peg, or sculpted into clefts and canals. Dendrites of the hygrosensitive neurons extend into the peg, whereas the thermosensitive dendrite is shorter. The two hygroreceptor cells function as an antagonist pair in which the moist receptor responds to a rise in humidity and the dry receptor to a fall in humidity. It is not entirely clear how hygroreceptors transduce humidity signals into neuronal pulses (Tichy and Loftus 1996; Tichy and Kallina 2010). Temperature sensors typically function as cold receptors, because the neurons fire more rapidly as temperature drops.

Hygroreceptors may seem superfluous for insects that live in the water, but they do occur and may be especially common in taxa or life stages inhabiting water bodies that dry periodically. Behavioural observations of various taxa in ephemeral environments suggest that some species first move into residual pockets of free water and then use humidity gradients (and perhaps temperature) to locate moist, cool refugia (alternative strategies include burrowing into the mud and entering a dormant phase). Larvae of the dragonfly *Libellula depressa* (Libellulidae), for example, typically live in small, shallow ponds and have this ability to seek out moist patches (Rebora et al. 2006; Piersanti et al. 2007). They have coeloconic sensilla on the tips of the antennae that function as hygroreceptors (Rebora et al. 2007) and may provide the sensory information larvae require to locate water during dry periods. For terrestrial adults, being able to detect different levels of humidity and temperature on a microclimatic scale may be enormously important to initiate behaviours that ensure the insect can avoid desiccation and to thermoregulate. The Odonata, for example, are well-known for their ability to thermoregulate behaviourally and physiologically (Section 4.2.1), and thermo-hygreceptors have been identified on the antennae (Rebora et al. 2007; Rebora and Gaino 2008; Piersanti et al. 2010; Piersanti et al. 2011).

7.5 Function of chemoreception and chemical communication

In general, chemoreceptors function to provide insects with information about their environment, including communication with con- or heterospecifics, and some functions have been mentioned already. Insect responses will vary with the kind of chemical or the simultaneous combination of multiple chemicals. Responses to specific chemicals are not always exactly the same and can depend on the environmental context or the physiological state of the individual. For example, sensitivity to chemicals that stimulate feeding often decreases with gut fullness and sensilla may become unresponsive after a meal. Although the details of chemical detection and behavioural response are complex and many details are unknown, some discussion of general classes of response is possible).

7.5.1 Perception of the environment

Chemical signals play an important role in food choice, at long and short distances. Many insects find food by following chemical trails or plumes detected by olfactory receptors (Section 7.5.2). Once food has been located, the initiation and cessation of feeding behaviours are also often strongly influenced by chemical signals, and contact chemorecep-

tors are particularly important in this context (Mitchell et al. 1999). Accordingly, contact receptors are abundant on mouthparts, although contact chemoreceptors on the tarsi and antennae may also be important in identifying suitable food resources. Sensilla on the maxillary palps and the galea are often considered to be especially important in assessing food quality, as shown in some larval caddisflies (Spänhoff et al. 2003; Spänhoff et al. 2005). Stimulation by a specific chemical may initiate feeding and, similarly, other chemicals, such as plant secondary compounds, may inhibit feeding.

Oviposition site selection may also be strongly influenced by chemoreception. Olfactory cues may help to locate potential oviposition sites from a distance, whereas contact chemoreception may be used to assess site suitability. Contact chemoreception in this context has perhaps been investigated most for terrestrial species that oviposit on specific host plants (e.g. Banga et al. 2003), but some aquatic taxa also use chemical cues in the water to select oviposition sites (this will be discussed in more detail in Section 11.2). Females may select water bodies based on physico-chemical characteristics of the water that reflect habitat suitability for larvae (i.e. potentially food-rich) or the presence of conspecifics (potentially resulting in local aggregations). Alternatively, they may avoid water bodies with chemical cues that signal the presence of potential predators or that are chemically unsuitable in some way.

7.5.2 Communication via chemical cues

Many insects communicate via chemical signals; this is common in terrestrial life stages, and occurs to a lesser extent in aquatic habitats. Chemical compounds used in communication are known as semiochemicals and may be used between individuals of the same species (pheromones) or different species (allelochemicals).

Perhaps the most well-known use of pheromones is in sexual communication (Section 10.5.3), in which attraction pheromones operate over long distances to bring together members of the same species, and courtship pheromones operate over shorter distances for mate pairing. Sex pheromones are typically produced by adult insects and are airborne—although some surface-dwelling hemipterans may communicate using surface-dispersible chemicals (Cheng and Roussis 1998; Tsoukatou et al. 2001). Pheromones can, however, serve several other functions including aggregation (for functions other than sex), spacing (avoidance), trail-marking, and signalling alarm. Most of the examples of pheromone communication not related to reproduction are from terrestrial insects. Exceptions include some adult gyrinid beetles in which chemicals released at the water surface from pygidial glands may be used as alarm signals to conspecifics (Henrikson and Stenson 1993).

Allelochemicals may be classified as kairomones, allomones, or synomones, according to the relative benefit or cost to the producer and the recipient. Kairomones benefit the receiver and disadvantage the producer (Dodson et al. 1994), and there are two main mechanisms: either insects detect potential food sources via chemical cues (receiver of the chemical eats, the producer gets eaten) (e.g. Evans 1982; Formanowicz 1987), or prey detect predators from their chemical odours and take evasive action (receiver of the chemical escapes, the producer goes hungry). Many larval aquatic insects alter their behaviour in response to chemical detection of predators, although the nature of the escape behaviour will vary with prey and predator species (Forrester 1994; McIntosh and Peckarsky 1999). Behavioural responses may include the short-term benefit of avoiding predation or induced morphological defences (Dahl and Peckarsky 2002), but there may also be longer-term consequences for individual fitness and the dynamics of whole populations (Peckarsky et al. 2002). In contrast, allomones modify the behaviour of the receiver in a way that benefits the producer. Common examples of allomones are repellent chemicals that signal distastefulness and deter would-be predators (producer of the chemical escapes predation, the receiver forages elsewhere), and are well documented for some

gyrinid beetles and their putative fish predators (Benfield 1972; Eisner and Aneshansley 2000; Harlin 2005), but also documented for a range of aquatic beetles and bugs (Scrimshaw and Kerfoot 1987). A similar defensive function against invertebrate predators has been suggested for chemicals produced by adult caddisflies in the genus *Pycnopsyche* (Duffield et al. 1977). Finally, synomones benefit both the producer and the receiver, and may operate in mutualistic or commensal relationships, but we are unaware of any examples within the aquatic insects.

In order to communicate via chemical signals, aquatic insects must also have the potential to produce chemicals. In the case of adventitious signalling, such as when predators detect prey, the chemical cues released by the prey may be by-products of some different biological process (e.g. faeces) or simply intended for a quite different purpose (sex pheromones intended for suitors, not predators). The most detailed, but still scarce, information on sex pheromones in aquatic insects comes from work on some Trichoptera and Diptera (Section 10.5.3). Many insect groups, however, have glandular epidermal cells or aggregations of secretory cells that form glands (often on the abdomen) that produce chemicals that may be released into the environment. These are perhaps best described for some adult Hemiptera and Coleoptera (Scrimshaw and Kerfoot 1987), and the molecular structure of the chemicals produced is remarkably diverse. Chemical secretions may be synthesized *de novo* from smaller molecules or by modifying specific chemicals derived from food. Functionally, such chemicals are commonly used for defence from predators (i.e. allomones). Alternatively, chemical secretions may be spread over the body surface during grooming behaviours to maintain water repellence of hydrofuge structures, such as those associated with gas gills (Kovac and Maschwitz 1989).

CHAPTER 8

Locomotion in and on water

8.1 Introduction

No insects are truly sessile throughout the life cycle and all have some ability to move. Indeed, movement is often essential in order to satisfy the fundamental needs of acquiring food, avoiding predators, finding mates, and many other autecological processes. Aquatic insects have complex life cycles, with both terrestrial and aquatic life stages. Movements can occur in both environments and different life stages may have quite different modes of locomotion. This chapter discusses and provides examples of the diversity of means of locomotion in and on water, and considers some of the physical principles involved. Movement in the terrestrial environments will be discussed in Chapter 9. Because the insects evolved on land and many groups secondarily invaded aquatic environments, there is enormous diversity in the adaptations for movement in and on water.

As should be apparent from any discussion of the physical properties of water (Chapter 5), moving in water is likely to present special challenges to insects, challenges that can be quite different to those faced by insects in the terrestrial environment. The size of the insect as well as the size of its appendages has an effect upon the Reynolds number at which it lives, which in turn determines whether the insect deals primarily with viscous or inertial forces. Such aspects create profound differences between taxa and instars in the equipment they need to move about in the aquatic environment, as well as how they deploy that equipment. Segmented legs, for example, can be used in many ways, including various forms of walking and swimming. Some insects move without the use of any appendages, but their seemingly erratic wiggling is in fact a much more complex and efficient form of propulsion than it appears. A few aquatic insects exploit other animals to supply movement, typically over longer distances than they can move themselves. Particularly fascinating are the ways in which insects have evolved a means for standing and moving around on the top of water using surface films. Grasping how insects achieve this lies in a basic understanding of the properties of surface films and water repellency of the insect integument. In addition, the movement of water creates challenges as well as opportunities, especially for insects that live in running waters. It is important to note that any one species may use more than one mode of locomotion depending, for example, on motivation (escaping predators versus foraging) or water movement (fast versus slow flows).

Many forms of locomotion involve the legs, so we start with a description of the basic leg structure common to many insects, and how those legs can be moved (Section 8.2). The subsequent sections are organized around the general places where insects move: in the water column via various forms of self-propulsion (Section 8.3), in the water column and taking advantage of water current (Section 8.4), on and in substrates (Section 8.5), on the water surface (Section 8.6), and attached to other animals (Section 8.7). Within each of these general areas, we consider the movement of insects with and without appendages. We also consider some aspects of behaviour, as behavioural decisions during locomotion have profound effects on the direction and distances that insects move.

8.2 Basic leg structure and movement

The basic body plan of adult insects includes three pairs of legs attached to the thorax, one pair per thoracic segment (Chapter 1). Legs evolved for walking

Aquatic Entomology. First Edition. Jill Lancaster & Barbara J. Downes.
© Jill Lancaster & Barbara J. Downes 2013. Published 2013 by Oxford University Press.

and running on hard surfaces, and the basic mechanisms may work equally well in terrestrial and aquatic environments. Legs have undoubtedly been modified secondarily to assist swimming and these adaptations are generally restricted to the aquatic insects.

Like many parts of the insect body, the legs are covered with exocuticle, a tanned cuticle that forms a hard exoskeleton and that provides rigidity against which muscles can work. Muscles for the legs may have one end inserted into the thorax ('extrinsic' muscles) or both ends inserted within the leg segments ('intrinsic'). The large, important muscles that are critical in leg movement are those that move the coxa and these are typically extrinsic, whereas muscles that move leg segments relative to each other are usually intrinsic. At joints, the hard exocuticle is absent so the cuticle is both soft and flexible, and this permits both articulation and movement. Each leg comprises six sections or segments: the coxa, trochanter, femur, tibia, tarsus, and pretarsus (Figure 1.4). The tarsus is usually subdivided into between two and five tarsomeres. Although often called tarsal segments, tarsomeres are not true segments because they have no muscle attachments and are connected only by flexible membranes, which permit movement. In adult insects, the pretarsus typically comprises a pair of tarsal claws and a lobe in the middle called the arolium (Figure 1.5); larvae typically lack an arolium and may have one or two claws.

Between the leg segments lies a membrane (the corium) and, where the segments are articulated, there is a joint. Some adjacent segments are fused together, e.g. the femur and trochanter are typically fused. Joints are of two general kinds. Monocondylic joints have one point of articulation, which means the joint has the capacity to be rotary and these joints are almost as good as ball-and-socket joints (Figure 8.1a, b). Diconodylic joints have two points of articulation, which results in movement only in a single plane, like a hinge (Figure 8.1a, b). Diconodylic joints can differ depending on whether the articulation points are at the tip of a segment or further back. The overall range of movement of legs thus depends on the locations and types of joints, but some arrangements are common to many insects (Figure 8.1c). Starting with the most proximal

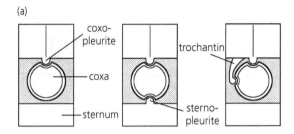

(a)

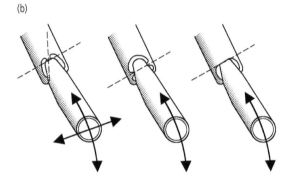

(b)

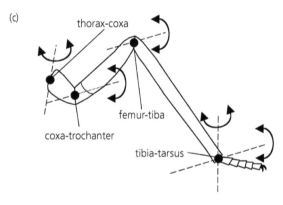

(c)

Figure 8.1 Types of joints in arthropod legs. (a) Articulation of the coxa with the thorax, illustrating a monocondylic joint with a single articulation point on the coxopleurite (left) and two diconodylic joints in which the two articulation points are opposite (pivot joint, middle) or more adjacent (hinge joint, right). (b) Hypothetical joints, corresponding to those in (a) and viewed from the inside angle, illustrating movement directions of adjacent segments. Dashed lines indicate the axes of rotation, solid arrows indicate the direction of movement around each axis. (c) Simple model of an entire leg showing how the different segments move relative to one anther (see text for description). Movement directions between adjacent tarsomeres have been omitted to reduce clutter.

segment, the coxa is typically articulated with the thorax via a diconodylic joint. The trochanter is articulated with the coxa also with a diconodylic joint, but is fused to the femur. Both these dicono-dylic joints thus operate only in one plane but, because they are typically set at a 90° angle to one other, they allow the base of the leg to be moved in three dimensions relative to the thorax. The tibia is articulated with the femur also with a diconodylic joint, but articulated with the tarsus by a mono-condylic joint. Basic leg structure and movement has been studied in relatively few types of insects, but in some species (e.g. adult dragonflies) the tho-racico-coxal joint has only a single point that acts as a fulcrum (monocondylic joint), so that movement is almost rotary (Frantsevich and Wang 2009).

The legs are attached to the side of the thorax, not to the ventral surface, so that the body typically hangs suspended between the legs, although this may be less obvious in bodies with strong dorso-ventral flattening. Forces on the legs thus comprise downward force caused by supporting the weight of the body (i.e. gravitational force) coupled with horizontally directed forces occurring during pro-pulsion, or by water moving past a stationary insect. Because water is much denser than air, it provides some support to the body and aquatic insects typi-cally experience smaller downward forces than their terrestrial counterparts. Horizontal forces can occur in both the longitudinal (front to back) and lateral (side to side) planes. Because of these orthog-onal forces, the legs play different roles in support-ing the body and in generating movement. The forelegs have a very wide range of movement, much greater than either the mid- or hind legs, due to their position at the front of the thorax and the angle at which they protrude from the body. This range of movement means they can produce high leverage, but the forelegs actually tend to impede forward motion in most positions. Much of the forward pro-pulsion comes from the mid- and hind legs, which also bear most of the animal's weight. In fact, the forelegs are often used predominantly for purposes other than locomotion or supporting the body, such as the raptorial forelegs some insects use for feeding (Chapter 13).

When walking, forward movement is attained because the legs can be *flexed* (i.e. the angles between segments are reduced and the leg folds up), which means the tips can be lifted clear from the ground allowing the leg to be swung forwards. The leg can

also be *extended* (extension is the opposite of flexure as angles between segments are increased) and this can produce thrust. How flexure and extension pro-duce movement, however, differs between legs. Because the forelegs have such a wide range of movement, they can be swung well to the front and extended. After the tip is placed on the ground, flex-ure of the foreleg *pulls* the body forward (Figure 8.2) until the leg is perpendicular to the body; subse-quent extension then produces thrust and the body is *pushed* further forwards. A similar process oper-ates with the mid- and hind legs, but, because they have much narrower ranges of movement, leg extension and thrust (i.e. pushing) are more impor-tant than pulling (Figure 8.2), and the greatest thrust overall is generated by the hind legs. This sequence of movements produces forward and lateral motion, and this is counterbalanced by legs on opposite sides of the body. Transverse movements created by the fore- and hind legs on one side are counterbal-anced by transverse movement of the mid leg on the other side. Insects thus have at least three points of contact with the ground and this 'tripod' ensures balance. At low speeds, only one leg is raised at a

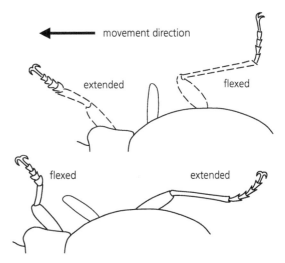

Figure 8.2 Diagram illustrating the roles of flexure and extension of the fore and hind legs of a walking insect (dorsal view of a hypothetical beetle). In top panel, the foreleg is fully extended and the hind leg fully flexed in preparation to move. Both legs are anchored by the tarsal claws at the tips of the legs. In the bottom panel, the foreleg is flexed to pull the body forward, at the same time the hind leg is extended producing thrust to push the body forward. The middle leg has been omitted to reduce clutter.

time and a typical gait would see the hind-, mid-, and foreleg on one side moved sequentially, followed by the hind-, mid-, and foreleg on the other side. At high speeds, the mid leg on one side and the fore- and hind legs on the other side are on the ground at the same time, thus achieving the tripod gait mentioned above.

8.3 Movement in the water column—self-propelled

Masters of moving through water, aquatic insects display an impressive array of mechanisms for propulsion. Most common are various forms of swimming with jointed legs (via rowing and flapping) and body undulations, but a few are capable of jet propulsion and of manipulating buoyancy to move vertically through the water column. Such movements, however, also need to be controlled, i.e. animals must be able to maintain stability in body orientation and manoeuvre at will. Unsurprisingly, there have been strong selection gradients to maximize energy efficiency of movement, e.g. by maximizing thrust and minimizing drag along the body as it moves forward.

8.3.1 Swimming using jointed appendages

Jointed legs are used by many aquatic insects to row or swim through the water column and at the surface (movement on top of the surface film is discussed in Section 8.4). Some are very competent swimmers, others much less so. As highlighted in Chapter 5, understanding movement through water means first considering the Reynolds number (Re) at which the limbs act in order to understand the conditions under which they create thrust. This section considers insects that swim at sufficiently high Re that they experience water as turbulent and where pressure drag dominates. This is also where most research has been carried out. For movement at low Re, see Section 8.3.5.

When swimming at high Re, legs row and act as paddles in a form of drag-based propulsion. Put simply, thrust is generated by drag on the paddle as it pushes against the water, which moves the insect in an opposite direction (i.e. momentum is conserved). How quickly the body moves will depend

on many factors including the shape of the paddle and the various drag forces along the body (e.g. pressure drag, skin friction, formation of vortices). Drag on the paddle creates turbulence in the water (in much the same way that turbulence is created as water moves past a stationary object, Chapter 5), and typically involves the creation of vortices, which are spinning, often turbulent, flow patterns. Such vortices will be familiar to anyone who has watched the oars on a row boat.

The sequence or order in which legs are moved affects swimming efficiency. In synchronized-leg swimming (also called perfect transverse phase synchrony), each leg within a pair that is used to generate thrust moves in perfect synchrony (Figure 8.3). This form of synchrony ensures that the lateral forces produced by leg motion are perfectly counterbalanced so that the animal moves straight ahead—just as a row boat does when propelled by a competent sculler. Most proficient swimmers use synchronized-leg swimming, including some adult beetles (Dytiscidae and Gyrinidae) and several families of Hemiptera (Notonectidae, Corixidae, Nepidae, Pleidae, and Belostommatidae). In contrast, alternate-leg swimming (also called diagonal phase

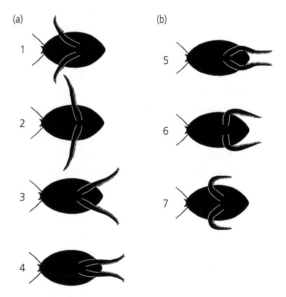

Figure 8.3 Illustration of perfect transverse phase synchrony. The stroke phases of the hind legs of a dytiscid beetle, showing the sequence of leg positions during (a) the power stroke, and (b) the recovery stroke.

Source: Adapted from Nachtigall (1985).

synchrony), where legs on the same side of the body and opposite legs within pairs are 0.5 cycles out of phase, produces a distinct side-to-side movement of the body. This less proficient form of swimming is employed by many adult beetles in the families Hydrophilidae, Haliplidae, and Curculionidae, and some larval Dytiscidae (Barr and Smith 1980). Alternate-leg swimming is essentially a tripod gait (Section 8.2) and may represent a more plesiomorphic state, whereas synchronized-leg swimming is likely to be a specialization for swimming and, indeed, it is may be a relatively inefficient way of moving on land.

Body shape and limb structure have been modified in several different ways to aid energy-efficient swimming. Efficiency involves minimizing drag, but also involves stability, manoeuvrability (i.e. ability to make turns), the ability to swim up and down in the water column, and the attendant problems of buoyancy. Stability involves ensuring that the insect does not 'roll' about the long axis of the body, 'pitch' up and down around the lateral axis perpendicular to the body, or 'yaw' about the vertical axis in the horizontal plane. Adult dytiscid beetles, in particular, have minimized drag and achieved excellent stability. Often, their bodies are dorso-ventrally flattened and streamlined, and have sharp lateral edges along the elytra; these help to reduce roll and pitch, and thus provide stability. These beetles swim at a small angle of attack (the angle of the body to the surface of the water), which they are able to maintain at all speeds. Adult dytiscids are air breathers and gas exchange is via an air bubble under the elytra, and the dorsal position of this gas store also helps them to stay upright. In contrast, the bodies and legs of adult Gyrinidae (whirligig beetles) produce superior manoeuvrability, but at a cost to stability. They are unable to swim at a consistent angle of attack without constant correction during propulsion (Nachtigall 1985), but 84 per cent of energy in swimming is converted to thrust that achieves forward movement and this is certainly among the highest (if not the highest) in the animal kingdom. Their hind legs can beat at 60 times per second (Nachtigall 1961), and their distinctive, rapid turns are achieved by using the legs in an asymmetrical way in which outboard legs paddle faster than inboard legs. The foreleg may be extended and used as a brake, allowing the animal to pivot, and in some cases these beetles may even extend a wing on one side and use it to scull (Fish and Nicastro 2003).

Swimming legs may be modified in various ways, the most common being flattened segments and the provision of swimming hairs to increase the effective surface area used to create thrust. Segments of swimming legs may also be elongated, as in the notonectids and corixids that have long, flattened hind legs. Similarly, the mid legs of the weevil *Hypera eximia* have significantly longer tibias and longer and wider femurs than their non-swimming congeners (Cline et al. 2002).

During the power stroke, swimming legs are deployed at an angle that maximizes the surface area used to create thrust, from both the leg segments and the swimming hairs. During the recovery stroke, the leg is rotated through 90° and presents a much smaller surface area in the direction of movement, thereby reducing backwards thrust. This is exactly analogous to orientation of an oar in the drive and recovery strokes used when rowing a boat. In adult gyrinids, the flattened tarsomeres, which can move relative to each other, may fan out during the power stroke and then collapse together like a pack of cards during recovery, again reducing surface area. The legs may be held close to the ventral surface of the body during the recovery stroke, which keeps them inside the boundary layer at the body surface (Figure 8.3b). Deploying the legs at different angles means that they can act as brakes or diving planes, and turning can be achieved by either using different frequencies of beating on each side or by beating the legs out of phase.

Swimming hairs are often the most significant part of creating thrust and can produce more thrust than the leg segments. In the dytiscid beetle *Acilius*, for example, swimming hairs produce 80 per cent of the surface area (Figure 8.4) and 68 per cent of the drag that creates forward thrust (Nachtigall 1961). Swimming hairs are typically attached to the tibia or tarsus and, in most species, simple water pressure during the power stroke ensures that they spread out into the water column and create drag. During the recovery stroke, the hairs have mechanisms ensuring they lay flat against the leg. In the dytiscids and gyrinids, swimming hairs have a 'snapping mechanism' that ensures they flip back

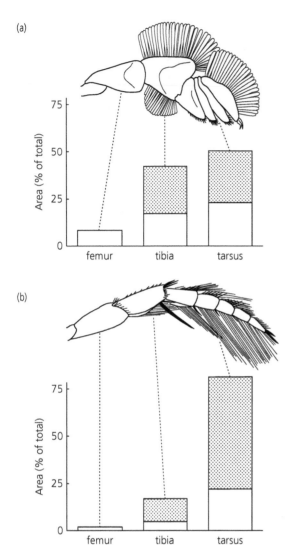

Figure 8.4 Diagrams of the hind legs of two aquatic beetles, (a) *Gyrinus* and (b) *Acilius*, illustrating the modifications for swimming. The graphs illustrate the percentage surface area accounted for by the hairs (stippled areas of bars) and leg segments (open areas).

Source: From Nachtigall (1961). With kind permission from Springer Science and Business Media.

against the leg during recovery. Rowing legs of the backswimmer, *Corixa punctata*, have about 5000 swimming hairs, many of which are extremely fine at only 2.5 μm diameter. Each flexible hair is attached via a basal ring that contains a groove toward the outer edge of the leg, such that the hair swivels towards the flattened edge of the leg during the recovery stroke (Nachtigall 1985).

Swimming at the water surface, as is characteristic of adult gyrinids, can limit swimming performance through an additional component of drag on the body called wave drag or wave resistance. Wave resistance relates to the energy transferred to the water when the body and legs create surface waves during movement (see also Section 8.5.1). *Gyrinus substriatus* has different swimming speeds depending on the pattern and frequency of leg use, and travel speed is closely associated with minimizing overall fluid resistance. At slow speeds (when they may deploy only the mid legs), the beetles travel at speeds less than 23 cm s^{-1}. Coincidentally, there is a minimum speed at which waves can travel at the air–water interface, and this minimum speed is 23 cm s^{-1}. Consequently, an object must travel at speeds of at least 23 cm s^{-1} to create surface waves in water (Denny 2004). Consequently, by employing a gait that delivers speeds of less than 23 cm s^{-1}, the beetles avoid creating surface waves with their bodies (if they travel in a straight line at more or less constant speed), which means they do not suffer the loss of forward momentum caused by wave drag (Voise and Casas 2010). Gyrinids can swim quickly when necessary and at fast speeds the hind legs generate most thrust and the mid legs create little forward momentum. It is possible that the mid legs are needed instead to maintain stability, because gyrinids do not have a keel on the bottom of the body that can reduce yaw (Nachtigall 1961). Once gyrinids exceed 23 cm s^{-1}, wave resistance is immediately at a maximum, but declines as the animal speeds up (Figure 8.5). On the other hand, fluid resistance (i.e. drag and energy lost through creation of vortices by the legs) increases with speed, at all speeds. Using the fast gait, the typical travel speed is approximately 40–50 cm s^{-1}, which minimizes both forms of fluid resistance (Figure 8.5). At very high speeds (up to 70 cm s^{-1}), gyrinids lose a lot of energy in fluid resistance, so these speeds may be used only in escaping from predators or other life-threatening situations (Voise and Casas 2010). Whirligig beetles are also unusual in that they can occur in swarms, often with hundreds or even thousands of beetles per aggregation. There is an energetic advantage to being at the rear of the pack, where individuals can take advantage of the reduced pressure created by moving individuals that are

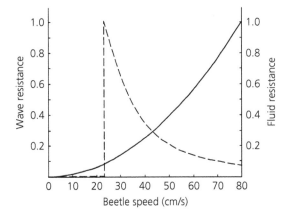

Figure 8.5 Illustration of the relation between the speed at which *Gyrinus substriatus* moves and the resultant wave (dashed line) and fluid (solid line) resistance. At low speeds, the beetle suffers little fluid resistance and no wave resistance by travelling at < 23 cm s⁻¹. Above 23 cm s⁻¹, wave resistance is initially maximal then declines rapidly with increasing speed, but fluid resistance increases. Speeds of 40–50 cm s⁻¹ produce the least overall resistance for fast travel.

Source: Republished with permission of the Royal Society, from Voise, J. and Casas, J. (2010). The management of fluid and wave resistances by whirligig beetles. *Journal of the Royal Society Interface*, 7, 343–352, Figure 7.

ahead of them (which is called 'drafting' or 'slip-streaming'), much like the benefit reaped by cyclists in packs in road races (Romey and Galbraith 2008).

Despite the energetic disadvantages of creating surface waves, there are situations in which gyrinids apparently create waves for communication purposes. Mechanosensory perception of surface waves, by use of Johnston's organ (Section 7.2.2) is a crucial way that these beetles receive information about the surrounding environment. Gyrinids are often seen chasing one another (presumably for mate location given that males are inevitably the chasers) and this behaviour typically involves travel speeds above 23 cm s⁻¹ and inevitably involves swimming 'in circles' (Bendele 1986). As mentioned above, a speed of 23 cm s⁻¹ is the minimum speed required to create surface waves but beetles travelling more slowly can generate capillary waves (Section 8.6.1), especially if they turn frequently because acceleration is inherent in circling motion (Voise and Casas 2010). Similarly, when gyrinid beetles travel in 'trains' during the formation and break up of rafts, the trains move in straight lines, but beetles within the train swim in a zig-zag fashion (Heinrich and Vogt 1980). Theoretically, this swimming motion may create waves

required for individuals to communicate with one another and stay in line. Thus, the whirling swimming behaviour of these aptly named whirligig beetles may function in energy-efficient communication as well as to confuse would-be predators.

Not all aquatic beetles are proficient swimmers. Many aquatic weevils, for example, live associated with vascular plants, some are semi-aquatic and others are fully aquatic, and this is reflected in a wide range of swimming abilities. For example, of weevils in the genus *Hypera*, *H. eximia* is a reasonably proficient swimmer, using all three pairs of legs and with most thrust generated by the mid- and hind legs, which are deployed in transverse phase synchrony. In contrast, both *H. postica* and *H. nigrirostris* are unable to swim and sink if placed in water (Cline et al. 2002). The rice weevil, *Lissorhoptrus oryzophilus*, also uses transverse phase synchrony but creates thrust using only the mid-legs. It is a capable diver and swimmer (Hix et al. 2000). Several weevil species (e.g. *Ochetina uniformis*, *Ludovix fasciatus*) swim at the surface with some of the body and legs submerged, but other parts above the surface—a similar orientation to adult gyrinids, but they are much less efficient. All three pairs of legs are used in propulsion, although the mid-legs provide most thrust (de Sousa et al. 2007).

Some seemingly un-streamlined insects are quite proficient swimmers, including some cased caddisflies. Swimming with a case produces some unique problems, but swimming is common in the Leptoceridae and some species appear to swim more often than walk. These caddisflies often have dense setae or swimming hairs and exceptionally long legs, particularly the hind legs, which protrude forward out of the case. Larval *Triaenodes* (Leptoceridae) swim with their legs in transverse phase synchrony with the legs fully extended during the power stroke and folded on recovery (Tindall 1964; Gall et al. 2011). The articulation between the coxa and the femur permits the legs to be swept through a very wide angle from front to rear. Nevertheless, the swimming motion is quite jerky and the case has to be dragged along. Given its spiral shape, the case has an asymmetrical projecting edge at the front that likely causes the case to spin around its axis, which must be offset by the swimming legs and by the larva gripping the case with its anal hooks. Although potentially inelegant, it is clearly an effective mode of swimming, as evidenced by records that *T. tardus* can swim up to 1.47 cm s^{-1} whilst carrying a case almost twice the mass of the larva, and larvae can swim continuously for up to six minutes (Gall et al. 2011).

8.3.2 Swimming using undulations of the body

Moving through the water column can also be achieved by undulating the body, either from side-to-side (lateral undulation) or up and down (dorsoventral undulation or 'cetacean' swimming), and this occurs in insects with and without jointed appendages. The basic principles of creating thrust with body undulations is essentially the same as for paddles (Section 8.3.1). Bending the body generates forces against the water and propels the animal in the opposite direction, typically creating vortices at the same time.

The damselfly *Enallagma cyathigerum* (Coenagrionidae) swims with its legs when moving slowly, but also with lateral undulations of the body. These undulations stay in phase with the movement of the legs, which use the typical, slow hexapodal gait described above (i.e. hind-, mid-, and forelegs move sequentially on either side of the body) and all three

pairs of legs are employed in a rowing motion. When the animal needs to move quickly, perhaps in response to a perceived threat from a predator, it uses only body undulations, and the legs are trailed toward the rear to reduce their drag. The caudal lamellae or 'fins' of these larvae are critical to creating thrust; they comprise three, movable plates, one of which is median and dorsal, whereas the other two are ventral and lateral. During undulations, water pressure causes the two ventral plates to come together and press against the median plate, and this enables the plates collectively to act like the caudal fin of a fish. Vortices are created on each side of the body for each half-stroke and, while about one-third of the energy is lost to the wake, about two-thirds create forward motion of the body (Brackenbury 2002). The mayfly *Cloeon dipterum* (Baetidae) also uses undulations for movement, and has a tail fin comprising three, long bristles covered in hairs and held in the horizontal plane. *Cloeon* uses cetacean swimming, but because it often swims on its side, it appears superficially to undulate from side to side. Vortices are shed above and below the larva and travel away (Figure 8.6), but efficiency is lower than that of the damselfly larva with only about 14–17 per cent of energy translated into forward motion. Nevertheless, this mayfly can move much more quickly than the damselfly and as quickly as dytiscid beetles,

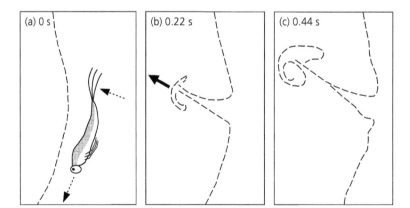

Figure 8.6 Illustration of a discrete ring vortex shed from the tail fin of a swimming mayfly larva, *Cloeon dipterum* (Baetidae), when viewed from above. The larva encounters a dye streamer (dashed line) when swimming in open water; *C. dipterum* swims using dorsal-ventral undulations, but swims on its side. In (a), the larva is swimming downwards in the field of view; the tail fin is moving to the left and is within 20 ms of the completion of its dorsal half-stroke. To the left of the larva is a curved streamer of dye. Frames (b) and (c), show the progressive outlining of the ring vortex shed from the tail as a wake element penetrates the dye streamer and propagates outward, in the direction of the arrow in (b).

Source: Adapted from Brackenbury (2004).

possibly due to the minimization of drag on the body (Brackenbury 2004).

Undulations in the dorso-ventral plane are commonly used by larval mayflies. For many baetids, such as *Baetis tricaudatus,* large larvae (body length > 4 mm) swim exclusively via body undulations and, like the damselfly above, the legs are trailed behind to reduce their drag. This may seem like an ungainly form of movement but these mayflies can achieve instantaneous swimming speeds of 16 cm s^{-1} (Kutash and Craig 1998). Small *Baetis* (≤ 3 mm) use a combination of undulations and rowing with the legs, which appears to reflect the constraints on movement at low *Re* (Section 8.3.4), and the fastest instantaneous swimming speeds are much slower at approx 3 cm s^{-1}.

Movement via body undulations is perhaps the main form of swimming in aquatic insects that lack jointed appendages and have either only stumpy prolegs or no legs at all. Many are larval Diptera, with bodies that are typically long and somewhat worm-like. Larval ceratopogonids, for example, move via sinusoidal undulations of the whole body that is moved rapidly from side to side. The amplitude of these undulations increases toward the rear and this pushes the body forward, although not very quickly because the insect only moves about one-fifth of a body length forward per oscillation (Nachtigall 1985). Similar undulations occur in some chironomid pupae, such as *Chironomus plumosus,* and this movement seems designed to achieve rapid ascent prior to adult emergence from the water surface (Brackenbury 2000). Side-to-side flexure, sometimes called body curling, is adopted by a variety of dipterans (chironomid, culicid, and corethrid larvae, and also pupae in some cases: Brackenbury 2001a). In body curling, the head is flexed toward the abdomen on one side then the other, so that the body coils and uncoils repeatedly. At the same time, the body is rotated clockwise and then counter-clockwise, with the net result that the larva moves forward at a speed of about two body lengths per second, with the head and abdomen moving through figures of eight (Brackenbury 2003). The mechanism of gaining forward momentum is similar to other body undulations in that a vortex is shed down the body just below the horizontal, from each half-stroke, as forward thrust is

created. Movement may also be aided by fans of swimming hairs or blades on the abdomen (Figure 8.7), which are spread fully to increase drag when the abdomen is in thrust phase and closed when the abdomen is moving forward (Brackenbury 2000). Similar, rapid body flexure and somersaulting is used by dipteran pupae (*Chironomus, Culex, Aedes*), and this is a mechanism to escape predation (Nachtigall 1985; Brackenbury 1999a, 2000).

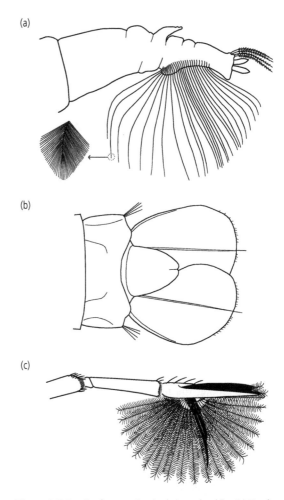

(a)

(b)

(c)

Figure 8.7 Details of some swimming hairs and paddles. (a) Tip of abdomen and tail fan of a larval phantom midge, *Chaoborus vagus*. Inset shows enlarges tip of a single hair to show branching that occurs on all fan hairs and increases the effective surface area. (b) Tip of abdomen and paddles of a mosquito pupa, *Culex*. (c) Tip of middle leg of the velid *Rhagovelia* with expanded swimming fan.

Sources: (a) From Colless (1986). Reproduced with permission from CSIRO Publishing. (c) Republished with permission, from Anderson (1976).

8.3.3 Jet propulsion

Jet propulsion is quite specialized and spectacular, and occurs in some larval dragonflies (e.g. *Anax*, *Aeschna*) and some mayflies (*Chloeon*). Slow walking is the normal mode of travel for most larval dragonflies, but some can move much faster in emergencies (Corbet 2004). To achieve rapid propulsion, water is drawn into the hindgut (often during gas exchange, Chapter 3) and ejected sharply out of the anus by contraction of dorsal and longitudinal muscles. The velocity of this water jet can be up to 2.5 m s^{-1}, which shoots the larvae forwards, allowing them to cover up to 30 cm and achieve peaks speeds of 50 cm s^{-1} with jet frequencies of up to three per second (Hughes 1958; Mill and Pickard 1972).

8.3.4 Vertical movement using changes in buoyancy

Insects that are able to control or manipulate buoyancy can move up and down in the water column using little or no energy. Air-breathing aquatic insects that carry a gas store have positive buoyancy, which can aid the insect in rising through the water column, but can also create difficulties during dives. These problems of controlling buoyancy are further complicated by changes in size of the air store as the insect uses up oxygen (Chapter 3). For many, the only solution is either to swim vigorously to counter positive buoyancy or cling to objects under the water. Some backswimmers, however, have a fascinating strategy to maintain neutral buoyancy over long periods. Species of *Anisops* and *Buenoa* (Notonectidae) are among the few insects that have haemoglobin. During dives, the O$_2$ removed from the gas store during respiration is replaced by releasing O$_2$ bound to the haemoglobin into the air store, thereby preventing the loss of nitrogen and maintaining an air store of constant volume. In *Anisops*, O$_2$ from the air store is supplied to the swimming muscles via the thoracic spiracles, but the haemoglobin lies in tracheal cells of the abdomen. Oxygen stored in the abdomen does not move via the tracheal system to replenish the gas store, but follows an external route. Using scraping movements of the hind legs, the backswimmer pushes the gas store backwards from its normal location over the thoracic spiracles, so

that it lies briefly over the abdominal spiracles communicating with the tracheal cells containing haemoglobin. Oxygen enters the bubble, which is then flicked anteriorly. Manipulation of the bubble in this manner can occur 20 or more times during a single dive (Wells et al. 1981; Matthews and Seymour 2006) and allows the insect to move freely around in the water column for long periods (Matthews and Seymour 2008).

Phantom midges (Chaoboridae) are not air breathers and use a different mechanism to control buoyancy and vertical movements through the water column. Some lake-dwelling species undertake an impressive diel migration, emerging from the mud at lake bottoms at sunset and rising many metres into the water column at night (possibly to evade predation by fish, Juday 1921). As described in Section 5.3.1, the larvae are able to change their buoyancy by manipulating the walls of tracheal sacs, which allows them to move up and down or hang in the water column.

8.3.5 Movement at low *Re*

The animals described in previous sections all swim at Reynolds numbers where they deal primarily with inertial rather than viscous forces. Even small chironomids undergoing body curling operate at a Reynolds number of about 350, meaning they create turbulence as they move forward (Brackenbury 2003). For small organisms moving at low *Re*, viscous forces predominate, flow is laminar, and the main form of drag is skin friction. For these organisms, appendages equipped with hairs can be effective means of moving, but whether such appendages act as paddles and create thrust (because water does not pass between the hairs) or act more like sieves depends critically on the distance between hairs and the *Re* at which the limb is deployed. In addition, whether an appendage should be rowed or flapped to create thrust can change over a very narrow range in *Re* (Sensenig et al. 2009). Consequently, it is difficult to generalize and detailed research is needed if we are to be confident of understanding how insects move at low *Re*.

The few studies of insect movement at low *Re* suggest that the mechanisms are as diverse as at high *Re*. As discussed in Section 8.3.2, the mayfly

Baetis uses only body undulations to move through the water column, whereas small larvae use a combination of rowing and undulations. The *Re* for small larvae was estimated at around 30, sufficiently small to suggest that viscous forces will dominate and hence make body undulations alone an ineffective means for creating thrust. As larvae grow larger, they gradually reduce the use of legs for rowing until they have transitioned to body undulations completely (Kutash and Craig 1998). In contrast, mosquito larvae (*Culex pipiens*) living in still water use body flexing to move rapidly, but when undisturbed they can glide through the water column keeping the body straight. This is achieved using a pair of mouth brushes, which can be extended so that the hairs spread out and form a domed surface, or are flexed closed. The brushes flex and extend repeatedly at a rate of just over 10 strokes s^{-1}. The action of the brushes creates a current in front of the head that jets down the body, pulling food particles in the water column toward the larva's mouth as well as causing the body to glide gently through the water column. The mouth brushes thus act as paddles (rather than sieves) and a calculation of their *Re* of operation of 0.2–0.4 is consistent with this behaviour. It is a relatively efficient way of being a filter-feeder in still water because it allows the larva to move between patches of food (Brackenbury 2001b).

8.4 Movement in the water column— exploiting water currents

Stream-dwelling aquatic insects can make use of downstream currents to move, a phenomenon called drift. Drift is often considered to be a passive form of movement, but most species actually have a great deal of control over drift, suggesting that it is far from passive (Lancaster 2008). Some insects can choose when to enter the drift (behavioural drift); most can also control the distance they travel, and also the point at which they escape currents and return to the substrate. For many species, drift shows a marked diel periodicity, with the greatest abundance of animals drifting at night-time and, for some species, particularly just after sunset. There has been considerable debate about whether such diel periodicity is caused by aquatic insects choosing to enter

the drift deliberately at night (e.g. to avoid visually feeding predators) or whether they tend to be dislodged or swept off the tops of stones at times of day when they forage most frequently (Waters 1972; Brittain and Eikeland 1988). Although some animals are certainly dislodged by accident, others exhibit behaviour that is clearly deliberate entry into the drift. For example, larvae of the mayfly *Baetis* (Baetidae) use the drift to leave stones when their capacity to gather food falls below a threshold rate (Kohler 1985), and several mayfly species drift to escape predators (Lancaster 1990; Forrester 1994). Changes in stream discharge can also increase drift of mayflies through behavioural mechanisms (*Baetis, Epeorus longimanus*) or passively (*Paraleptophlebia heteronea, Ephemerella infrequens*) (Poff et al. 1991). Stream discharge can also decrease drift in some taxa (Downes and Lancaster 2010). Blackfly hatchlings can also control entry into the drift and are more likely to leave the natal site if water velocities are too slow for filter-feeding or to escape high densities of siblings (Fonseca and Hart 1996).

Once in the water column, insects need to control the distance travelled and to be able to regain the bottom within suitable habitat. If not, they risk being transported into unsuitable habitats, injury, reduced fitness, or death—time in the drift is time insects cannot feed and may be vulnerable to predation. Generally, most drifting insects probably do not travel very far, except perhaps during catastrophic floods and spates (Elliott 1971a; Lancaster et al. 1996; Lancaster et al. 2011). Control of drift distance typically involves moving up or down in the water column either by swimming or changing body posture to promote sinking (or not). These simple vertical movements will have a profound effect on the distances individuals travel because of the increasing speed of water at greater distances from the substrate. The logistic difficulties, however, of observing individuals closely under natural conditions means there is relatively little detailed information, other than in flumes. Species that are capable swimmers, such as the mayflies *Baetis* (Baetidae) and *Cinygmus* (Heptageniidae), can swim downwards to regain the substrate (Allan and Feifarek 1989; Oldmeadow et al. 2010). Few studies have examined this behaviour using the natural topography of stream beds, which creates turbulence patterns that are different from simple, flat-bottom flumes. The replication of natural or near-natural bed topography in flumes can be very revealing. In fast flows, mayfly larvae of *Baetis rhodani* adopt a 'parachute' posture with the legs splayed out, which prolongs drift, but when they encounter the turbulence created by water flowing over abrupt steps (natural features in many rocky streams), they swim, even upstream and through very turbulent flow, to regain the bottom (Oldmeadow et al. 2010). In contrast, larvae of another mayfly, *Ecdyonurus torrentis* (Heptageniidae), also adopt a parachute posture to prolong drift, but are unable to maintain this posture or swim in high turbulence. Instead, they tumble and drift beyond zones of high turbulence before regaining the bottom. These observations match the distribution of both mayflies species in natural streams, where *Baetis* are frequently found in hydraulically complex habitats, whereas *Ecdyonurus* are more often found in hydraulically smooth flows (Oldmeadow et al. 2010).

An unusual mechanism to control drift distance is through the use of silk threads, as seen in some blackflies (Simuliidae). Many blackfly larvae are filter-feeders and consequently need to exit the drift in relatively fast-flowing places. They can extrude silk from their mouths in thin strands or lines, and these silk lines may aid their transport in the drift if, by chance, they end up in slow-moving sections of streams, which are unsuitable for filter-feeding. Silk lines reduce the velocity at which larvae sink and increase their drag, both of which help larvae to move through slow-flowing sections of streams or to regain the drift if they are deposited in pools (Fingerut et al. 2009).

Exiting the drift, especially from fast flows, may require specific adaptations. Species that have jointed legs may simply attach themselves to the bottom using their tarsal claws when within reach of the substrate, but this may depend on the roughness of the substrate surface (Lancaster and Mole 1999). Blackfly larvae, which lack true legs, may once again use silk. In slow-moving water, larvae of *Simulium vittatum* (Simuliidae) can re-attach themselves by 'biting' the substrate with their mouthparts. This is difficult in fast-moving water, but if they can contact the substrate even briefly with their mouthparts, they can attach a silk drag line to the substrate and then 'reel'

themselves in like rope-climbers, although even this is impossible in very fast flows (Fonseca 1999). *Simulium tribulatum* behaves similarly and silk lines up to 13 mm long increase the settlement chances of larvae passing over rough substrate by fortyfold (Fingerut et al. 2006).

8.5 Movement over the substrate surface

It is perhaps unsurprising that the primary means of moving around on a substrate surface are broadly similar for aquatic and terrestrial insects. Those movements, however, are often affected strongly by the aquatic environment, especially in flowing water, and the way the surface topography interacts with flow. There are also some surprises, for example, in the ways that silk can be used to aid locomotion.

8.5.1 Walking and crawling

The basic gait used by insects walking along the substrate under water is much the same as for terrestrial insects, i.e. a tripod gait, Section 8.2, but because the body has only slight negative buoyancy, there is less downward force on the legs. Instead, the legs must cope with loads that are distributed along and across the limbs, plus the torsion created by drag on both the body and legs as they are moved through a medium with much greater viscosity than air. Rough terrain can make the gait irregular, given the types and arrangements of joints and the consequent angles over which the legs can be extended and flexed are numerous. Detailed studies of how aquatic insects move over the substrate under water are scarce compared, for example, with crustaceans (Wootton 1999). Remember that insects that walk are not necessarily restricted to walking, and many species move around using a mix of walking, swimming, and drifting (e.g. hydropsychid caddisflies: Downes et al. 2005; Downes and Lancaster 2010).

Behaviour during walking or crawling is an important aspect of locomotion, and substrate topography, water velocity, and food levels can all affect travel speed and direction. Field collections suggest, for example, that some species may avoid or seek places with particular speeds of flow, with some species being fast-flow specialists and others occurring most often in slower flows. Some walk-ing or crawling is in an upstream direction in streams, potentially offsetting downstream move-ment from the drift (Elliott 1971b; Söderström 1987; Williams and Williams 1993). Food abun-dance, substrate topography, and water velocity can all affect the directions of movement of aquatic insects, and therefore the locations in which they are likely to occur (Lancaster 1999; Hoffmann et al. 2006). Most experimental flumes have a flat sur-face or an artificial arrangement of pebbles, but one unusual study of the cased caddisfly *Pota-mophylax latipennis* (Limnephilidae) used a cast that faithfully replicated stream-bed topography (Lancaster et al. 2006; Rice et al. 2008). Larval *P. latipennis* crawled more slowly with increasing discharge and near-bed flow, and also changed direction depending on the arrangement of rocks such that junctions between adjacent rocks were travelled most frequently. Such behaviour in response to topography and near-bed flow cer-tainly affects both the directions and distances individuals will travel, a finding supported by field studies on marked caddisflies (Jackson et al. 1999) and stoneflies (Freilich 1991).

Insects without jointed legs can move about the substrate and often colonize new substrates remarkably quickly. The chironomid *Chironomus plumosus*, for example, is able to crawl along the bottom using its abdominal and thoracic prolegs to attach alternatively to the substrate and by flexing its body in a looping motion similar to the move-ment of caterpillars (Brackenbury 2000). The cham-pions of unique crawling styles are surely larval blepharicerids. These dipterans have a row of suckers on the ventral surface (Chapter 5) with which they are able to hang on to the substrate. When moving, the animal detaches one or more suckers, moves part of its body and then replaces them, resulting in forwards, backwards or side-ways movement depending on whether the animal is foraging for algae, dispersing, or escaping pred-ators (Frutiger 1998).

Before leaving the topic of walking and crawling, mention should be made of aquatic life stages with capacity to walk on land. Insects left stranded after a sudden drop in water level are often capable of crawl-ing and wriggling back into the water, perhaps using

gravity as a directional cue. Female mosquitoes of some species, such as *Anopheles gambiae*, lay their eggs on the edges of puddles, so that the larvae hatch out of water. The larvae use sinusoidal undulations to move over wet mud, and, on dry mud, move like caterpillars: segments are lifted in waves that travel from the abdomen to the head and move the animal forward. Their capacity to move across the substrate outside of water means the species can occupy a wide range of environments (Miller et al. 2007).

8.5.2 Using silk

Silk can be used by insects to move over the substrate as well as to aid movement in stream currents (Section 8.4). The caddisfly *Agapetus boulderensis* (Glossosoma-tidae) crawls in slow water velocities using all three pairs of thoracic legs but, at high velocities, it also uses silk and a pivoting movement (Olden et al. 2004). The larva, which occupies a stony case that is open at both ends (Figure 8.8), faces upstream and anchors the ven-tral surface of the upstream end of its case to the sub-strate with silk. It then turns around inside the case, protrudes its head and legs from the other opening, grasps the substrate and, using the silk attachment point, pivots the case broadside to the flow, manoeu-vring it to face upstream once more. More silk is used to bind the case to the substrate, and the larva then once again reverses position in order to sever the first silk binding. In this way, the larva is able to move upstream a case length at a time. Larvae make use of pivoting particularly in fast flows, meaning that water velocities and topographic refuges from the current in the substrate are important in determining the type of movement mode and direction taken.

Another group to make use of silk to move around on hard substrates is the blackflies, which adopt a behaviour called looping to move short distances. Blackfly larvae attach to hard substrates by spin-ning a small pad of silk into which they insert their abdominal hooks, enabling them to lean into the current to filter-feed. In looping, larvae lean over and spin a new pad of silk to which they attach with their mouthparts. The abdomen is then released from its silk pad and attached to the new one in a movement that looks superficially similar to that of inchworms. Water velocity may affect the extent of looping and distance travelled (Kiel 2001).

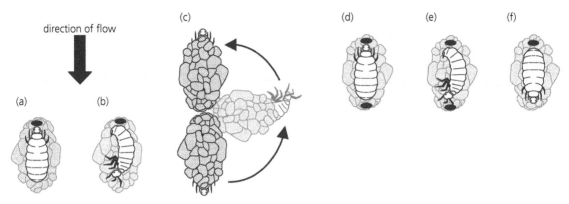

Figure 8.8 Illustration of the sequence of movements and probable behaviours during pivoting motion of the caddisfly *Agapetus boulderensis* (Glossosomatidae). Black ovals represent silk anchors. (a) The larva attaches the ventral, upstream surface of the case to the substrate. (b) The larva then reverses position within the case (so it faces downstream) and then, (c) using its thoracic legs, moves the case broadside to the current and then into an upstream position, using the silk attachment as a pivot. (d) The larva binds the new upstream, ventral surface to the substrate with more silk, (e) then reverses position, and finally (f) removes the previous silk attachment point.

Source: Adapted from (Olden et al. 2004).

8.5.3 Burrowing

Although most aquatic insects are found on the surfaces of substrates, or in the large interstitial spaces of objects close to the surface, a surprising array of insects burrow and move around in fine gravels and soft sediments. Insects may not move very far once inside burrows, but how they create these burrows is perhaps best considered in the context of locomotion. Burrowing implies that these insects are able to move sediment particles, and the viscosity of water may make this somewhat easier than burrowing is for terrestrial insects. Characteristic of burrowing animals, many of these insects have long and narrow bodies that minimize the volume of excavated sediments and facilitate in-burrow manoeuvring, and include a variety of Diptera, Ephemeroptera, and Plecoptera. Among the larval dipterans, some chironomids and tipulids (e.g. *Pedicia*) burrow into silty sand or gravels. The tipulids sometimes have 'inflatable' or bulbous segments near the tip of the abdomen with which they can anchor the rear of the body whilst digging with the front. Some burrowing chironomids (e.g. *Chironomus* spp.) line their burrows and spend most of the larval life in the sediment. The mayfly *Baetis pumulus* (Baetidae) and stoneflies *Chloroperla* (Chloroperlidae) and *Leuctra* (Leuctridae) have bodies that are more worm-like than many other members of these orders presumably an aid to their burrowing habits.

The so-called 'burrowing' mayflies, representatives of six families in the superfamily Ephemeroidea, do indeed create burrows. Some live in relative simple U-shaped tunnels dug into the sediment, others in more complex tunnel networks, and some burrow into wood (e.g. some Polymitarcyidae). Typical movements involve digging with the forelegs and mandibles, which may be modified into tusks, and driving the body forward with dorso-ventral undulation of the body, especially the abdomen (Lyman 1943; Keltner and McCafferty 1986). Once a tunnel begins to form, the mid- and hind legs may be used to anchor the body to the tunnel sides. Body undulations provide forward thrust into the sediment surface and also help to displace loosened sediment in a backwards direction.

8.6 Movement on the water's surface

Being able to stand on the water surface may be considered a somewhat remarkable feat in itself; being able to move often quite rapidly on an apparently slippery water surface is quite extraordinary. The hydrodynamics underlying the surface locomotion of semi-aquatic insects is still poorly understood, although there have been some recent breakthroughs. The physics of surface films is complex and as usual, the physical challenges to movement vary with insect body size. We will attempt brief descriptions of the various phenomena, by way of an introduction, but these are by no means detailed descriptions.

8.6.1 Surface walking, rowing, running, and jumping

Water walkers have a variety of gaits. They can gallop, row, and walk, but all tend to move at relatively high *Re*. The most efficient water walkers are gerrids (Hemiptera), which row using the middle pair of legs by quickly sweeping them backward over the water's surface. Until fairly recently, it was thought that these animals moved using capillary waves, which are waves with very short wavelengths that are created as their legs interact with the surface tension as they move. Capillary waves move due to an interaction between mass and surface tension (as distinct from the more familiar gravity waves, which result from an interaction between inertial mass and gravity) and it was thought that the momentum they generate as they move away from the insect propelled the latter forward. However, as mentioned earlier (Section 8.3.1), calculations demonstrated that the minimum speed at which surface waves will travel is 23 cm s⁻¹, which means that the organism must be able to move at least this fast. Early-instar gerrids have short middle legs—only about 2 mm long—and would have to swing them at an angular velocity of 115 radians per second to achieve a speed of 23 cm s⁻¹. Clearly gerrids cannot do this and yet they can walk readily on water, a problem called 'Denny's paradox' (Denny 2004). It appears, however, that gerrids are able to generate thrust by rowing and using their legs as oars. The water-repellent state of the legs means that the legs deform—but do not penetrate—the water surface (Hu et al. 2003; Hu and Bush 2010). Inevitably, such rowing creates surface waves, but they do not play a significant role in propulsion. Instead, the gerrids generate momentum by pushing against the surface and producing from each leg hemispherical, dipolar vortices, which are at the surface and travel backwards at about 4 cm s⁻¹ (Figure 8.9). Calculations show that the energy produced by moving relatively large amounts of water via slow-moving vortices is translated into forward moving energy with an efficiency of 96 per cent. This explains how these animals are able to move at

speeds of over 1 m s⁻¹ (Hu et al. 2003; Denny 2004). Gerrids can also leap off the water's surface by changing the driving angle of the rowing legs and rowing very vigorously, which allows them to leap up to 10 body lengths away. They land with legs splayed to minimize penetration into the water on landing, but the hydrophobic nature of their integument means they do not get wetted (Hu and Bush 2010).

Alternatively, the broad-shouldered water strider *Microvelia*, the water treader *Mesovelia*, and the water measurer *Hydrometria* all use a typical tripod gait (Section 8.2). Both *Mesovelia* and *Hydrometria* generate the same dipolar vortices as described for gerrids (Figure 8.9). The marine springtail *Anurida maritima* also uses a tripod gait, but these animals walk on the ends of their claws and move relatively inefficiently. They generate a single vortex through the collective efforts of all legs together, rather than a vortex from each leg (Hu and Bush 2010). Springtails can also jump. As mentioned in Chapter 5 (Section 5.4) these animals have a non-wetting, ventral tube. Penetration and then removal of the wetting tube from the surface film, while the animal simul-

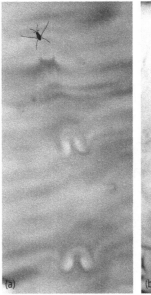

Figure 8.9 Photographs illustrating the production of surface vortices (visualized by Thymol Blue) in the wake of (a) an infant gerrid and (b) an adult gerrid.

Source: From Hu, D. L. and Bush, J. W. M. (2010). The hydrodynamics of water-walking arthropods. *Journal of Fluid Mechanics*, 644, 5–33. Reproduced with permission from Cambridge University Press.

taneously pushes down at the front and back ends, propels the organism upwards (Figure 8.10), and this is likely to aid escape from potential predators (Bush and Hu 2006).

8.6.2 Skimming, sailing, and skating

Surface skimming and sailing are employed by some adult stoneflies and mayflies that are poor fliers or incapable of flight, and there is evidence that skimming is an ancestral condition, which provides a potentially interesting perspective on the evolution of flight in aquatic insects (Kramer and Marden 1997; Thomas et al. 2000). Sitting on the water surface (without sinking) conforms to the same physical principles followed by gerrids and other water walkers (Section 5.4), but movement of these insects is powered by the wings flapping or harnessing breezes like sail boats, and sometimes rowing with the legs.

In skimming, the insect flaps its wings to produce uplift and forward thrust, but maintains contact with the water, which provides support as the animal skims across the surface. Skimmers may keep all six legs in contact with the water surface, the mid- and hind legs only, or only the hind legs, and some even use body undulations and essentially swim along the surface (Marden et al. 2000). The

stonefly *Taeniopteryx burksi* (Taeniopterygidae) uses six-leg skimming as well as body undulations to cross open water to reach land after individuals emerge as adults; they have not been observed to fly (Marden and Kramer 1994). In contrast, stoneflies in the genus *Leuctra* use hind-leg skimming, and may occasionally fly. The body is held upright in an S-shape, which allows the insect to beat its wings through much greater angles and attain greater speeds than six-leg skimmers. Interestingly, the stonefly *Diamphipnopsis samali* (Diamphipnoidae) can fly, but also uses its forewings to row across water (Marden and Thomas 2003). In contrast, the mayfly *Cheirogenesia* (Palingeniidae) is incapable of flight, and both sexes skim over the water, where mating and oviposition occur (Ruffieux et al. 1998). This species has relatively small wings, but can achieve fast speeds. Their wings are shaped like those of bats and are covered with unique scales that stand upright and provide water repellency (Figure 8.11).

Alternatively, sailing is a method used by adults that cannot flap their wings. Instead they stand on the water's surface, hold their wings up and

(a)

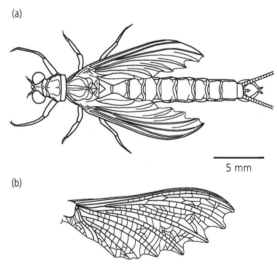

(b)

Figure 8.11 Illustration of (a) a flightless, male mayfly *Cheirogenesia decaryi* and (b) its bat-like wing, which may be an adaptation for skimming on the water surface rather than flying.

Source: From Ruffieux, L., Elouard, J. M., and Sartori, M. (1998). Flightlessness in mayflies and its relevance to hypotheses on the origin of insect flight. *Proceedings of the Royal Society of London Series B*, 265 (1410), 2135–2140, by permission of the Royal Society.

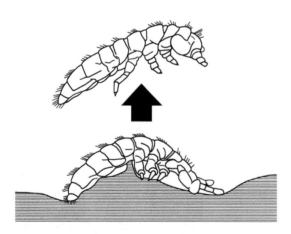

Figure 8.10 The water-walking springtail *Anurida maritima* is able to make use of surface films to escape predators by pulling up with a central wetting tube at the same time as it pushes down at the head and tail (which are non-wetting). The resultant forces become unbalanced when the animal releases the wetting tube from the surface film, causing the animal to be propelled into the air.

Source: Adapted, in part, from Bush and Hu (2006).

allow the wind to blow them across the water until they reach the banks or shore. For example, sailing by the stonefly *Allocapnia vivipara* (Capnidae) produces speeds that are greater than could be attained by gliding in air (Marden and Kramer 1995).

'Skating' is a form of motion in which insects hang downwards from the surface of the water film by their dorsal, tracheal gills, which can be used as a 'foot' to skate along the water–air interface. The dixid *Dixella aestivalis* and hydrophilid beetle *Hydrobius fuscipes* both use side-to-side swimming movements while anchoring the dorsal gill against the surface film, and this enables them to skate along at speeds reaching comparability with skimming (Brackenbury 1999b).

8.6.3 Meniscus climbing

To small insects, the surface of even still water is topographically rough, because wherever the surface film encounters an object (floating plants, leaves, twigs, land), it creates a meniscus (i.e. the water forms a curved surface). Larger water-walking insects can step over such menisci, but smaller individuals cannot and may instead use meniscus climbing. In meniscus climbing, the insect exploits a physical property of surface films in which distortions in the film can create lateral forces of attraction that move objects toward each other. The physics of meniscus climbing is complex and readers wishing detailed explanations are referred to other sources (Hu and Bush 2005; Bush and Hu 2006). Briefly, meniscus-climbing insects have the ability to curve the surface film and create forces that propel the insect up the meniscus. Wetted insects, such as larvae of the chrysomelid beetle *Pyrrhalta nymphaeae*, have part of the body below the water's surface. Arching of the back produces a distortion of the surface film and they are impelled up the meniscus by capillary forces. Non-wetting insects can use their retractable wetting claws to grasp the surface and pull it upwards. The water treader *Mesovelia* pulls upwards with the front and hind legs while pushing down with the middle pair (Figure 8.12). The front and hind legs are attracted toward the background meniscus while the middle pair are repelled, and this propels the insect up the meniscus with the body held in a static position (Hu and Bush 2005; Bush and Hu 2006).

8.6.4 Marangoni propulsion

Many insects can exude a surfactant, a substance that reduces surface tension (e.g. soap), and this reduction in surface tension causes the insect to be propelled forwards, a phenomenon called Marangoni propulsion. Examples in the Hemiptera include *Microvelia*, which is able to travel at 17 cm s^{-1}, twice as fast as it could achieve by walking at top speed (Bush and Hu 2006). *Velia* squirt the surfactant backwards out of the proboscis in order to generate forward motion (Nachtigall 1985). Semi-aquatic rove beetles such as *Stenus* (Staphylinidae) also use surfactants, reaching speeds of 45–70 cm s^{-1}, about 30 times faster than they can run or swim (Nachtigall 1985).

8.7 Hitching a ride: phoretic relationships

Some aquatic insects attach themselves to other animals, and may be dispersed in a truly passive manner. It is difficult to gather evidence that can differentiate between true phoresy (where an organism hitches a ride and potentially attains a benefit without any cost to the 'host', so the relationship is commensal) and other sorts of relationships, which can range from mutualism (where both partners benefit) to true parasitism (where the host suffers negative effects on fitness). There are certain to be many incidental or occasional hitch-hiking events, but true phoresy implies much more frequent and systematic movements. Empirical evidence is rare, but evidence suggests this is common in the aquatic medium for some Diptera (movement in terrestrial environments is discussed in Chapter 9).

Larval chironomids are well-known for attaching to a variety of animals including many other insects (e.g. mayflies, caddisflies, hemipterans, stoneflies, dipterans, megalopterans, dragonflies) as well as bivalves and fish (Tokeshi 1993). Many of these associations are thought to be commensal and in which the main benefit is dispersal of the chironomid larvae, but most associations have not been sufficiently well-studied to determine the nature of the relationship. Particularly interesting examples are marine chironomid larvae (*Clunio* and *Pontomyia*)

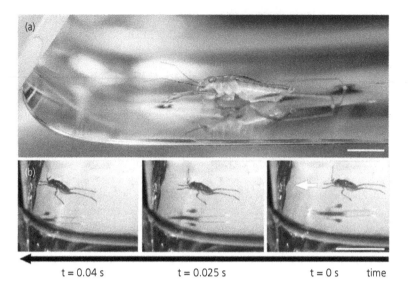

t = 0.04 s   t = 0.025 s   t = 0 s   time

Figure 8.12 Meniscus climbing by the water treader *Mesovelia*. (a) *Mesovelia* approaches a meniscus moving from right to left. (b) High-speed video images of an ascent, illustrated from right to left. Lighting from above reveals the surface deformation produced. In pulling up, the insect generates a meniscus that focuses the light into a bright spot; in pushing down, it generates a meniscus through which light is diffused, casting a dark spot. Characteristic speeds are 1–10 cm s^{-1}. Scale bars, 3 mm.

Source: From Hu and Bush (2005). Reproduced with permission from Nature Publishing Group.

that have been found attached to hawksbill turtles, and this association is almost certain to be a dispersal mechanism for the chironomids, because the females are flightless and essentially legless (Schärer and Epler 2007).

Blackflies (Simuliidae) are another group with larvae that commonly attach to a variety of animals, including other insects. For example, larvae of a species of *Simulium* attach themselves to larvae of the mayfly *Afronurus peringueyi* (Heptageniidae). The blackfly larva is always found at the base of the coxa of the mayfly's hind leg, with the head and body projected posteriorly over the mayfly's gills. When the mayfly moults, the blackfly detaches itself and re-attaches after moulting is complete. The blackfly even pupates on the mayfly, forming the pupal cocoon when the mayfly goes through its last moult before becoming a subimago, so that both individuals emerge as adults within hours of each other (van Someren and McMahon 1950). Almost half of the mayflies collected in that study had blackflies attached, a species that was not found anywhere else in streams, suggesting an obligate relationship. In other work, species of *Simulium*, both larvae and pupae, have been found attached to freshwater crabs (van Someren and McMahon 1950; Disney 1971a) and larvae have also been found on atyid prawns (Disney 1971b), where, as filter feeders, they may get the benefits of the currents these animals use to aerate gills as well as dispersal. In a particularly interesting twist, blackfly eggs have been found attached to dragonfly larvae (Lewis et al. 1960).

CHAPTER 9

Dispersal in the terrestrial environment

9.1 Introduction

Movement of aquatic insects in the terrestrial environment occurs largely through flight, flight capabilities vary enormously among taxa, and this has interesting implications for many aspects of their ecology and biology. If dispersal is highly restricted (e.g. poor flight ability), populations can be completely isolated from each other over quite short distances, and this has consequences for both the genetic make-up and extinction probabilities of those populations. Even when dispersal is less restricted, knowing how often and over what distances adults move is important to population ecology. An aspect unique to stream species is that some may drift frequently as juveniles and theoretically could be translocated considerable distances downstream before they emerge as adults (Section 8.3.6). For these species, it has been hypothesized that adults should fly preferentially upstream to 'compensate' for this downstream translocation—a model called the colonization cycle (Müller 1982). It is thus of considerable interest to know how well adults can fly or otherwise disperse, whether they routinely travel away from natal sites to other water bodies, the directions and distances they travel, and the interactive roles played by various environmental factors.

Understanding why species differ in flight capability requires some information about the structure and movement of wings, and the basic principles of flight, i.e. aerodynamics. These are enormous topics that are well-covered in most entomology texts and other references about flight. As such, we provide only a basic discussion of wing structure and movement (Section 9.2), and of how wings can create lift and

thrust (Section 9.3). This material then allows us to discuss how adult dispersal varies with wing morphology, the distances and direction travelled, the cues and barriers to dispersal (Section 9.4). Among the most spectacular feats of flying involve long-distance migration, and some dragonflies commonly migrate hundreds to thousands of kilometres, often over the open ocean (Section 9.5). Flight is energetically expensive and some aquatic insects have flightless adults, or have a proportion of individuals that are flightless (flight polymorphisms), which presents interesting challenges and trade-offs (Section 9.6). Even less energetically demanding is dispersal by other animal vectors, as when hitching a ride on other animals as a means of dispersal (Section 9.7). We do not discuss topics that are generic to the insects more broadly, such as the neural control of flight, details of the movement and control of wing muscles, or metabolism associated with flight. Readers are referred to any good entomology text for such information.

9.2 Wing structure and movement

In the basic insect body plan, adults have two pairs of wings (the forewings and hind wings) attached to the dorso-lateral surface of the thorax (meso- and meta-thorax, Figure 1.3) and comprising two, thin membranes of integument that are in close juxtaposition. The wings contain veins (Figure 9.1), which are places where the layers of integument are not closely apposed, the cuticle is thickened and where haemolymph, trachea, and nerves may be present. Veins arise at the base of the wing and branch toward the tip. The complexity of branching patterns varies greatly between insects from different

Aquatic Entomology. First Edition. Jill Lancaster & Barbara J. Downes.
© Jill Lancaster & Barbara J. Downes 2013. Published 2013 by Oxford University Press.

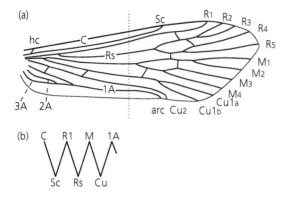

Figure 9.1 Forewing of a hypothetical caddisfly. (a) Illustration of wing venation and the associated nomenclature. C, costa; Sc, subcosta; R, radius; Rs, radial sector; M, media; Cu, cubitus; 1A, 2A, 3A, anal veins; hc, humeral cross-vein; arc, arculus (where 1A terminates at the wing margin). Branches of these veins are distinguished using subscripts 1, 2, etc. (b) A cross-section through the dotted line in (a), showing those veins that are convex versus those that are concave (not to scale).

taxonomic groups, but the basic arrangement and order of the major veins is highly conserved, and there is an accepted nomenclature for veins and the areas of the wing they enclose (Figure 9.1a). The veins on the leading edge of the wing are rigid and provide support yet resist bending and twisting. Cross-veins, which run across the wing and link different veins, may be flexible and bend in multiple directions. Thus, wings are not flat, but are quite three-dimensional and appear corrugated or pleated in cross-section (Figure 9.1b). Collectively, the veins ensure the wing is rigid during the downstroke, but can flex at particular points along the wing during the upstroke, which is necessary for flight (Section 9.3). Flexion points are critical because insect wings do not contain any musculature allowing the wings to be directly flexed (as in birds and bats). Tipulids (Diptera) even appear to have 'crumple zones' on the wing tips, which may allow them to hit objects while flying without sustaining serious damage (Wootton 1992). Numbers of veins and cross-veins can vary between and within orders, and over evolutionary time. Stoneflies known only from fossils have relatively few costal cross-veins and their wings were probably poorly adapted for flight. Among the extant Nemouroidae, there appears to have been an evolutionary reduction in cross-veins, and a correlated reduction in flight proficiency

(Illies 1965), and many of these species use their wings for surface skimming rather than flight (Section 8.6.2). This wing morphology (i.e. few costal cross-veins) suggests that skimming is an ancestral condition, with increases in cross-veins associated with flight developing later in other, derived families (Thomas et al. 2000).

Wings are often delicate but essential structures, and most insects have the ability to fold their wings longitudinally or transversely over the back to protect them when resting or seeking concealment in cracks and crevices. Fold lines and points of flexion delineate different wing areas that are recognizable in different insects, although the sizes of these areas vary greatly with wing shape, and so do wing angles and lengths of wing margins (Figure 9.2a). Wings may also have a pigmented spot called the pterostigma on the leading edge (Figure 9.2a). The mass of this spot is greater than a similarly sized area elsewhere on the wing, and this increased mass may improve flight. For example, small insects may be able to achieve greater control over the angle at which the wings are held during flapping, and for Odonata it may help prevent wings fluttering during gliding flight.

The process of wing folding is somewhat analogous to sorting a handful of playing cards and occurs in all insects except the Ephemeroptera and the dragonflies in the Odonata (modern damselflies can fold their wings). In order to fold the wings, the insect must be able to move different veins or areas of the wing relative to one another and, because there are no muscles in the wings themselves, this is typically achieved at the wing base. Each wing is attached to the thorax by a membranous and flexible basal area, which contains several articular sclerites. The shapes and sizes of these sclerites, and the associated musculature, differ among orders and this influences how, or whether, the wings are folded. There is some evidence of homology between the orders that cannot fold wings (e.g. mayflies and dragonflies) and all other orders, which suggests that the design of articular sclerites is probably ancient (Willkommen and Hörnschemeyer 2007). Perhaps the simplest arrangement is seen in the dragonflies, which do not fold their wings and have two large plates: an anterior humeral plate attached to the costal vein, and a posterior axillary plate (composed of several smaller

(a)

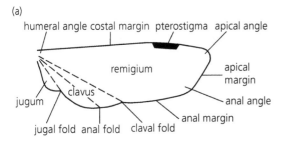

(b)

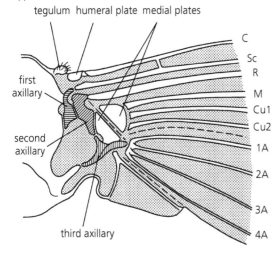

(c)

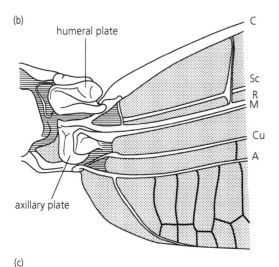

Figure 9.2 Schematic illustrations showing (a) nomenclature of edges, angles, folds, and wing areas; the articulation of the wing in (b) a dragonfly that does not flex its wings, and (c) a generic insect that does. Lettering for veins, on the right, as in Figure 9.1 Dashed lines in (a) and (c) are major fold lines.

Source: (b) and (c) adapted from Snodgrass (1935).

sclerites fused together) attached to the four post-costal veins (Figure 9.2b). Insects with folding wings have a more complicated arrangement, typically with multiple small plates: the humeral plate is associated with the costal vein; three axillary sclerites are associated with the subcostal, radial, and anal veins, respectively; and two less definite medial plates in the mediocubital area (Figure 9.2c).

Flapping wings move up and down, but also rotate and twist (Section 9.3), and articulation with the thorax is profoundly important in setting the degrees of movement. Three sorts of muscles are involved in wing movement. Muscles inserted into the base of the wings move the wings directly. A second group of muscles move the wings indirectly by distorting the thorax; these distortions are transferred to the wings via the articular sclerites. The final group of muscles involved in wing movement is the accessory muscles that change the shape or elastic properties of the thorax in a lateral direction. Such elasticity is an important source of energy that can be transferred back to the wings. In addition, many insects store elastic energy accumulated from the upstroke in a pad of resilin (a flexible rubbery protein) that forms the wing hinge, and over 95 per cent of this energy is transferred back to the wing during the downstroke. Such sources of energy are important because flying is energetically expensive, and flying insects must invest significant amounts of food intake to build and fuel the flight muscle mass. The size of this energetic investment is demonstrated by the fact that some insects histolyse flight muscles when flight is no longer required and the released calories can significantly increase reproductive effort (Section 9.6.2).

The two pairs of wings are typically coupled together, whereby the fore- and hind wings on each side are linked together and beat as a single unit. Wing coupling occurs in all aquatic orders except the Odonata (and Diptera where only one pair of wings is used for flight) and involves lobes or spines near the bases of the wings. There are many different mechanisms and within the Trichoptera, for example, wing coupling may have evolved independently several times (Stocks 2010a, b). In some taxa, the jugal lobe of the forewing is modified to connect with setae on the hind wing. In others, the jugal lobe is reduced or absent, and wing coupling involves curved setae, spines or microtrichia on the hind wings that link with projections on the forewings (Figure 9.3). These setae or spines may occur in patches or along the margins of wings, and their specific shapes and method of locking the wings together vary among families. Some other caddis-flies have incomplete coupling such that the wings are linked together on the downstroke, but not on the upstroke.

Wing shape, relative size, and function vary greatly among insect groups, which is reflected in a wide range of flight capabilities (Section 9.4). The power for flight is often generated by the hind wings, which are usually larger than the forewings, except in the Ephemeroptera, which have tiny hind wings. Some wings are highly modified, as in the beetles for example, where the forewings (elytra) are heavily sclerotized covers that lock together with a tongue and groove mechanism and protect the hind wings when they are folded. In the Diptera, the forewings are used for flight, while the hind wings are modified to form structures called hal-teres. These sense organs vibrate in time with the wings and help with stability during flight. Halteres may be of different relative lengths: > 12 per cent of forewing length in Tipulidae, but < 6 per cent in mosquitoes. Finally, some insects have both pairs of wings reduced, a condition called brachyptery or micropatery, and others are completely wingless (apterous)—a topic for Section 9.6.

Attached to the wing surface may be a range of structures; many have a sensory function, but the function of many spines and papillae are still poorly understood. Most insects have sensilla along the wing veins, which are probably used to detect air flow, but may be involved in chemoreception (some Diptera). The wings of Lepidoptera are coated with scales, whereas trichopteran wings are usually cov-ered with large microtrichia (non-innervated spines), which gives them a 'hairy-winged' appear-ance. Damselfly wings often have pigment spots (e.g. *Calopteryx*, Calopterygidae) that play a role in thermoregulation (Outomuro and Ocharan 2011). Wings that are coloured or patterned, either by scales, hairs, or pigment, may also have behavioural functions in defence or display.

9.3 The principles of flight—lift and thrust

In order to understand how differences in wing morphology and deployment affect insects' capac-ity to fly, it is necessary to explain some basic prin-ciples of aerodynamics. As in earlier chapters

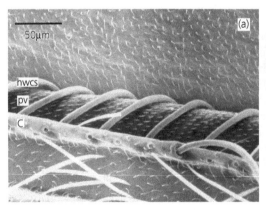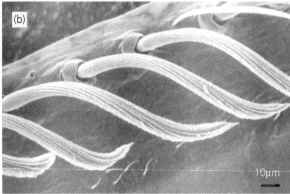

Figure 9.3 Wing coupling in the caddisfly *Molana ulmerina* (Molannidae). (a) A ventral view of the wings coupled together. The costa (C) of the hind wing contains a series of curved setae (hwcs) that project from its dorsal surface. These setae hook over a pseudovein (pv) on the trailing edge of the forewing with the curvature of the setae matching that of the pseudovein to ensure a snug fit. The pseudovein has microtrichia that project distally and help to prevent slipping. (b) The hind wing setae, showing longitudinal grooves that are slightly twisted such that the grooves attain a spiral contour, which is thought to engage with the microtrichia on the forewing and aid locking.

Source: (a) From Stocks (2010a). Reproduced with permission from John Wiley & Sons. (b) Republished with permission, from (Stocks 2010b).

touching on physical principles (Chapters 5, 8), the presentation of this information is kept to the minimum needed to explain some of the differences in flight capability between insects and consequently is fairly basic. In addition, understanding the aerodynamics of flapping flight is horrendously complicated and our description is somewhat superficial. Readers interested in the physics underlying flight should consult alternative sources.

A simple introduction to the principles of aerodynamic lift is provided by an airfoil, which is a curved surface with the convex side facing upwards, like the wing of a modern aircraft (Figure 9.4a). Wind coming toward an airfoil separates and moves both above and below the airfoil, but the path to the trailing edge is longer along the top than the bottom. For reasons we will not explain, the velocity of the air is faster along the top than the bottom; this creates lower pressure along the top than along the bottom, which results in lift. The amount of lift, calculated as a lift coefficient, is proportional to the area of the airfoil in plan view (Figure 9.4b), the density of air, and the square of the free-stream velocity. Lift coefficients share some similarity with drag coefficients (Section 5.4) in that their values depend on the shape of the object, its orientation with respect to air flow, and the ever-important Reynolds number, Re (Section 5.4). Airfoils, like any object immersed in a moving fluid (or moving through a fluid), experience skin and pressure drag, and this will influence the net amount of lift. The conditions in which the ratio of the coefficient of lift to the coefficient of drag is maximal are particularly interesting. One relevant factor is the angle of attack, or the angle of the airfoil relative to the oncoming wind direction (Figure 9.4c). For an airfoil, both lift and drag increase with increasing angle of attack, but after a certain point, lift declines and eventually stall occurs (lift drops very rapidly), yet drag continues to increase. Therefore, there is one angle of attack that produces the greatest lift, given attendant drag. To complicate matters, airfoils suffer another form of drag called induced drag. When airfoils create lift, they tip upwards at the front, and consequently the lift vector is tipped backwards (Figure 9.4d), causing a loss of lift and increased drag. However, induced drag is actually lower at higher speeds because a lower angle of attack can produce a lift vector closer to

vertical. At slow speeds, the angle of attack must be far higher in order to generate sufficient lift, and induced drag can become high, which explains why flying slowly is actually more difficult and energetically expensive than flying quickly.

The shape of an airfoil will strongly influence flight and, similarly, an insect's ability to fly will be related to wing shape. Shape is described by the aspect ratio and calculated as the square of the wing span divided by its area. High aspect ratios indicate long, skinny wings, whereas low ratios indicate short, stumpy wings. Long, skinny wings are advan-

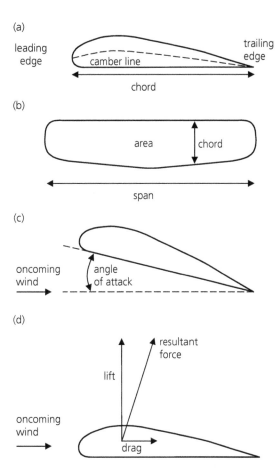

Figure 9.4 Lift-producing airfoils have (a) curved surfaces with the convex surface facing upwards. The chord is the distance from the leading to trailing edge. (b) In plan view, the span is the length at right angles to the chord. The aspect ratio is equal to (span)^2/area. (c) In cross section, the angle of attack is the angle of the airfoil relative to oncoming wind direction. (d) Oncoming wind creates lift and drag, such that the resulting lift vector is tipped backwards.

tageous in two ways. They create lift with less over-all drag and hence overcome the cost of staying aloft more easily, plus they are less affected by the vortices that are shed from each wing tip (wing-tip vortices are created during lift, but are undesirable because they are a major component of wake turbulence and are associated with higher drag). Short, stumpy wings suffer more from tip vortices and also have lower lift-to-drag ratios. Simple measures of insect wing shape, such as aspect ratio, are potentially useful, albeit indirect, measures of dispersal capability, which may be linked to species-specific distribution patterns. Evidence does indeed suggest that relatively larger-winged species may occupy more places across landscapes (e.g. caddisflies in Sweden: Hoffsten 2004) or have greater overall ranges (stoneflies and mayflies in Sweden: Malmqvist 2000; damselflies: Rundle et al. 2007), than smaller-winged species. Similarly the wing shape of dragonflies are related to whether they are migratory and whether they carry out mate-guarding flights (Johansson et al. 2009).

To understand differences in flight capabilities further, we need to consider the sizes and shapes of insects, because their bodies create 'parasite' or extra-to-wing drag that must be overcome. Small insects create relatively more drag (because they have greater surface-to-volume ratios and relatively more frontal area), but have less mass to lift. As Vogel (1994) puts it, it is relatively easier for small insects to get into the air than large ones, but much harder for them to move around. Drag-to-weight ratios illustrate this point, e.g. drag is about 18 per cent of body weight for fruit flies, but about 2–4 per cent for larger insects. Of course, Re plays a role here too given that small insects operate at smaller Re than large ones, and, at low Re, drag on wings rises markedly and becomes much greater than induced drag.

At this point, we can consider—very briefly!—the additional complexity created when wings are flapped rather than held stationary. Flapping needs to create forward motion as well as flight, and so wing strokes need to generate thrust as well as lift. During wing flapping, an up-and-down motion is accompanied by rotation. The downstroke creates mostly lift and some thrust because the wings are moved forward as well as downwards; conversely, the upstroke creates mostly thrust with some lift because wings are moved backward as well as upward (Figure 9.5). Importantly, the wings also twist during flight and rotate through dual figure-of-eight motions, so that the entire animal moves up and down as well as forwards. (Hovering is an exception, where flapping wings move forwards and backwards in an almost horizontal plane.) By twisting the wings, the angle of attack can be changed throughout the stroke, which is advantageous because lift is improved (and drag reduced) by having a high angle of attack during the downstroke and a lower angle of attack on the upstroke. The ability of the wings to distort along flexion lines and change various aspects of shape are the keys to their capacity to provide lift and thrust at the right moments. Wootton (1990) likened insect wings to sails in their capacity to deform and flex. One final aspect of wing flapping that deserves mention is the clap-and-fling (or clap and peel) method, in which the wings actually touch (or nearly touch) at the top of the upstroke. As the wings peel apart for the downstroke, air moves in to fill the gap and produces local circulation that helps impart lift in addition to the lift generated by the wings themselves.

The flight efficiency trade-offs between wing size, shape, and body size are tricky. Large-bodied organisms need bigger wings to carry the greater load. Smaller wings, however, are less effort to use (less

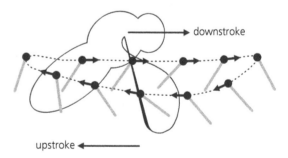

Figure 9.5 Side view of an insect body in outline and with one wing, which is shown twisted such that the leading edge is facing backwards and about to start the upstroke. The grey lines indicate the orientation of the wing's chord and the black circles the leading edge as the wing travels along the dotted line. As the wing moves into the upstroke, it travels both backwards and upwards, then downwards and forwards on the downstroke. The grey lines indicate how the wing twists during flight.

Source: Republished with permission from the *Journal of Experimental Biology*, from Ishihara et al. (2009).

area means less drag) and may be able to beat faster than larger wings, providing more lift—but of course—things are much more complicated than this. Because wings can flex, insects can change effective wing area during particular strokes and speeds, and therefore do not have to alter the angle of attack constantly (with its attendant problems of variable drag and the dangers of stall). Additionally, wing shape and *Re* play significant roles. While birds, bats, and large insects operate at high *Re* of roughly 5000 to 50,000, small insects can fall below 1000. At low *Re*, the maximum possible lift coefficient is small, the maximum possible lift-to-drag ratio is also small, and stall occurs at a smaller angle of attack. Flight looks impossible, but of course there are exceptionally small insects capable of flight! Once again, the issue is that life at low *Re* is not intuitive or straightforward, and indeed the wings of insects flying at low *Re* can look fairly odd, e.g. some are very thin with protruding veins. At low *Re*, wing corrugations (Figure 9.1b) may operate as if they are filled (i.e. no different from a smooth surface). In some small insects, the 'wings' are collections of bristles, rather than a continuous surface, and sheets of bristles, with their thick boundary layers at small *Re*, can act like a solid plate (Section 8.3.4). Some insects may have wings capable of changing area drastically in ways allowing them to overcome the problems of flight at low *Re*. It is possible that these insects use wings in ways analogous to the generation of thrust by swimming beetles, which have swimming hairs that collapse on the recovery stroke, drastically reducing area and minimizing reverse thrust (Section 8.3.1), but that is entirely hypothetical (Vogel 1994). Indeed, some suggestions that small insects 'row' through the air have proven to be incorrect (Usherwood and Ellington 2002), and this is clearly a poorly understood topic.

9.4 Dispersal by flying

Now that some of the basic principles of wing structure and flight have been established, albeit rather simply, we can begin to consider dispersal of adult aquatic insects more closely. Species differ enormously in their flight capabilities, and this is strongly linked to their overall sizes and shapes, and especially

their wing morphologies. From an ecological perspective, the distances and directions adults fly are important, but it is logistically difficult to gather sound empirical information. Consequently, this topic is relative rich in examples, but generalizations are tentative and may be subject to methodological biases. Given that flight is energetically expensive it is worthwhile considering what cues insects use to motivate and direct movement, and what constitute potential barriers to dispersal.

9.4.1 Wing morphology and flight capability

While many insects can fly, the majority of research has focused on odonates, particularly dragonflies (Anisoptera), which are masters of the air. This creates an unavoidable bias in our discussion, but also provides an ideal example to illustrate how complex flight can be. Wing morphology and flight capability of other taxa is likely to be similar to the odonates in at least some respects, but they have received comparatively little attention, except perhaps in the context of wing evolution.

Dragonflies are among the fastest flyers and also have tremendous manoeuvrability, including the ability to glide, hover, fly slowly, fly backwards, and fly in tandem during mating (review: Corbet 2004). This aerial agility is possible due to various adaptations in the wings themselves. Their resilin-constructed wing vein-joints provide superb bending capacity (Donoughe et al. 2011). The wings contain two very special structures, the nodus and basal complex, which provide superior capacity to twist and flex in directions allowing great manoeuvrability during flight (Wootton et al. 1998). The nodus (Figure 9.6), present in Anisoptera and Zygoptera, comprises bracket-like cross-veins that link three veins along the wing's leading edge (the leading-edge spar), and divide the spar into two parts: the proximal section is rigid and the distal section is more flexible. The nodus allows the wing to twist in response to inertial forces created when the leading edge is upwards, but restricts such twisting when the leading edge is downwards. This allows superior manoeuvrability without the wings bending during parts of the wing stroke when they need to remain rigid. The basal triangle–supratriangle complex, diagnostic of the Anisoptera, is an area where particular veins intersect and produce a highly angular and three-dimensional area (Figure 9.6). Functionally, it holds the trailing edge down and improves the camber of the wing during lift. Aspects of such 'smart engineering' are found also in fossil odonates from the mid-Carboniferous, indicating that versatile flight was developed some 80–100 million years ago (Wootton et al. 1998). This illustrates nicely that the arrangement of veins and cross-veins on insect wings may reflect phylogenetic relationships, but we should also expect strong links between form and function.

Unlike most other insects, the fore- and hind wings of odonates are not coupled together and can be moved independently, leading to considerable complexity in how wing structure and function provide lift, acceleration, and so forth. An interesting paradox is that four wings should generate less lift than two wings coupled together, which seems at odds with the superior flight capabilities of odonates. In practice, dragonflies take advantage of the interactions between the fore- and hind wings in much the same way that the two rotors on helicopters work, with the hind wing able to recover energy created by 'swirl' from the forewing to reduce the power needed to fly by 22 per cent (Usherwood and Lehman 2008). Dragonflies have greater wing velocity and can accelerate and fly faster than damselflies, which generally have a slower wing beat and a greater stroke amplitude (Rüppell 1989), but this is not always straightforward. For example, the damselfly *Calopteryx splendens* (Calopterygidae) is indeed slower overall than a dragonfly of comparable size,

Sympetrum sanguineum (Libellulidae), but it can get higher acceleration and fly further per wing beat (Wakeling and Ellington 1997a, b). This damselfly can also achieve greater lift, partly due to using clap-and-fling (or partial clap-and-fling) and partly because it is able to beat its wings more synchronously than the dragonfly.

Beating the fore- and hind wings independently means odonates can use various patterns, which include counterstroking (hind and forewings beat in opposition such that the downstroke of one pair coincides with the upstroke of the other), parallel stroking (hind and forewings beat in virtual synchrony), and phase-shifted (fore- and hind wings are out of phase). Angles of attack and stroke plane also vary and are associated with particular manoeuvres. For example, *Megaloprepus coerulatus* (Pseudostigmatidae) can take off backwards by beating its wings forwards with a large angle of attack, then immediately doing the exact reverse to begin forward flight. Odonates also alter the angle of attack in concert with shifting the phase of wing-beating, depending on whether they are attacking, carrying prey, or courting prospective mates (Rüppell 1989).

Like some other aquatic insects, many odonates are parasitized or exploited by organisms that attach to their thoraxes to 'hitch a ride' to other water bodies (Section 9.7). These organisms can add to the 'parasite' or extra-to-wing drag and consequently affect the host's flight capacity. For example, the damselfly *Nehalennia speciosa* (Coenagrionidae) is small (abdomen 20–25 mm long) and is frequently parasitized by water mites (genus *Arrenurus*, Arrenuridae), with individual damselflies carrying up to eight mites. Damselflies carrying many mites flew significantly shorter distances than unparasitized individuals, although in this case the mites created problems by penetrating the flight muscles rather than simply adding weight (Reinhardt 1996). Other organisms may attach to the wings or elytra, causing potentially serious problems. For example, freshwater limpets have been found on the hemielytra of water bugs (Walther et al. 2008).

The problem of carrying extra weight is encountered by many non-odonate taxa that mate on the wing. In some caddisfly species, females fly into swarms of males, with copulation beginning in the air and the female hanging upside down from the

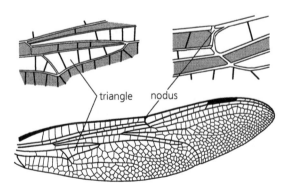

Figure 9.6 Forewing of *Hemianax papuensis* (Aeshnidae) with enlarged insets of the basal triangle–supratriangle complex and the nodus. Stippled versus white areas are angled in different directions as a result of wing corrugations.

male. Males must be able to carry their mates and tend to have longer wings than females, whereas in other species, male and female pairs fly in tandem (both individuals can flap their wings) and there are no strong differences in wing size (Gullefors and Petersson 1993). (Mating in odonates also involves flying in tandem so there is typically no sexual dimorphism in wing aspect ratio.) In some situations, males within the same species may differ in size and in flight capability. One such example is the caddisfly *Athripsodes cinereus* (Leptoceridae), where copulation begins in the air with the male carrying the female, but small males can end up in the water with their prospective mates because they are unable to carry the load (Petersson 1995). Larger males are able to carry females successfully to the ground where copulation is completed.

Unusual among the insects, dragonflies can glide, which is a more energy-efficient form of locomotion than flapping wings. Their wings have some of the highest aspect ratios of any insects and this may be the key to their gliding ability. For example, the migratory dragonfly *Pantala flavescens* (Libellulidae) (discussed further in Section 9.5) can glide for 10–15 s at 15 m s^{-1} depending on wind speeds (Hankin 1921), species of *Aeshna* (Aeshnidae) can glide for 30 s or more without losing altitude, and the dragonfly *Sympetrum sanguineum* (Libellulidae) has been clocked gliding at 2 m s^{-1}. Dragonfly wings can generate greater lift in steady-state (gliding) conditions than all other insects except desert locusts, a feat not solely due to wing size or high aspect ratio but due to superior construction that results in low drag at low angles of attack (Wakeling and Ellington 1997c). The surfaces of dragonfly wings are highly corrugated (Figures 9.1b, 9.7) due to the large number of cross-veins; patterns of corrugations also change along the wing length (Kesel 2000, Figure 9.6). Theoretically, the gliding performance of such rough surfaces ought to be lower than smooth surfaces because of increased drag, but gliding dragonflies can be moving at relatively low Re (< 500). In fact, dragonfly wings perform as well or better than smooth airfoils over a wide range of Re and angles of attack. Increased pressure drag caused by wing corrugations is more than offset by a drop in skin friction, because air circulates within the pockets created by wing veins (Kesel 2000; Vargas et al. 2008).

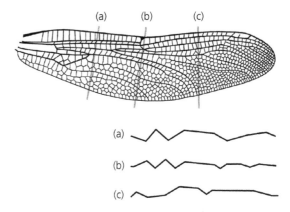

Figure 9.7 A drawing of the forewing of *Aeshna cyanea* (Aeshnidae) The three grey lines indicate the approximate positions of the three cross-sectional profiles (below the wing) that were measured at (a) 0.3, (b) 0.5, and (c) 0.7 distance from the base of the wing and show the corrugated nature of the upper surface.

Source: Cross-sectional profiles republished with permission from the *Journal of Experimental Biology*, from Kesel (2000).

9.4.2 Flight directions and distances

How far and in what directions do flying insects disperse? Many ecological questions require some information about flight direction and distance. Flight capability of various taxa is clearly related to their morphology, but whether they actually use that capability to go anywhere is a different matter. This will depend on many factors, including adult longevity (short-lived species are unlikely to have time to travel very far); air temperature (flight performance is constrained by temperature; see Chapter 4); wind, which can either suppress flight behaviourally or passively disperse individuals long distance in particular directions; motivation for movement (i.e. foraging, between-habitat dispersal, migration); landscape characteristics (there are many potential barriers to dispersal including bridges, mountains, trees, etc.). It should be immediately obvious that this is a multifaceted research area riddled with logistical challenges. Unsurprisingly, collecting empirical information on flight distance (and direction) is difficult and many estimates are potentially subject to biases imposed by the methods used. There are many ways to catch an insect, but it is much more difficult to determine where it came from and where it was going. Consequently, generalizations about movement patterns must be tempered with caution, and must be considered in the context of the methodological limitations. What follows is a brief description of some common methods, their strengths and weaknesses, and then an attempt to make some generalizations about flight direction and distance. It may read like a catalogue of examples, but the number of good examples is surprisingly small so it may be worth considering many. (Note: migration, or mass movements over very long distances, will be discussed in Section 9.5.)

There are three methodological approaches commonly used in work on flying insects, including interception traps, mark-recapture, and genetic techniques. Interception traps (e.g. malaise traps, sticky traps, light traps) tell you where an insect was at the time it was trapped, but provide no information on where the insect came from or where it was going. Although comparatively simple to implement, evidence suggests that interception traps can be misleading with respect to estimates of directionality (Macneale et al. 2004). Labelling insects with tags unique to each individual, yet small enough that they do not interfere with flight, is possible but technically difficult, especially given the typically low recapture rates (approximately 5 per cent or less) and constraints on the number of individuals that can be followed by radar trackers (Riley et al. 1996; Roland et al. 1996; Boiteau et al. 2011). An alternative is to label insects en masse with, for example, stable or radio isotope labels, thereby increasing the likelihood of recapturing a reasonable number of individuals. Although mass-marking methods inevitably rely on interception traps to recapture individuals, they can provide unequivocal information about movement distance and direction from the marking point, if not the route travelled. Genetic methods for inferring movement and population structures are now common and this field is developing rapidly with new techniques being implemented all the time. The challenge here is to ensure that the results allow inferences about movement on ecologically relevant scales.

The adults of many stream-dwelling insect species disperse predominantly along stream corridors and for some species there appears to be a bias for upstream movement, as predicted by the colonization cycle, i.e. the suggestion that upstream flight of

adults would be required to compensate for downstream displacement of aquatic larvae (Müller 1982). Using interception traps, some studies have shown an upstream bias (Winterbourn and Crowe 2001), but others have not (Wagner 2003). Remember, however, that directionality estimates based on interception traps can be misleading (Macneale et al. 2004) and reveal little about distances travelled. Somewhat more convincing data of upstream movement and distance comes from mass-marking studies. Thousands of larval *Baetis* (Baetidae) tagged with stable isotopes (^{15}N) in an Arctic river drifted downstream and emerged as adults. About 30–50 per cent of these individuals flew about 2 km back upstream, suggesting strong upstream flight (Hershey et al. 1993). Similarly, adult stoneflies tagged with ^{15}N flew predominantly upstream along stream corridors an average of some 200 m, with a few individuals travelling greater than 500 m (Macneale et al. 2005). Coutant (1982) took advantage of juvenile caddisflies marked with radioactive ^{65}Zn in water discharged from a nuclear reactor. Adult caddisflies were collected and those with radioactive signals were detected both downstream and upstream of the reactor, with substantial numbers found upstream, to a distance of 16 km. In summary, an upstream bias in flight direction seems likely for some species, but it is more difficult to know whether this is a significant proportion of all stream species, or whether most have no directional flight bias or, indeed, are not displaced downstream as larvae (Chapter 8).

Within populations there is much variation in individual flight distances, but the majority of individuals do not travel far from the water body occupied by larvae of the same population. How one defines 'far' depends, of course, on species-specific flight capabilities and longevities. Much of this information comes from studies using interception traps placed at various distances from water bodies and assume that trapped adults did indeed originate from that water body. For example, in New Zealand, 20 per cent of all caddisflies were trapped within 20 m from the stream edge, although some were trapped up to 70 m away, and there was great variability between species (Collier and Smith 1998). Similarly, in Scandinavia, some stoneflies (*Leuctra*, Leuctridae; *Nemoura*, Nemouridae; and *Isoperla*,

Perlodidae) travelled up to 60 m (the maximum distance examined), but others were trapped only ≈ 1 m from the stream (Kuusela and Huusko 1996). Other studies have also trapped a large proportion of adults only a metre from stream banks, including caddisflies (Sode and Wiberg-Larsen 1993) and stoneflies (Winterbourn 2005). At least some adults, however, can travel substantial distances away from water. For example, mayflies and hydropsychid caddisflies in Canada were trapped up to 5 km away from the nearest water body. The average dispersal distances away from water of five species of hydropsychids varied between 600–1600 m while that of a mayfly species (*Hexagenia*, Ephemeridae) was 1200 m, illustrating good dispersal capacity in all taxa (Kovats et al. 1996). Nevertheless, as with other studies, the majority of individuals were caught in traps placed relatively close to banks (< 100 m). Interestingly, similar patterns are found in many odonates, even though they are strong flyers and large enough for mark-recapture studies in which individuals are each labelled with a unique tag. A study of over 1600 dragonflies across seven species found that most individuals remained near their natal ponds. Individuals of some species dispersed between neighbouring ponds, but only a few individuals travelled significant distances (hundreds of metres) to distant ponds (Conrad et al. 1999). A similar study of the rare damselfly *Coenagrion mercuriale* (Coenagrionidae) found individuals travelled only a median distance of 32 m during their lifetimes, and two-thirds moved less than 50 m. The longest distance recorded was only 1.8 km, suggesting that this species is quite sedentary (Rouquette and Thompson 2007). Mass marking of larval stoneflies, *Leuctra inermis* (Leuctridae), showed that 90 per cent of adults moved < 11 m, although some moved up to 1 km between catchments (Briers et al. 2004). However, the travel directions were aligned with prevailing winds, suggesting that long-distance travel for these relatively weak fliers was largely wind-assisted. Other mass-marking studies also indicate that many adult mayflies, stoneflies, and caddisflies do not travel more than a few metres away from streams (Francis et al. 2006; Petersen et al. 2006). In summary, regardless of the capacity of aquatic insects to move long distances, it appears that most individuals do not travel far. An important, but as yet unsolved,

problem of this apparent pattern is how to disentangle the distance travelled from the time an individual spends in a particular location, i.e. exactly the same results described above would arise if all individuals made a brief, long-distance journey away from water and then spent a long time close to water. Additionally, dispersal ability can differ between sexes (Caudill 2003), but there is insufficient information to generalize about sex-specific movement patterns.

For those few individuals that do travel far, some may travel very far from their larval home, often travelling many kilometres and crossing catchment boundaries. For example, adults of species with larvae that are sensitive to acid waters were trapped near acidified streams where their larvae had not been caught in over 20 years; this could only arise if the adults had dispersed from catchments up to several kilometres away (Masters et al. 2007). In a mass mark-recapture study, larval blackflies (mostly *Simulium venustum*) were exposed to radioactive ^{32}P and X-ray film was placed inside traps to catch adults spread out over 2300 km^2 around the tagging point—radioactive flies hitting the film left a mark. The results showed that blackflies travelled an average of 9.3–14 km and some were trapped about 35 km away two days after emergence. Most blackflies travelled either west or south of their release point, directions that coincided with prevailing winds (Baldwin et al. 1975). Genetic studies have found similar patterns and again have found marked variation both between and within taxonomic groups. The caddisfly *Mesophylax aspersus* (Limnephilidae) living on the Canary Islands occasionally moves long distances between islands, which involves flight across 20–50 km of open ocean (Kelly et al. 2001). In comparison, an upland net-winged midge from South Africa (*Elporia barnardi*, Blephariceridae) rarely dispersed even across catchment boundaries (Wishart and Hughes 2001). It is possible that upland (montane) species have more restricted dispersal than lowland species, but such generalizations are risky with the available information and even lowland species may show disparate dispersal patterns. For example, two species of mayflies (*Atalophlebia* spp., Leptophlebiidae) inhabiting the same lowland region showed different patterns, with one showing evidence of isolation by distance at < 40 km whilst the other was panmictic at ≤ 160 km (Baggiano et al. 2011).

At the opposite extreme, there is good evidence that populations of some flight-capable species have very restricted dispersal, in at least some circumstances. A fascinating example is that of the mosquito *Culex pipiens*, which colonized the London Underground railway system many decades ago (Byrne and Nichols 1999). In the Underground, mosquitoes breed using pools, flooded shafts, and sumps, and adults feed on train passengers. (Records show people taking refuge in the Underground from bombardment during the Second World War suffered greatly from mosquito bites.) Genetic analysis shows there are no exchanges of mosquitoes between the Underground and surface populations. In addition, there are strong genetic differences between mosquitoes from different train lines. This probably results from trains creating draughts that push mosquitoes strongly up-and-down particular lines, but with dispersal throughout the tunnel network being otherwise quite limited. Not all *C. pipiens* live in the London Underground, however, and populations in other locales may have different dispersal patterns, which can make it difficult to generalize about dispersal even at the species level!

One last interesting aspect of dispersal is provided by insects that live in ponds or streams that dry either seasonally or occasionally. Such species may have originally colonized these water bodies by dispersing from permanent waters nearby, with their offspring emerging and dispersing before the water dries up (Williams 1997). For example, the caddisfly *Hydropsyche siltalai* (Hydropsychidae) is widely distributed across Europe and occurs in both permanent and temporary streams in the Mediterranean. An analysis showed no significant genetic differentiation across four river basins (an area of over 9800 km² with individuals potentially dispersing up to 80 km) or between temporary and permanent waters (Múrria et al. 2010), a result that has been found also for caddisflies living in temporary waters elsewhere (Shama et al. 2011).

9.4.3 Cues, attractions, and barriers

Different aspects of the environment can serve either to attract or repel insects, or act as barriers that reduce dispersal. They may trigger movement in particular directions or simply increase movement activity in general. Attractants and repellents are typically sensory cues or signals in the environment and they only operate if insects have the requisite sensory structures to detect the cues (Chapters 6 and 7). Barriers to movement may break up a sensory cue such that insects cannot follow the cue, or they may be physical barriers that exceed an insect's flight capability.

Attractants to adult aquatic insects dispersing in the terrestrial environment can involve various cues, but chemical and visual cues are perhaps most prevalent. Pheromones are chemical cues produced by insects and are instrumental in altering various movements, including mate location for some taxa (see Chapter 10). Among the cues not produced by insects, plant chemicals may also act as attractants to herbivores, even though they are perhaps commonly considered in the context of toxins or digestibility-reducing compounds. Such attraction is common in terrestrial insects where airborne volatiles of plants can attract insects over long distances (Bruce et al. 2005). In water, however, the persistence and mobility of plant compounds are likely to be rather different and close-contact 'taste', not long-distance 'smell', may be more important for aquatic life stages (Chapter 7). Nevertheless, many aquatic plants have emergent parts (e.g. flowers, leaves) and the adults of most aquatic insects have a terrestrial stage with diverse sensory receptors, so it is entirely plausible that adults may be attracted to plants that are suitable food sources for larvae. Indeed, the North American milfoil weevil *Euhrychiopsis lecontei* is a specialist aquatic herbivore that feeds (larvae and adults), oviposits, and mates on the invasive freshwater macrophyte *Myriophyllum spicatum*, and adult weevils appear to be attracted to high concentrations of glycerol and uracil released by *M. spicatum* during rapid growth (Solarz and Newman 1996; Marko et al. 2005). Similarly, the volatile metabolites released by mats of cyanobacteria act as a habitat-finding cue for dispersing adult beetles of *Bembidion obtuscidens* (Carabidae), which inhabit the littoral margins of saline lakes (Evans 1982), and for ovipositing mosquitoes of *Anopheles albimanus* (Culicidae) (Rejmankova et al. 2000).

Visual clues can also be important attractants, especially if the result is that adult insects are attracted to water, where females must eventually oviposit. Many insects are attracted toward horizontally polarized light, which is reflected off water surfaces (Chapter 6). Refilling of ephemeral ponds and/or the presence of aquatic plants can thus act as a significant attractant to dispersing insects (Boix et al. 2011), which may also be attracted to dark ponds (e.g. brown water containing leaves: Williams et al. 2007) or ones with an open canopy (Binckley and Resetarits Jr 2007), illustrating a variety of visual cues. Unfortunately, insects can be attracted to human-made surfaces that reflect polarized light and, to insects, can look like water. For example, building windows, cars, asphalt, cemetery grave markers, and solar panels can attract adults in many groups of aquatic insects that sometimes end up mistakenly laying eggs on these artificial surfaces (e.g. Smith et al. 2009; Horváth et al. 2010). Street lighting can also attract aquatic insects and be a significant source of mortality (Ohba and Takagi 2005).

Some cues may act as repellents that trigger animals to disperse away from a water body. Aquatic plants, for example, can trigger dispersal by predaceous diving beetles (*Rhantus*, Dytiscidae), with low plant densities producing a higher probability of dispersal away from ponds (Yee et al. 2009). Food densities can also affect the likelihood that insects will disperse (e.g. giant water bugs: Ohba and Takagi 2005) and the presence of predators, like fish, can deter colonization (Kraus and Vonesh 2010). High salt concentrations can trigger dispersal of the water strider *Aquarius paludum*, with adults swept downstream from fresh to brackish water being more likely to fly back upstream (Kishi et al. 2007). Torrential rain falling onto stream surfaces causes dispersal away from streams (over 20 m) by the giant water bug *Abedus herberti* (Belostomatidae), which helps individuals avoid being swept away and possibly killed by flash floods (Lytle 1999).

Potential physical barriers in the landscape are numerous and include topographic features (e.g. mountains) and extensive areas of particular vegetation types. Examples of mountains as barriers are common, such as in the case of the caddisfly *Rhyacophila pubescens* (Rhyacophilidae), which lives

only in the upland areas of tufa spring brooks and, in the European Alps, adults do not disperse away from individual springs (Engelhardt et al. 2011). Other montane caddisflies also rarely disperse between stream valleys (Kubow et al. 2010; Lehrian et al. 2010). Alternatively, a comparative study in Chile found that while mountains were a barrier to dispersal by a widespread caddisfly (*Smicridea annulicornis*, Hydropsychidae), they were not to a co-occurring mayfly (*Andesiops torrens*, Ameletopsidae) (Sabando et al. 2011). Evidence that vegetation types can be barriers are less well documented, but convincing nevertheless. A mark-recapture study of dragonflies (*Leucorrhinia hudsonica*, Libellulidae) living in Canadian peatlands showed that individuals, particularly males, moved hundreds of metres between areas of peatland, but this depended on the nature of the intervening landscape. When peatlands were separated by long distances (> 700 m), forested environments deterred movement compared to more open territory, but the reverse was true over shorter distances (< 700 m). This outcome probably reflects differences in the behaviours underpinning deliberate long-distance dispersal compared to foraging movements (Chin and Taylor 2009). The impacts of human beings on landscapes are accompanied by a depressingly large catalogue of examples of altered insect dispersal. Chironomids (*Echinocladius*) inhabiting upland streams in patches of rainforest in the Australian wet tropics rarely disperse between streams < 1 km apart, and this is likely due to fragmentation of the riparian rainforest vegetation (Krosch et al. 2009). Deforestation limits the dispersal capacity of some forest-dwelling insects to move between catchments (Smith and Smith 2009; Alexander et al. 2011) and can even affect species living in highly disturbed systems like urban creeks (Smith et al. 2009). On much smaller scales, some human-made structures such as road culverts can deter movement by caddisflies or increase their risks of mortality from enterprising spiders that build their nets across the culverts (Blakely et al. 2006). Bridges across streams can also deter movement with one study showing that 86 per cent of mayflies, *Palingenia longicauda* (Palingeniidae), did not fly upstream past bridges. Given that the mayflies did not make contact with the bridge, the likely mechanism is that bridges disrupt the polarization of light from the water surface, which mayflies use as a visual guide during dispersal (Málnás et al. 2011).

9.5 Migration

Some species of insects undertake extraordinary migrations across hundreds or thousands of kilometres. A formal definition of what constitutes 'migration' is difficult because most migrating insects do not undertake return trips, as do migrating birds, for example. Instead, some insects may migrate only in one direction, with their offspring doing the return journey, and so gathering evidence of true migration is exceedingly difficult.

Of aquatic insects known to migrate, the clear champions are the dragonflies, which is not unexpected given their extraordinary flight capabilities. At least 25–50 species out of a world total of some 5000 are thought to undertake migrations (Russell et al. 1998), but this could be an underestimate given the difficulty of studying migration (May and Matthews 2008). Migrations may involve movements between continents and also seasonal 'swarms' in which large numbers of dragonflies may move long distances over relatively short periods of time. Such swarms commonly follow landmarks such as coastlines, cliffs, and other topographical features. Moving swarms are often associated with weather fronts, where insects can take advantage of winds (Srygley and Dudley 2008).

Dragonflies often undertake swarm migrations, with densities reaching a par with that of locust swarms. For example, a swarm migration of *Anax junius* (Aeshnidae) travelling through Chicago at tree-top height was estimated to be ≈ 850 m wide (at its peak) and to contain about 1.2 million dragonflies (Russell et al. 1998). Although it is difficult to know whether swarming dragonflies actually travel great distances, at least some do. A study in which individual *A. junius* were tagged with transmitters and followed by a Cessna aircraft showed that these dragonflies migrated southward during the colder months, dispersing approximately every three days and moving > 58 km in each six-day period. Individuals only moved after two consecutive nights of successively colder temperatures, patterns that are very similar to those of migrating songbirds (Wikel-

ski et al. 2006). Dragonfly swarms also occur on other continents, including the South American species *Aeshna bonariens* (Aeshnidae), which swarms in regions of Pampas ahead of cold fronts typically containing thunderstorms (Russell et al. 1998).

The truly amazing migrants, however, are those that travel across oceans. A common, global migrant is the wandering glider, *Pantala flavescens* (Libellulidae), which can track dense aggregations of aerial plankton caused by weather patterns. It can be found 1000 km out to sea (Russell et al. 1998) and on atolls where there are no resident populations (e.g. Buden 2010). The Maldive Islands, for example, lack much surface fresh water (being coral cays) and yet millions of dragonflies arrive there in October each year, 98 per cent of which are *P. flavescens*. It is speculated that this species regularly migrates across the Indian Ocean, from southern India to East Africa, across 3500 km of open ocean (Anderson 2009) and uses the Maldives as a stopover or resting location. These dragonflies fly at high altitudes (1000 m), which means they can take advantage of north-easterly tail-winds delivered by the Inter-tropical Convergence Zone (ITCZ), which moves in a southerly direction in October. They take further advantage of the ITCZ, because it also delivers rains on their arrival in Africa and hence provides reproductive opportunities for this species, which uses ephemeral pools for oviposition and larval life. A reappearance of dragonflies in the Maldives during May, when the weather patterns are reversed, suggests there may be a return migration of offspring from Africa to India (Anderson 2009). *P. flavescens* has also been detected migrating at high altitudes in China, where it was estimated individuals might be flying 150–400 km in a single night, again possibly taking advantage of movements of the ITCZ to travel to areas of China at times when land is flooded to grow rice (Feng et al. 2006).

9.6 Flight polymorphisms

While flight is the predominant way adult aquatic insects disperse, it is energetically very expensive. Wing muscles can comprise a substantial proportion of the body mass (50–60 per cent of total body mass in some male dragonflies, Marden 2008) and a large proportion of the thoracic cavity (e.g. 70 per cent in gerrids: Roff 1990). Apart from the energy required to build flight muscle, flying itself consumes significant amounts of energy, and this can be a significant drain on resources. For example, the mayfly *Siphlonurus aestivalis* (Siphlonuridae) is typical of mayflies in that it does not feed as an adult, instead relying on energy accumulated during the juvenile stage. These mayflies undertake significant flights while looking for mates, and incur significant weight loss as they deplete fat reserves. Males, for example, lost 52 per cent of body fat during imaginal stages and this reflected the relatively long distances flown in search of females (Sartori et al. 1992).

This energetic cost is probably one of the ultimate reasons why some species of adult insects have nonfunctional wings and are restricted to walking. Other species show so-called flight polymorphisms, where some individuals within populations have reduced wings (wing polymorphism) or reduced flight muscle mass (flight muscle polymorphism) that preclude flight (Zera and Denno 1997). While these strategies may help solve the energy problem, there may be consequences for other life-history traits, such as dispersal.

9.6.1 Wing polymorphism

Wing polymorphisms describe insects where there is marked inter-individual variation in the sizes of the wings and flight muscles, such that some individuals have full flight capability (macropter or alate) while others cannot fly at all ('brachypter' meaning short wings, or 'apter' meaning without wings). Wing polymorphisms may be sex-linked, as in many stoneflies where males are brachypterous yet females have fully functional wings, and some caddisflies where females are wingless yet males have functional wings.

The proportions of flight-capable versus flight-incapable individuals can vary between populations and between species. For example, the Amazonian waterstrider *Potamobates williamsi* (Gerridae) has both winged and wingless individuals in reasonable numbers whereas the congeneric *P. sumaco* had only wingless individuals, which meant greater restrictions on its dispersal (Galacatos et al. 2002). In the northern hemisphere waterstrider

Aquarius najas, wingless individuals comprise 99 per cent of individuals in high-latitude populations, whereas low-latitude populations contain many more winged individuals. Wingless *A. najas* move only about 300 m during their lifetimes, meaning that high-latitude populations are likely to be far more isolated from each other due to limited dispersal than low-latitude populations (Ahlroth et al. 2010).

Flight capacity, or lack of, is often traded off against reproductive success in many insects having wing polymorphism. In females, this trade-off often means that individuals capable of flight have lower fecundity than females that cannot fly, while for males the trade-off may mean that flight-capable males have lower mating success than flight-incapable individuals (Zera and Denno 1997; Guerra 2011). For example, wingless individuals of the gerrid *Limnoporus canaliculatus* had significantly faster growth rates and laid more eggs than winged individuals (Zera 1984). In a study of Veliidae, fertility was higher in wingless than winged individuals across three different species of *Microvelia* (Muraji and Nakasuji 1988). However, flight capability may mean reduced life span rather than reduced fertility, as observed for some gerrids (Kaitala 1991), and in some species wing form (i.e. long or short wings) can be determined mostly by environmental factors such as photoperiod or temperature during critical life stages (Fairbairn 1988; Inoue and Harada 1997).

9.6.2 Flight muscle polymorphism

In species with flight muscle polymorphism, all individuals have wings, but some have reduced flight muscle mass and are incapable of flight (Zera and Denno 1997). Nearly every order of insects has species in which there is polymorphism in respect of flight muscle development (Marden 2000). Studies on aquatic insects are relatively few, but some aquatic hemipterans have individuals that are fully winged but with non-functional flight muscles (e.g. Scudder 1971). In some corixids, the development of functional flight muscles may depend on temperature (Scudder and Meredith 1972).

Some insects that are initially able to fly have the ability to histolyse wing muscles after dispersal (Young 1965; Harrison 1980; Spence and Andersen

1994). This renders them flightless, but the energy accrued from breaking down muscles can be invested into reproduction, an arrangement termed the oogenesis-flight syndrome. This syndrome occurs among aquatic insects, particularly beetles and bugs (Bilton et al. 2001). For example, three congeneric species of gerrids histolysed flight muscles directly after the dispersal period with flight-incapable females laying more eggs than females still capable of flight (Kaitala and Huldén 1990). Direct evidence that histolysis of wing muscles resulted in greater development of eggs in female mosquitoes has also been observed (Hocking 1952). Flight-muscle histolysis can be promoted if days are long and halted if feeding conditions are poor, again suggesting fairly sophisticated responses by some species to environmental cues affecting reproductive success (Harada and Nishimoto 2007).

9.6.3 Flightlessness

Flightless species occur in every order of insects where the species are otherwise typically winged as adults. Flightless species have no wings at all or stubby projections that are largely inoperable as wings, and some of these species are highly specialized. Flightlessness in stoneflies is often associated with low temperatures and, for example, 25 of the 104 New Zealand and subantarctic species are wingless or have reduced wings. As described in Section 8.5.2, many stoneflies and at least a few mayflies have reduced wings suitable only for sailing or skimming, although some skimmers have the capacity to fly weakly. Some species of hydrophiline beetles are completely flightless (unusual for this subfamily) and are restricted to wet, headwater gulches on the sides of particular mountains in Hawaii (Short and Liebherr 2007). Marine chironomids in the genera *Pontomyia* and *Clunio* typically have males with fleshy wings and females that lack both wings and legs. These tiny insects live in coastal lagoons or rock pools as larvae. The pupae float to the surface and emergence occurs on nights with a full or new moon, with males emerging first shortly after sunset and females about 20 minutes later. Males move by using their fleshy wings as oars and are able to glide quickly over the water's surface at speeds of up to 15 cm s^{-1} to locate females, which

typically have no means of locomotion. Mating and egg-laying takes place quickly as these insects live no more than one to three hours, one of the shortest adult longevities on record (Neumann 1986; Huang and Cheng 2011). As noted in Section 8.6, some species in these genera live out in the open ocean, but are attached to and dispersed by hawksbill sea turtles. Another wingless, marine chironomid is *Telmatogeton amphibius*, a subantarctic, intertidal species that lives on islands in the Southern Ocean and is not only wingless but also parthenogenetic (Nondula et al. 2004).

9.7 Dispersal by vectors

Some species disperse by attaching themselves to other highly mobile animals that then take them to other locations, potentially allowing much longer-distance dispersal than species may be capable of themselves. Such phoresy or parasitism was discussed in Section 8.6 in the context of dispersal in the aquatic environment. Obviously, some vectors are capable of also carrying attached individuals across land and may disperse them between water bodies. For example, dragonflies occasionally act as vectors for other aquatic animals and, as discussed above, they are capable flyers. Dispersal of aquatic invertebrates across terrestrial environments by other types of organisms, such as birds, has been only poorly studied and much of the information is largely anecdotal or phenomenological. Some invertebrates (particularly crustaceans) can survive ingestion and are defecated alive by birds, however there is little information on whether eggs and larvae of aquatic insects ingested by waterfowl might be transported between water bodies (Green and Figuerola 2005). One study found intact, corixid eggs in the faeces of waterbirds (ducks and coots) (Figuerola et al. 2003) although the viability of the eggs was not ascertained.

There are instances of adult insects being transported by other, flying animals. Perhaps the most astonishing of these is a report of chironomids, *Chironomus salinarius*, being transported in the guts of black-tailed godwits, waterfowl that can travel up 20 km between nesting and feeding sites and also migrate between Europe and northern Africa during the northern hemisphere winter. The chironomids are eaten by godwits, but a significant proportion survive passage through the gut and are deposited alive with the faeces, meaning that this chironomid species can be potentially transported great distances (Green and Sánchez 2006).

These examples of phoresy are fascinating, but it is unclear to what extent these are relatively rare and 'accidental' movements, or whether they are frequent and ecologically important elements of species' ecologies. There is likely to be a bias in the reporting of these incidents in favour of the successful episodes, with little or no information on unsuccessful events. If dispersal by vectors through the terrestrial environment is an integral part of species' ecologies, it might be advantageous if these insects could identify or select vectors that will travel to suitable locations. Now that would be truly amazing.

PART 4

Population Dynamics and Population Persistence

This next set of chapters considers aspects of aquatic insect biology that are especially pertinent to questions posed in population ecology. Ecologists who investigate populations focus their research on identifying factors that drive population numbers; where, when, and how these factors operate; and ultimately how populations are regulated. This kind of research also underpins much applied ecology and species conservation, because it allows us to understand what is likely to make some populations particularly vulnerable to human impacts, such as climate change, and the implications for species distribution patterns. To do this kind of research means gathering information on rates of births, deaths, and migration, because these are the basic parameters that drive population increase and decline. Dispersal that may move insects across population boundaries (i.e. migration) was considered in Chapters 8 and 9. In this section we focus on births and prospective sources of mortality as insects grow and develop. The three chapters examine how reproduction occurs; the production, deposition, and hatching of eggs; and the growth and development of juveniles as they recruit into the next generation of adults.

To study the population dynamics of insects, with their complex life cycles, is no easy task. Like their terrestrial counterparts, aquatic insects have multiple developmental stages, but aquatic insects may also move between aquatic and terrestrial environments. Each life-cycle stage is exposed to particular physico-chemical stressors and potential food shortages, as well as prospective competitors, predators, parasites, and disease-causing organisms. All of these factors may change at each developmental stage, and yet we need to know what sets limits at each stage because demographic bottlenecks may occur at any point in the life cycle. To conduct a comprehensive population study of an aquatic insect requires details of the production and survivorship, growth, and development of each stage in its natural environment, and these details may vary enormously between species. We need to understand how adult males and females find one another to mate, and whether they use particular environmental cues to locate each other. Fertilization and oviposition of eggs may take place immediately after mating or be delayed, and females of different species show great variety in oviposition behaviour and habitat. Some species simply broadcast their eggs into the water, whereas others use specialized behaviours to locate specific habitats in which to lay their eggs. The hatchlings of these species begin life distributed across the aquatic environment in very different ways, with potential implications for population numbers if mortality of hatchlings is high and spatially variable. Species pass through variable numbers of larval instars and spend different lengths of time in the larval stage (days to years), partly due to innate differences in larval longevities, and partly due to environmental variation. Cold temperatures, for example, can slow the growth of individuals, meaning that the same species can exhibit a wide range of larval sizes and development in different places. Mortality rates of larvae may vary considerably as a result. Additionally, some aquatic insects pass through a complete metamorphosis during a pupal stage before they emerge as adults. Huge variety is seen in the construction, location, and prospective survivorship of pupae, again adding a layer of prospective sources of mortality that may limit recruitment.

The intrepid population ecologist who chooses to work on insects must therefore be armed with considerable, biological information. When and where do mating and oviposition occur? Can eggs be counted and identified and can we detect and count those that hatch successfully? How fast do larvae grow and how many generations are there per year? Where and when do pupae occur, and can they be successfully identified and counted? Can successful production of adults be measured? All of these questions must be addressed, and the information needed to begin answering them is covered in Part 4.

CHAPTER 10

Reproduction and mating behaviour

10.1 Introduction

Reproduction is essential for populations to persist and for species to remain extant. It is a simple word, but it entails many successive processes, each of which must occur in the correct order. Perhaps most fundamentally, reproduction requires the production of gametes by the reproductive organs: males produce sperm; females produce eggs. The reproductive organs of insects are diverse in form, but all conform to the same basic design and function in similar ways. Like humans, fertilization is internal in the insects so females typically retain eggs internally, and sperm must be transferred from the male to the female reproductive tract. In order to mate and transfer sperm, males and females must first locate one another and correctly identify members of their own species. A necessary first step involves the formation of pairs or aggregations and communication between individuals to identify conspecifics, members of the opposite sex, and individuals that are ready to mate. Copulation (sperm transfer) is typically the final step of the mating process. Fertilization does not necessarily occur during or immediately after copulation and sperm is often stored until, for example, the female is fully sexually mature. Many aquatic insects reproduce only once in a lifetime (semelparous species), and this is particularly common in species with short-lived adults. Multiple bouts of reproduction (iteroparous species) are more common in long-lived adults, such as some aquatic Coleoptera and Hemiptera.

Reproductive events up to, and including mating, typically take place in the adult stage and in the terrestrial environment. Once mating has occurred, females must then lay eggs (oviposition), the embryo develops within the egg (embryogenesis), and the neonate or first instar larva hatches out. At some point during the processes of oviposition or egg hatch, members of the next generation must enter the aquatic environment. Aquatic insects have diverse strategies for making this transition from terrestrial to aquatic environments, but we will defer that discussion until Chapters 11 and 12. This chapter will discuss events up to the stage of egg fertilization, beginning with a brief description of the male and female internal reproductive organs and the formation of eggs and sperm (Sections 10.2, 10.3). Some species are ready to mate almost immediately after they emerge as adults, whereas others require a period of maturation (Section 10.4). How many offspring each female produces (fecundity) may also depend on events during the maturation period and also during the larval stage. Inter-individual communication is very important in finding a suitable mate. The structure and function of sensory organs have been discussed in Chapters 6 and 7; Section 10.5 will discuss how they operate in the context of reproductive processes, to bring males and females together. Once a mate has been located, the pair must copulate to transfer sperm (Section 10.6). For most species, the male and female separate immediately after copulation, but mating pairs of a few species, especially the Odonata, remain together in post-copulatory mate guarding (Section 10.7). Finally, a few species (primarily Ephemeroptera) have done away with males and mating altogether, and the females reproduce parthenogenetically (Section 10.8).

10.2 Female reproductive organs and egg formation

The main functions of the female reproductive system are to produce eggs (the process of oogenesis),

Aquatic Entomology. First Edition. Jill Lancaster & Barbara J. Downes.
© Jill Lancaster & Barbara J. Downes 2013. Published 2013 by Oxford University Press.

which have yolk inside and a chorion (egg shell) on the outside, to receive and store sperm from the male, and to coordinate events that lead to fertilization and oviposition.

10.2.1 Internal organs

The female internal reproductive organs typically include a pair of ovaries that release mature oocytes (eggs) into the lateral oviducts via the calyses (singular calyx), which are cup-like expansions of the oviducts (Figure 10.1). The lateral oviducts fuse to form the common or median oviduct, which typically enters a sac-like structure called the genital chamber (or vagina) via the gonopore. The external opening of the genital chamber is the vulva. The Ephemeroptera are the exception in that the lateral oviducts remain separate and open to the exterior independently. During mating, the genital chamber (often called the bursa copulatrix) serves as a copulatory pouch in which spermatophores or seminal fluid are deposited initially. Connected to the genital chamber are the accessory glands, which may have a variety of functions. Also connected to the genital chamber is the slender spermatheca, in which sperm are stored prior to fertilization. The spermatheca may have an associated tubular spermathecal gland, the secretions of which may provide nourishment for the sperm during storage. Although most female insects have a spermatheca or some facility to store sperm, there are exceptions. Most Ephemeroptera, for example, cannot store sperm and eggs must be fertilized at the time of mating.

The ovaries each comprise several tubular ovarioles, typically ensheathed by connective tissue that is embedded with tracheoles and muscles. In the Plecoptera, however, the ovarioles do not occur in bundles and remain separate from one another. The number of ovarioles varies among species, but is generally constant within species. Each ovariole consists of a terminal filament at the distal tip (Figure 10.1), followed by the germarium, vitellarium, and finally the pedicel or ovariole stalk where it joins the calyx (Figure 10.2). The terminal filaments may be fused together and anchor the ovaries to the body wall or dorsal diaphragm. Within the germarium, mitosis of germ cells gives rise to oogonia (the primary oocytes) and, in some types of ovarioles, to nutritive cells. The oocytes typically form a linear sequence along the ovariole as they mature. Upon entering the vitellarium, they become enclosed in a one-cell thick layer of follicular epithelium. Oocytes accumulate yolk in the vitellarium (a process known as vitellogenesis) and produce an eggshell (choriogenesis). There are two main types of ovariole (Figure 10.2), depending on the presence or absence of special nutritive or nurse cells (trophocytes) that contribute to the nutrition of the developing oocytes. Trophocytes are absent in a panoistic ovariole (e.g. Ephemeroptera, Odonata, Plecoptera), and the oocytes obtain nutrients from the haemolymph via the follicular epithelium (Figure 10.2a). Trophocytes are present in meroistic ovarioles, which can be further divided into two sub-types. In polytrophic ovarioles (e.g. Trichoptera, Diptera, Hymenoptera, Lepidoptera), trophocytes are enclosed in each follicle along with the oocyte, and are directly connected to the oocyte via short cytoplasmic bridges, through which they pump nutrients into the oocyte (Figure 10.2b). In telotrophic ovarioles (e.g. Hemiptera, Coleoptera), trophocytes are clustered in the germarium and are connected to each oocyte by an ever-lengthening trophic cord (Figure 10.2c). The number of mature egg chambers per ovariole can vary among species. For example, among the caddisflies, *Brachycentrus incanus* (Brachycentridae) and *Sericostoma galeatum*

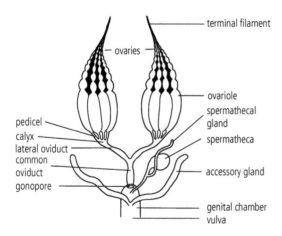

Figure 10.1 Schematic illustration of the female internal reproductive organs.
Source: Adapted from Snodgrass (1935).

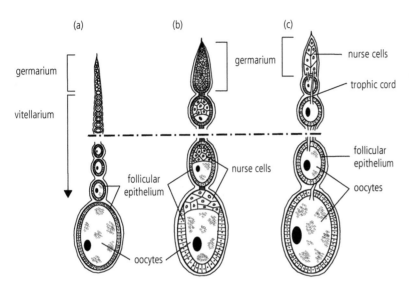

(a) (b) (c)

germarium

vitellarium

germarium

nurse cells

trophic cord

follicular epithelium

oocytes

follicular epithelium

nurse cells

follicular epithelium

oocytes

oocytes

Figure 10.2 Types of ovarioles. (a) A panoistic ovariole with no specialized nurse cells to support oocyte development. Meroistic ovarioles are of two types: (b) a polytrophic ovariole with nurse cells grouped near each oocyte: (c) a telotrophic ovariole with a cluster of nurse cells in the germarium. The upper portion of each ovariole (above the dashed line) is enlarged relative to the lower portion.

(Sericostomatidae) have only one mature egg capsule, whereas *Goera pilosa* (Goeridae) and *Plectrocnemia conspersa* (Polycentropodidae) have several (Denis and Le Lannic 1977; Conn and Quinn 1995). Mature eggs are typically held in the oviducts prior to fertilization and oviposition.

Various accessory glands may be present and usually open into the genital chamber. Not all insects have accessory glands and, among the aquatic groups, they are absent in the Ephemeroptera and most Plecoptera. Female accessory glands can take various forms ranging from simple glands to compound, branched glands, and they may play a diverse range of pre- and post-copulatory functions (Kaulenas 1992; Gillott 2002). Commonly, these glands secrete materials that form a protective coating around the eggs or adhesives to attach the eggs to the substrate during oviposition, although such coatings may also be produced by follicle cells in the ovarioles, especially in taxa which lack accessory glands (Chapter 11).

10.2.2 The mature egg

Each egg is a single cell, the female gamete. Egg shape is highly variable between species and may be spherical, conical, sausage-shaped, barrel-shaped, or have elongate respiratory structures (Chapter 11). Although eggs vary in shape, once

mature, they generally conform to the same basic plan (Figure 10.3).

The outer-most layer of the egg is the chorion, or shell, and beneath the chorion is the vitelline envelope, which is an extracellular, proteinaceous layer (sometimes called the vitelline membrane, although it is not membranous in structure). The vitelline envelope surrounds the entire egg cell. The contents of the egg cell consists primarily of yolk, and the cell's cytoplasm, or periplasm, is usually distributed in a thin band just inside the vitelline envelope and in diffuse strands that run throughout the yolk, called the cytoplasmic reticulum. The serosal cuticle forms later in development, incorporating the vitelline envelope on the outside, a chitinous endocuticle with an epicuticle often having a second wax layer. The zygote nucleus usually lies dorsally and close to one end of the egg, which is considered the anterior pole.

10.2.3 Vitellogenesis and choriogenesis

Insect eggs typically contain enormous amounts of yolk, which is produced during vitellogenesis and results in a marked increase in the size of the oocytes. Most insect eggs are large relative to the size of the adult, with the smallest eggs generally found in parasitic species. Yolk is a mixture of various nutritive substances, to support the developing embryo until

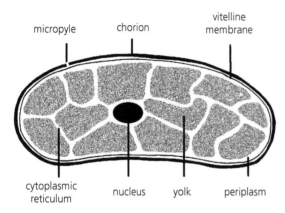

micropyle   chorion   vitelline membrane

cytoplasmic reticulum   nucleus   yolk   periplasm

Figure 10.3 Diagrammatic section through an egg at oviposition.

it can obtain its own food. Yolk is made up of round-ish granules or vacuoles (known as yolk spheres), which contain proteins, nucleic acids, lipids, and carbohydrates. Unfortunately, toxins accumulated by females during their larval life can be transferred to the egg yolk and may affect offspring development and survivorship (Standley et al. 1994).

Once vitellogenesis is complete, the vitelline envelope and later the chorion are formed, primarily by follicle cells in the ovariole. The chorion is rarely smooth (as in most bird eggs) and surface sculpturing of the chorion typically reflects the outline of the follicle cells. The time required for chorion formation varies greatly among species and is related to the size (thickness) of the final structure. The chorion comprises two main layers, an inner endochorion next to the vitelline membrane and an outer exochorion, although the chorion appears to be multi-layered in some species. The chorion is cuticle-like in nature and contains layers of protein and lipoprotein, some of which are tanned by polyphenolic substances released by the cells. Once chorion formation is complete, a waterproof layer of wax or lipid may be secreted between the chorion and the vitelline envelope. This waterproof layer is most common in insects with eggs that are laid terrestrially and at risk from desiccation, and it is characteristically absent from eggs that are laid aquatically.

10.2.4 Form and function of the chorion

The egg chorion is rarely produced as a uniform layer and it is often sculptured with microscopic grooves, ridges, and various complex structures that may be visible only under the higher magnification of an electron microscope. Some chorionic architecture may arise passively during choriogenesis (e.g. the outline of follicle cells is often imprinted on the chorion, as mentioned above), but many structures serve important functions in the reproductive process. The chorion serves several functions, which include allowing sperm entry, providing elasticity to allow oviposition through a narrow genital opening yet expansion of the egg as the embryo enlarges, protecting the embryo from environmental stresses (e.g. bacterial and fungal attack, predators, temperature fluctuations, wetting, and drying), attaching eggs to substrates, ensuring adequate gas exchange (oxygen supply) for the developing embryo, and finally allowing the larva to escape from the egg (hatch). Because species differ in their oviposition habits (Chapter 11), there is enormous variation among species in the structure of the chorion. There have been attempts to use chorionic architecture for taxonomic purposes, e.g. ootaxonomy (Koss and Edmunds Jr 1974; Hinton 1981; Dominguez and Cuezzo 2002), and eggs of closely related species may be very different (Morgan 1913; Zwick 1982; Yule and Jardel 1985). Indeed, illustrations of eggs now often accompany the taxonomic and systematic descriptions of new species. However, eggs from a single female may also vary in form (Gaino and Bongiovanni 1992), so caution is required when using eggs for taxonomic purposes. Some patterns in chorionic architecture appear to have a phylogenetic basis but, for our purposes, it is perhaps most logical to organize a discussion of chorionic architecture around the functional roles. Three of the general functions mentioned, gas exchange, protection, and attachment, are closely related to oviposition habits and will be discussed in Chapter 11, while another function, egg hatching, is related to development and will be discussed in Chapter 12.

During oviposition, as an egg moves past the opening to the spermatheca, a few sperm are released onto its surface. Sperm entry, and hence egg fertilization, is possible because there are channels through the chorion, called micropyles, through which sperm can enter the egg. Micropyles are created by some follicle cells in the ovarioles that have extra long microvilli, which span the width of the chorion during choriogenesis. When these microvilli are withdrawn after chorion formation, they leave behind open channels. The number of micropyles and their spatial arrangement varies between eggs of different species. Among the Odonata, for example, two micropyles per egg are typically found in the family Libellulidae, seven in the Cordulegastridae, and 12–14 in the Epiophlebiidae (Becnel and Dunkle 1990). In terms of spatial arrangement, micropyles typically occur at the anterior pole of dragonfly eggs, at the posterior pole of many stonefly eggs, but in mayflies, micropyles are arranged normal to the major egg axis and tend to be evenly spaced around the circumference. Micropyle structure also varies between species. In many Odonata and Megaloptera, the micropylar opening occurs at the tip of conical projections, whereas in many mayflies the micropylar opening lies in a depression in the chorion surface and sperm are presumably guided into the opening by this funnel-like arrangement. The shape of this depression or sperm guide also varies among mayfly species and in some Ephemerellidae, sperm guides may be oval, round, or teardrop-shaped (Ubero-Pascal and Piug 2009). Although some eggs have multiple micropyles, typically only one sperm (the first to reach its destination) fertilizes the egg.

10.3 Male reproductive organs

The main functions of the male reproductive organs are to produce, store, and deliver sperm to the female. Males may also produce substances that are transferred to the female during copulation and that may regulate female receptivity and fecundity, or provide protection for the fertilized egg. The basic male system includes paired testes (although these are fused in Lepidoptera), paired vas deferentia and seminal vesicles, a median ejaculatory duct and various accessory glands (Figure 10.4a). Each testis comprises a variable number of testicular tubes or follicles that produce the sperm, which are then released into the vas deferens. The testicular tubes are typically held together, as are the ovarioles of the ovary, but the tubes are free from one another in the Plecoptera (Figure 10.4b). As with the female ovarioles, various zones of sperm maturation can be iden-

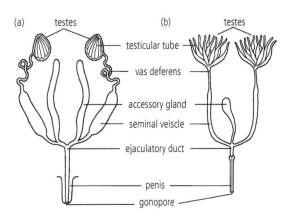

Figure 10.4 (a) Male reproductive system typical of many insects. (b) Male reproductive system typical of Plecoptera.
Source: Adapted from Snodgrass (1935).

tified within the male follicles. At the most distal end, the germarium, spermatogonia are produced from germ cells; they then divide and transform into spermatocytes. The spermatocytes then move into the zone of growth where they divide to form spermatids and each one becomes enclosed in a layer of somatic cells forming a cyst. They then travel into the maturation zone and finally to the zone of transformation where the cyst wall usually ruptures and spermatids differentiate into flagellated spermatozoa. Sperm are moved through the vas deferens to the seminal vesicles for storage. The seminal vesicles are dilations of the vas deferentia, but are well tracheated and often glandular, which may indicate a nutritive function. The paired vas deferentia join, and sperm are released into the ejaculatory duct. Ephemeroptera are the exceptions in that the ejaculatory duct is absent and each vas deferens opens directly to the exterior, i.e. male mayflies have two penes. Insect spermatozoa tend to be long and have a slender head region, probably as an adaptation to enter the micropyle. Within most insect orders (except Ephemeroptera and Diptera), some taxa will package sperm into a spermatophore. Within the Trichoptera, for example, some species produce two-part spermatophores consisting of a coagulated protein mass and a sperm sac, whereas others deliver free-swimming sperm (Khalifa 1949). In males, there may be spermatophore sacs associated with the seminal vesicles that function in spermatophore production. Spermatophore structure varies enormously among taxa and, accordingly, the female reproductive organs may vary in form depending on the structure of the spermatophore that they receive.

Tubular and often paired accessory glands are connected either to the lower end of the vas deferentia or the upper end of the ejaculatory duct. Their secretions commonly contribute to the seminal fluid and formation of the spermatophore but, as with females, the male accessory glands may play diverse roles in reproduction (Chen 1984; Kaulenas 1992; Gillott 2002). Accessory gland products of some species, when transferred to the female during copulation, cause increased egg production or alter female behaviour (decreased receptivity or willingness to mate with more than one male, or stimulate oviposition).

10.4 Sexual maturation and fecundity

The number of eggs carried by a female is often used as a measure of fecundity—this is actually a measure of 'potential' fecundity because not necessarily all the eggs a female produces will be fertilized, oviposited, or hatch. The best estimates of potential fecundity, however, are based on the number of mature eggs and, therefore, it is important to understand the processes that surround sexual maturation.

Newly emerged adults often lack fully developed reproductive organs, eggs, sperm, or accessory glands, and require a period of sexual maturation during which important morphological, physiological, and behavioural changes occur. In odonates, for example, adults may be in a pre-reproductive or maturation period for days or weeks, and females may take longer to mature than males (Corbet 2004). Individuals of many taxa move far from the emergence site, and often far from any water body, during the maturation stage and must find their way back to water before mating or egg laying. For other taxa (e.g. some Plecoptera, Ephemeroptera, Diptera), however, egg and sperm production may occur in the final larval or pupal instar, and mature sperm may be present in the male seminal vesicles at emergence, thus enabling mating and egg laying soon after emergence. Often males that are reproductively mature will mate with adult females that are not yet sexually mature, and females must store sperm for later fertilization while they complete oogenesis (Section 10.6). Indeed, mating may be required before maturation can be completed. Thus, maturation time may differ markedly between sexes of the same species. In other groups, the majority of oogenesis and spermatogenesis occur in the adult stage and such adults may be long-lived. Many maturation processes are controlled by the endocrine system (hormones) and various environmental stimuli may influence endocrine activity and, thus, influence the rate of maturation. The hormonal processes of aquatic and terrestrial insects are very similar and will not be discussed.

At the simplest level, maturation can be measured in terms of the time, after emergence, required to reach sexual maturity, i.e. the rate of maturation. The concept of maturation, however, is often inextricably linked to concepts of fecundity (number of eggs laid or offspring produced) and body size, given that fecundity is commonly correlated with body size in insects (Honěk 1993; Kingsolver and Huey 2008). These factors (maturation time, fecundity, body size) are enormously variable between and within species, and may depend on events in the adult, larval, and pupal stages. The causes and consequences of that variability is the subject of a wealth of studies on life-history strategies, plasticity, and trade-offs. We cannot possibly cover that literature and, instead, discuss briefly two variables that directly affect maturation rate during the adult stage: temperature and feeding versus stored energy reserves.

10.4.1 Temperature

Many environmental factors, including photoperiod, humidity, and temperature, affect maturation rates indirectly by modifying the activity of the endocrine system, which controls so much of the reproductive process. These variables are rarely independent and maturation may depend on particular combinations of environmental conditions.

Undoubtedly the most important abiotic variable influencing maturation time is temperature, simply because it has such a profound and direct effect on most biochemical and developmental processes (Chapter 4), and species differ in their optimal temperatures for reproductive performance (Kingsolver and Huey 2008). It may seem contradictory, but the amount of accumulated heat required for a particular organism's development is constant, but that cumulated heat is a combination of temperature (between thresholds) and time (i.e. the degree-day concept, Chapter 12), and this does vary spatially. Temperature extremes (beyond thresholds) may damage oocytes and sperm, so maturation may be impossible in some circumstances. Within species, differences in maturation rates may be driven by geographical, altitudinal, and latitudinal variations in degree days and, for species with more than one generation per year (multi-voltinism), by seasonal variations. The Corixidae of some northern temperate zones, for example, are multivoltine, and the rate of sexual maturity of adult males and females differs seasonally among generations (Jansson and Scudder 1974). Over-wintering adult males are sexually mature at the time of ice break-up, new generation males are mature at the time of emergence in early- to mid-summer, and late summer males remain immature until late in the autumn. Adult females show similar patterns of sexual maturity, except that over-wintering females mature only some time after ice break-up.

10.4.2 Feeding versus stored energy reserves

Energy is required for sexual maturation and for maintenance of the adult during maturation. For non-feeding adults (e.g. Ephemeroptera), all this energy must be accumulated during larval growth and the adults of such species are often short-lived, thereby ensuring that maximum energy reserves are directed to reproduction. For adults that do feed, the amount and kind of food consumed may have various consequences for reproduction and life-history characteristics. In the most extreme situation, starvation and lack of food may prevent sexual maturation (Jansson and Scudder 1974). The importance of adult feeding is perhaps best understood for the biting flies (e.g. Simuliidae and Culicidae), many of which need a blood meal for the eggs to mature. When food resources are scarce, however, adults of some species may draw on their own body tissues as an alternative energy source.

Although some of the most memorable blackflies and mosquitoes are those with adults that have

bloodsucking habits (anautogenous), many species within these groups do not feed as adults (autogenous). In any particular area, variations in the relative abundances of anautogenous and autogenous species may be related to the potential abundance of blood meals. For example, vertebrate hosts may be scarce in high latitude environments and the blackfly fauna of the Yukon (north-western Canada) has a very high proportion of autogenous species: 25 per cent compared with < 3 per cent in the Simuliidae as a whole (Danks et al. 1997). Autogenous females emerge with fully mature eggs, or mature their eggs exclusively from nutrients carried over from the larval stage and stored in fat bodies. In contrast, anautogenous females must feed in order to produce their eggs. The situation, however, is not quite that simple and some simuliid species may be obligate feeders, facultative feeders, or non-feeders, and food can include vertebrate blood meals (usually for females) and carbohydrates from flower nectar (males and females) (Anderson 1987). In facultative species, the larval diet determines whether the female will be able to develop eggs autogenously, or whether a blood meal will be required. In general, blood and nectar meals have different functions: the carbohydrates from nectar provide energy for flight and prolong life, whereas, blood meals provide the nutrients required for egg maturation. Inevitably, there are exceptions, such as the autogenous *Simulium ornatipes* (Simuliidae), which requires sugars to initiate oogenesis (Hunter 1977) and, similarly, sugar feeding is required for egg development in some facultative autogenous mosquitoes (Corbet 1967). In anautogenous female blackflies that have not yet taken a blood meal, oogenesis is usually halted before or at an early stage of vitellogenesis and proceeds only when blood is consumed. The number of eggs eventually produced is typically associated with the kind of host blood and the volume of blood ingested.

When food resources for adults are scarce and/or energy reserves acquired during the larval stage are inadequate, then the adults of some insects are able to mobilize energy locked up in their own body tissues and re-allocate it for reproductive purposes. The selective value of this process is that it enables the female to tolerate periods of low food and/or high stress. Histolysis of flight muscle in return for reproductive gain is common (e.g. in many water striders) and a longer term trade-off between flight and reproduction (the so-called oogenesis flight syndrome, Chapter 9) is apparent in various species with dispersal polymorphisms (Zera and Denno 1997). Tissues other than flight muscle may also be reabsorbed. When food is scarce, females may cannibalize their own recently laid eggs, allowing them to maintain their reproductive status, as observed in some Corixidae (Pajunen and Pajunen 1991) and Dytiscidae (Jackson 1958).

10.5 Aggregation and sexual communication

The first challenge faced by adults ready to mate is the problem of finding a mate that is of the same species, reproductively mature, and receptive to mating. The adult stages of most aquatic insects are terrestrial and, thus, mating generally occurs in the terrestrial environment. Common exceptions are many aquatic Coleoptera and Hemiptera that mate in the water, but mating follows the same general behavioural processes. Matchmaking may be facilitated by males and females being attracted independently to some feature in the environment and/or by signalling directly to one another. Such behaviour may result in dense, apparently organized aggregations (e.g. swarms) or more loosely arranged encounters.

Insects often form conspicuous, airborne swarms with many individuals (usually of the same species) aggregated in quasi-stationary flight at a particular place (flight station) for the purpose of mating (Downes 1969; Sullivan 1981). It is this criterion of aggregation about a particular station that distinguishes strict, station-holding, swarming behaviour from other forms of adult aggregation (often called dispersed swarms), such as those that occur when mature larvae are highly aggregated, emergence is synchronous, and newly emerged adults are reproductively mature. Although the word 'swarm' suggests many individuals, a true station-holding mating swarm can have only a single male. In practice, distinguishing between station-holding and dispersed swarms may not be straightforward. Mating swarms may be particularly effective for species with scattered larval habitats or low densities of larvae and long-lived, wide-ranging adults, such as some Diptera (Downes 1969; Yuval 2006). Swarms may be equally suitable for species with short-lived, non-feeding adults that swarm close to the emergence site and thus prevent further dispersal, such as many Ephemeroptera. Most swarms are composed predominantly of males (although female-only swarms occur in some species) that hang about waiting for receptive females to arrive and come close enough for detection and recognition. In some situations, recognition may not be perfect and a significant proportion of mating attempts may be between two males, for example one-third of mating grasps in the midge *Chironomus anthracinus* (Chironomidae) were between two males (Tokeshi and Reinhardt 1996). Even when a male grasps a female, mating does not necessarily occur every time and multiple attempts may be required before successful mating occurs. Females may join a mating swarm briefly if copulation occurs in the air, but leave the swarm once mated, whereas males may remain in the swarm with the possibility of a subsequent mating. Alternatively, a pair may leave the swarm and copulate some distance away, in the air or on the ground. After mating, males but not mated females may then rejoin the swarm. In general, males of swarming species may fly considerably more than females, which is likely to entail a substantial energetic cost and therefore males may require greater energy reserves (Sartori et al. 1992).

Swarming sites are identified by visual markers and are usually species-specific. Genetically distinct populations of the same species may even use different swarm markers (Savolainen et al. 1993). Markers are enormously variable and may include trees, tree branches, tall stands of vegetation, streams, hill tops, fence posts, buildings, yellow school buses, or large mammals in the case of biting flies (Yuval 2006). In general, males and females both respond to the same visual stimulus of the marker, which promotes swarming behaviour, and initially they do not respond to one another. The swarm has a definite position and boundaries relative to the marker. Within the swarm, however, individuals may hover and be relatively stationary within the swarm (e.g. Simuliidae), constantly fly in an up-and-down pattern (e.g. some Ephemeroptera, some Chironomidae), in a series of vertical and horizontal loops (e.g. mosquitoes, Gibson 1985), or a horizontal figure-of-eight pattern (e.g. some leptocerid caddisflies, Tozer et al. 1981).

Species in which males are territorial may display intense male–male competition and territories typically contain resources that are important to females, such as potential oviposition sites. Such behaviour is well-documented in the Odonata (Corbet 2004) and the rendezvous (where the male and female meet) is commonly centred on an oviposition site. Males search for females that might enter the territory, with searching carried out either by visually surveying whilst settled on a perch or whilst in flight, patrolling the territory. Any males entering the territory are usually met with aggressive behaviour from the incumbent.

As should now be apparent, these aggregation behaviours depend to a large extent on males and females being able to detect and respond to environmental cues, and to communicate with one another. Communication systems that put males and females in close proximity may operate on several spatial scales, with environmental cues generally operating at the largest scales (e.g. hilltops as swarm stations) and various behavioural signals at the smallest scales to bring males and females into physical contact. In addition to signals of attraction, isolation mechanisms, which ensure that only conspecifics are attracted to one another, are equally important forms of communication. Insects may communicate with one another using visual, vibrational, auditory, and chemical cues, and multiple forms of communication may be involved in any suite of mating behaviours. Sensory systems, how sensory receptors function, and sensory information is transmitted has been discussed in Chapters 6 and 7; here we focus on how they function in mating systems, i.e. the processes and behaviours related specifically to mating and courtship.

10.5.1 Vision

Vision is widely exploited as a means to recognize conspecifics and to distinguish between males and females, but may also be involved in more complex visual signals regarding receptiveness. Visual signals are perhaps most important in day-active species, and less important for taxa that mate at night.

In most odonates, for example, recognition of sex and species is based predominantly on visual cues such as body size, flight style, and colour. The range of potential responses to a visual signal are remarkably diverse and contingent upon the species, sex, and reproductive state of the two parties (Corbet 2004). Some encounters may lead into courtship, but many are met with refusal behaviours.

Eye morphology may be sexually dimorphic if visual mate recognition is more important for one sex than the other, and such dimorphism is common in species that swarm and mate in flight. In most Ephemeroptera and some dipteran families (Simuliidae, Blephariceridae, Tabanidae), for example, female eyes are of medium size and the facets are small and numerous (Downes 1969). The male eye, however, is often divided into two parts with a sharp line dividing the upper and lower areas: the lower is unmodified while the upper part is much enlarged and highly specialized with large facets, as discussed in Section 6.4. The male turbinate compound eyes of Ephemeroptera have upper ommatidia that are much enlarged and sometimes separated completely from the lower female-like part of the eye (Figure 6.11). The large upper elements of the male eyes presumably help the male to locate and capture a female. There are, of course, exceptions and these may prove the rule, e.g. males and females of some Simuliidae and Blephariceridae have similar eyes, but these species typically mate on the ground (Downes 1969).

10.5.2 Vibrational signalling

Communication through substrate-borne vibrations plays a crucial communication role for many insect groups (Virant-Doberlet and Čokl 2004). Vibrational signals may have many different functions, including various roles in mating. Substrate-borne signals are produced by diverse methods and may involve percussive actions on solid substrates (e.g. drumming on riparian vegetation), or by tremulation and creating oscillations that propagate along the substrate itself (usually waves and ripples along the water surface). Although percussive actions may create acoustic signals (as every drummer knows!), the important point is that many insects detect the transverse waves of substrate vibrations (often through sensors on the tarsi) and not the longitudinal waves of sounds (Section 7.3). Vibrations may allow communication over several metres and are generally considered to be short-range communication.

Male–female communication by substrate-borne vibrational signals has been reported in some Trichoptera (Ivanov and Rupprecht 1992; Wiggins 2004) and Megaloptera (Rupprecht 1975), but the best documented and perhaps most complex system is the so-called drumming of some adult Plecoptera, but only in the Arctoperlaria of the northern hemisphere (Stewart and Sandberg 2006). Although signals are species-specific, suggesting an isolation mechanism, regional dialects occur within some species (Rupprecht 1972; Stewart et al. 1982). The communication–search behaviour typically starts with a male call, a female ready to mate then replies and remains stationary while the male begins to search for her (Abbot and Stewart 1993). The pair establish a duet and continue calling to one another while the male locates the female (often via triangulation) and they then copulate. Unmated females typically respond to male calls and mated females may not answer, even though females may mate more than once. Drumming facilitates encounters between the two sexes, but is not an obligatory courtship ritual or a prerequisite for mating, because copulation without calling sometimes occurs in arctoperlarian species that normally drum, and all extant Antarctoperlaria manage to mate without drumming.

Substrate signalling on the water surface occurs among surface-dwelling insects (e.g. Belostomatidae, Gerridae, Veliidae) that create ripples or surface waves. Signals may be produced by vertical oscillations of the legs that remain in contact with the water (e.g. Gerridae) or by vertical oscillations of the entire body (e.g. Belostomatidae). The number and nature of these signals typically varies between species and sexes, and may play various roles in mating, as well as territoriality and food defence. When male belostomatids create a surface wave for mating purposes, females respond by swimming to the surface and then toward the signal source. Females are most likely to respond if they are ready to mate, and the likelihood of a male securing a mate increases with the time spent signalling (Kraus 1989). Similarly, surface wave signals generated by some male gerrids attract sexually receptive females in some species (Wilcox 1972). In other species, males can discriminate between sexes visually or by using surface signals (Wilcox 1979; Wilcox and Spence 1986), and males may use signals to repel potential competitors (males) from territories (Wilcox and Spence 1986).

10.5.3 Sound communication

Mating communication by sound is common in some terrestrial insects (e.g. crickets, cicadas), but less common in aquatic species. Some of the most interesting examples are the underwater sounds created during mating by the adult aquatic Hemiptera. The literature on underwater sound production dates back to the mid-19th century and was reviewed in the 1980s (Aiken 1985), but surprisingly little research has been carried out in the subsequent decades. As discussed in Chapter 7, underwater sounds are typically produced by stridulation using a file–plectrum system in which a sharp object (plectrum) is drawn across a series of pegs (file or pars stridens). The plectrum is often a rough surface made up of numerous, more or less parallel, ridges. The file and plectrum must be on body parts that can be rubbed against one another, and some species may have more than one stridulation system. The sounds produced typically differ between species, as one might expect if they play a role in mate finding.

In both male and female Corixinae (perhaps the most-studied aquatic insect group for sound production), fields of small pegs (the file) of the fore femora are rubbed against thickened flanges (the plectrum) on the maxillary plate of the head (Jansson 1972). Males and females have different songs: males may have several different songs yet females typically have only one (Jansson 1973; Aiken 1982; Theiss et al. 1983). The precise function of these call signals may differ among species, but there may be some general principles. The males' 'spontaneous' call results in an aggregation of the sexes; a subsequent male 'courtship' call elicits the 'agreement' call in a receptive female. The female remains stationary while the male approaches; in some species, 'mounting' and 'copulation' calls of the male may accompany copulation and a different stridulation mechanism may be used to produce these calls (a file on the femora of the middle leg drawn against the edges of the hemielytra).

Stridulation has been documented, or at least proposed, in several groups of Coleoptera, but there is still much confusion about the stridulation mechanisms and the function of these calls (Aiken 1985). Some calls, may be unrelated to reproduction, such as the buzzing of some *Rhantus* (Dytiscidae), which may serve to coordinate group emigration from degraded habitats (Smith 1973). Stridulation in the Hydrophilidae is perhaps the best documented of all the beetles and sound is produced by a file on the dorsal surface of the abdomen rubbing against a plectrum on the ventrolateral surface of the elytra. Although there is still some confusion about the function of these sounds, there is probably enough evidence that they play a courtship or copulatory role in some beetle species (Aiken 1985), such as *Tropisternus* spp. (Hydrophilidae) (Ryker 1976).

Sound communication in the air occurs during mating of some Diptera (Downes 1969; Clements 1999), with sound generated by the wing beats of females and detected by the males. The precise nature of the sound varies with wing beat frequency and wing size (thus sounds may be species-specific), and may also vary with age such that the sound produced by newly emerged, reproductively immature females is often below the auditory range of males. Hearing is via mechanosensory transducers on the antennae, and antennae of males are more conspicuous and elaborate organs compared with those of females (Figure 10.5), reflecting the fact that males use sound to locate females but not *vice versa*. Sounds produced by wing beats rarely travel far and detection typically occurs only when males are close to a female (up to tens of centimetres). Thus, long-distance cues generally bring males and females together into mating swarms, and short-range sound cues are used for males to locate females within the swarm.

10.5.4 Chemical communication

Chemical communication for the purpose of reproduction is common in several insect orders; among the aquatic insects this includes some Trichoptera and Diptera. Sex pheromones function generally to alter

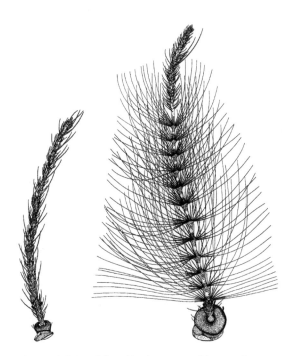

Figure 10.5 Sexual dimorphism in antennae of the mosquito *Anophelese maculipennis* (Culicidae). A female antenna on the left, a male antenna on the right.

Source: From Ismail (1962). Reproduced with permission from Elsevier.

the sexual behaviour of conspecifics, but not other species. Most sex pheromones are volatile, produced by glands that open to the external environment and the chemoreceptors for airborne pheromones are often located on the antennae. The range of behaviours associated with pheromones is diverse (Section 7.5) and there are two general categories of sex pheromones: long-distance attraction pheromones and short-distance courtship pheromones.

Sex pheromones are common in some families of Diptera and Trichoptera (reports are lacking for other groups of aquatic insects) and generally function as attraction pheromones that are released by females to lure males. In Trichoptera, aggregation pheromones may be produced by males of some species, such as some *Hydropsyche* (Hydropsychidae) (Löfstedt et al. 1994), which may be the stimulus for the formation of dispersed mating swarms. Pheromones are produced by abdominal glands and detected by olfactory receptors on the antennae (Resh and Wood 1985; Larsson and Hansson 1998). Chemically, sex pheromones are not a single, species-specific compound but mixtures of short-chain secondary alcohols and the corresponding ketones (Bjostad et al. 1996; Löfstedt et al. 2008). Chemical cocktails typically differ between species, even though the component chemicals may be shared by many species. The behaviours provoked by sex pheromones are, however, species-specific. It is this combination of species-specific cocktails and behaviours that contributes to reproductive isolation. For example, two congeneric caddisfly species, *Gumaga griseolai* and *G. nigricula* (Sericostomatidae), may have pheromones of similar chemical structure, but interspecific attraction does not occur because the species differ in the time of day that males are responsive to the female pheromone (Jackson and Resh 1991). Occasionally, mating caddisflies become so numerous that they are considered a nuisance to humans and synthesized pheromones may be used to bait pheromone traps, luring amorous individuals to a tragic and untimely death.

10.6 Copulation and sperm transfer

Ultimately, the function of mating is to fertilize eggs. Most insects have internal fertilization and sperm are transferred from male to female during copulation. Copulation may occur when pairs are airborne (e.g. many Diptera, Odonata, Ephemeroptera, Trichoptera), resting on the ground (e.g. Plecoptera, Megaloptera), resting on the water surface (e.g. Gerridae, Veliidae) or under water (e.g. some Hymenoptera, Coleoptera, Corixidae). Copulation is normally between adult males and females but, in a process called paedogenesis, females develop fully functional reproductive organs in the final larvae (or subimago) stage and adult males mate with larval females. Paedogenesis is well-documented in some mayflies where male imagos mate with female subimagos (Edmunds Jr and McCafferty 1988). Paedogenesis should not be confused with species that have wingless females, such as some Trichoptera and marine Chironomidae, but are otherwise mature adults with fully developed reproductive organs.

In the simplest form of sperm transfer, male and female genitalia make physical contact, part of the male's aedeagus (penis) is inserted into the female reproductive tract via the gonopore, and sperm is transferred (insemination) (see Chapter 1 for description of genitalia). Sperm are received in the female's genital chamber, directly into the spermatheca or its duct, and then stored in the spermatheca. The storage period varies and may be a matter of seconds or minutes if oviposition occurs soon after mating (e.g. some mayflies, caddisflies, and chironomids) or may be days or weeks (e.g. some blood-sucking Diptera). As mentioned earlier, rather than injecting sperm directly into the female, sperm are often packaged into spermatophores produced by the male accessory glands and the entire spermatophore is inserted into the female's genital chamber. Where a protein mass makes up part of the spermatophore, absorption of the protein by the female may take several days and this process may be largely independent of sperm transfer.

The transfer of spermatophores may be a simple process, but it may also involve more complex behavioural routines. For example, there are three distinct forms of spermatophore transfer within the Megaloptera. In all megalopterans, the narrow tube of the spermatophore is inserted into the female's genital opening, and the rest remains outside. (1) Male alderflies (Sialidae) attach a simple spermatophore externally to the female's genitalia, then the male departs, the sperm swim into the spermatheca, and females subsequently eat the spermatophore before searching for an oviposition site (Pritchard and Leischner 1973; Elliott 1977). (2) Male dobsonflies (Corydalidae, Corydalinae) produce large spermatophores that are attached externally to the female genitalia and that consist of two parts: the sperm ampulla, which contains the ejaculate, and the spermatophylax, which is a 'nuptial gift', a sperm-free gelatinous mass eaten by the female after mating (Hayashi 1993). While the female is occupied eating the spermatophylax (this could take more than two hours), the sperm make their way to the spermatheca and the female then eats the ampulla. Through this nuptial feeding, the male may make a nutritional investment in the offspring and/or simply protect the sperm by making sure the ampulla is not removed until all the ejaculate has been transferred. (3) Male fishflies (Corydalidae, Chauliodinae) have a simple spermatophore, but remain physically paired with the female (mate guarding, Section 10.7) for two to six hours after spermatophore attachment to prevent the female removing the spermatophore and to ensure complete ejaculate transfer (Hayashi 1996). Further, fishfly sperm are grouped into sperm bundles in which hundreds of sperm are joined together at the head (Hayashi 1996, 1998). These sperm bundles move faster than single sperm and presumably increase the likelihood that sperm will reach the spermatheca.

Multiple mating (with different partners) by females (polyandry) and males (polygyny) occurs in many species. Multiple matings usually occur sequentially, not all at once. However, mating balls have been reported in some Plecoptera, in which multiple males attempt to copulate with a single female (Tierno de Figueroa et al. 2006). Females may oviposit between mating events, as do some Plecoptera, or store spermatophores from multiple males for later fertilization. Polygyny is probably more common than polyandry, and post-copulatory behaviour of males may prevent or reduce the likelihood of polyandry (Section 10.7).

10.6.1 Copulation in Odonata

The act of copulation in odonates is remarkable and warrants special consideration because the primary genitalia of the two sexes do not come into actual physical contact. Further, the male's primary genital opening (where the sperm leaves the internal reproductive tract) and penis (for inseminating the female) are at opposite ends of the abdomen. As with most insects, the male's primary genital opening is at the tip of the abdomen, whereas the second and third abdominal segments bear the secondary genitalia, which includes the penis, a sperm storage reservoir and, in the Anisoptera, structures to grasp the female during copulation. Before copulation, the male transfers sperm to the secondary genitalia by bending the tip of the abdomen downwards to the base of the abdomen and fills the sperm reservoir. In the first step leading to copulation, a male and female join in tandem formation (Figure 10.6a). The male first uses his legs to grasp the female's thorax and head, then brings the tip of his abdomen forward between his legs and fastens his anal appendages to the back of her head, and finally releases his legs. Tandem formation may occur when the female is in flight or when perched.

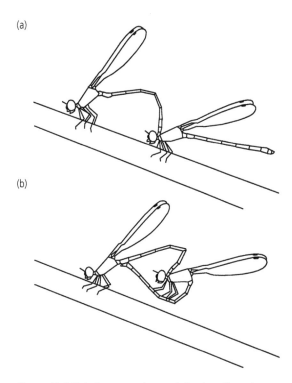

(a)

(b)

Figure 10.6 Typical sequence of stages during damselfly mating. The male is to the left of the female in both (a) tandem, and (b) copulation stages, as described in the text.

Second, the pair joins in the wheel formation (Figure 10.6b). The male lifts the female up and forward, while she curls her abdomen ventrally, pushes it forward to engage with the male's secondary genitalia and sperm are then passed from the male to the female. The wheel is terminated when the female withdraws her abdomen, although the pair may remain in tandem for some time (Section 10.7).

The Odonata are also unusual in that a male can remove sperm stored in a female, from a previous mating, before inseminating his own sperm (Waage 1979; Córdoba-Aguilar et al. 2003). Copulation appears to take place in two stages: first, the male repositions or removes sperm from the female's storage organ, and second, he transfers his own sperm. The precise mechanism for sperm removal varies among species and is strongly associated with the species-specific morphology of the genitalia (Waage 1986; Corbet 2004). The method, however, is effective and 88–100 per cent of sperm can be removed (Waage 1979). Generally, this behaviour

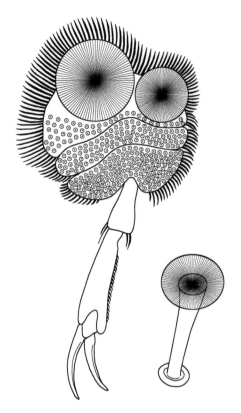

Figure 10.7 Male secondary sexual features. Foreleg (ventral aspect) of a male dytiscid beetle showing multiple 'suction cups' (adhesive seta) on the three, laterally expanded tarsal segments. A small cupule is enlarged on the right.

Source: Adapted from Miall (1903).

leads to sperm competition among males (when sperm from multiple males occur in a single female, they may compete to fertilize the egg), but also makes possible cryptic female choices where the female decides—after copulation—which male fertilizes her eggs (Córdoba-Aguilar and Cordero-Rivera 2008). The evolutionary implications of this behaviour are fascinating.

10.7 Post-copulation behaviour and sexual selection

After copulation, male insects often remain with the female in a behaviour known as mate guarding (post-insemination or post-copulatory mate guarding). The duration of copulation varies among species, lasting seconds to minutes, and the duration of mate guarding may also be variable, lasting minutes

to days. One function of mate guarding is to prevent the female from re-mating and obtaining sperm from rival males, and may be prevalent in species where males outnumber females (Alcock 1994). Various kinds of mate guarding include 'prolonged copulation' where genital contact is maintained long after insemination, 'mate grasping' where males remain attached to the female but the genitalia are disengaged, and 'mate monitoring' or 'non-contact guarding' where males remain close to the female but without grasping or physical restraint. Examples of all these different guarding behaviours can be found within the Odonata (Corbet 2004), and other taxonomic groups may be equally diverse in their behaviour.

Polyandry can also be prevented if the male applies a mating plug that physically prevents subsequent mating. For example, a male beetle of *Dytiscus alaskanus* (Dytiscidae) passes a spermatophore to the female, withdraws his genitalia, waits for some of the accessory gland material to seep from the female's gonopore, and then smears this material over the gonopore and terminal sternite to form a plug (Aiken 1992). Mating plugs are not necessarily formed at every mating, however. *Dytiscus alaskanus* have two mating periods a year (spring and autumn) and mating plugs are produced primarily in autumn, whereas post-copulatory mate guarding is more common in the spring. One possible explanation (Aiken 1992) is that autumn females are sexually immature, sperm are stored over the winter while the eggs mature, and the female becomes ready to oviposit the following spring. The male is able to guard autumn-mated females long enough for the plug to harden, but not for the entire winter, so a mating plug is a more efficient way of protecting paternity. Spring-mated females, however, are sexually mature with ovarioles filled with eggs. A brief period of mate guarding is adequate to allow the sperm to reach the spermatheca; fertilization and oviposition occur shortly afterwards.

Many elements of mate selection are viewed from a male perspective and presume that post-copulatory behaviours, for example, centre around a need to assure paternity. Not all choices about mating are made by males, however, because females may have some influence over which male to accept or reject, and females of many taxa will shake off amorous males if they are not ready to mate. Indeed, female choice is a critical part of various hypotheses regarding the diversification of male genitalia and secondary sexual characteristics. A good example of this is seen in the secondary sexual characteristics of many dytiscid beetles. The males have forelegs, and sometimes middle legs, with tarsi that are enlarged to form a rounded palette and that are equipped with modified adhesive setae (Figure 10.7). The setae are sucker-like, shallow saucers on narrow stalks and that can tilt about any axis perpendicular to the shaft. The large, primary and secondary setae are few, and the smaller tertiary setae are typically more numerous. There is little or no courtship behaviour in the Dytiscidae and the male simply attacks a passing female, using these adhesive setae to attach to a female's dorsal surface during an initial, intense struggle phase. If the male remains attached, he then initiates copulation attempts and holds the female in a variety of ways (Aiken 1992). Females of these same species often have sculptured or furrowed elytra and pronota (in contrast to the smooth dorsum of males) and it is sometimes said that the 'sticking power' of the male's forelegs is increased by these furrows. In fact, the opposite is more likely to be correct, as surface sculpting generally decreases the effectiveness of large suckers (Aiken and Khan 1992; Bergsten and Miller 2007), and female sculpturing is often viewed as an adaptation to interfere with the male grasping device, giving the female greater control over the decision of who to mate (Bergsten et al. 2001). Females may incur costs when mating (e.g. reduced swimming efficiency, predation risk, loss of opportunities for gas exchange), possibly related to prolonged duration of mating, which can last several hours in some *Dytiscus* (Aiken 1992). Thus, being able to reject potential mates may allow the female some control over the timing and cost of mating. To make it more interesting, females within a species or population may be dimorphic, with smooth or sculptured elytra (Bergsten et al. 2001), and there may also be variations among the forelegs of males that correspond to the different female morphs (Bilton et al. 2007). The modified forelegs of males are certainly for grasping females and female sculpturing reduces male attachment, and the variations in the

sculpturing of female elytra is now generally interpreted as evidence of intersexual arms races and sexual conflict (Miller 2003; Bergsten and Miller 2007). Analogous arms races over secondary sexual characteristic also occurs in some Gerridae where males have complex appendages for grasping females and females have traits for dislodging males (Arnqvist and Rowe 1995).

A particularly unusual system of mate guarding and nuptial gifts from the female to the male occurs in Zeus bugs of the genus *Phoreticovelia* (Veliidae) that inhabit backwaters and slowly flowing areas of tropical rivers (Arnqvist et al. 2003, 2006, 2007). Like many semi-aquatic Heteroptera, females are typically larger than males, and there are winged (macropterous) and non-winged (apterous) morphs of Zeus bug, but the two morphs have different mating systems. Winged morphs mate like most semi-aquatic Heteroptera, with a brief copulation period, little or no mate-guarding, and a likelihood of polygyny. Mating of the apterous morphs, however, is quite different and males ride on the backs of females for long periods of time, e.g. longer than a week (Arnqvist et al. 2003). The apterous population is sex-biased with more males than females and an adult male will ride on an immature female in a form of pre-mating guarding, i.e. to ensure that he gets a mate once the female matures. Adult apterous females have dorsal glands that produce wax-like secretions that are consumed by the riding male, i.e. sex-role-reversed nuptial feeding (in arthropods, nuptial gifts are predominantly from the male to the female). Monogamy may be common in this morph and females may benefit from reduced interference and kleptoparasitism from males (Arnqvist et al. 2006).

10.8 Parthenogenesis

Most of the discussion thus far has focused on sexual reproduction with mating between adult females and males, and this is the most common mode of reproduction among aquatic insects. In contrast, parthenogenesis is a form of asexual reproduction in which females produce unfertilized eggs that produce viable offspring. Mating, therefore, is not required. Parthenogenesis is widespread in the insects and may be obligate or facultative, in which all (obligate) or only some individuals (facultative) reproduce parthenogenetically. There are several forms of parthenogenesis depending on the sex of the eggs produced: thelytokous, arrhenotokous, and amphitokous or deuterotokous. Only female eggs are produced in thelytoky (characteristic of obligate parthenogenesis); only male eggs in arrhenotoky; both male and female eggs in amphitoky or deuterotoky. In most insects, the female is the homogametic sex (i.e. having two X sex chromosomes), and the male is heterogametic (i.e. having XY chromosomes). The Y chromosome has been lost in the Odonata, so that the male is XO. In the Trichoptera and Lepidoptera, however, the female is heterogametic (ZW) and the male homogametic (ZZ). Which and how many sex chromosomes occur in parthenogenetic offspring, however, are variable. The genetic mechanisms by which the different chromosomal patterns lead to gender determination are variable and poorly known for all except a few terrestrial insects, and environmental factors (e.g. temperature) can also play a role.

Some form of parthenogenesis occurs in most insect orders and, among the aquatic orders, it is common and perhaps best-documented in the Ephemeroptera (Funk et al. 2010) where unmated females of many species will lay eggs in the normal manner and at least some eggs hatch successfully. A few species are obligate parthenogens, such as the thelytokous *Cloeon triangulifer*, Baetidae (Gibbs 1977), and micropyles may be present in eggs of this species even though mating is not required (Koss and Edmunds Jr 1974). Males are typically absent from populations where parthenogenesis is obligate, and sex ratios may be skewed (usually more females than males) in facultative populations (Harker 1997; Funk et al. 2008). Facultative parthenogenesis is widespread among mayfly species and families, as in the Baetidae (Bergman and Hilsenhoff 1978; Harker 1997), Heptageniidae (Humpesch 1980), Polymitarcyidae (Tojo et al. 2006), and Ephemerellidae (Sweeney and Vannote 1987). Geographic parthenogenesis occurs when sexual and parthenogenetic populations of the same species exist in the same geographical area (Sweeney and Vannote 1987; Tojo et al. 2006). In tychoparthenogenesis, both sexual and parthenogenetic reproduction occurs within the same population (Ball 2002). The ecological and evolutionary consequences of these different forms of parthenogenesis for population dynamics are fascinating and, as yet, not fully understood.

CHAPTER 11

Oviposition and eggs

11.1 Introduction

Oviposition is the term used to describe laying of eggs. Once an egg is laid, the developing embryo requires time to develop—this could be as short as a few days or as long as several months if eggs enter diapause. Throughout this period, eggs and their embryos are potentially vulnerable to predation, disease, and physiological stress from enumerable abiotic factors. For species to remain extant, therefore, eggs must be placed in locations that will allow the embryo to develop and the neonate eventually to emerge into a location that is suitable for larval life. Accordingly, morphological characteristics of the eggs and behaviours of the female associated with selecting a suitable oviposition site are likely to have been subject to strong selection pressures.

Not surprisingly, oviposition is a complex event that may be split into two different sets of behaviours: pre-oviposition (Section 11.2), which incorporates all the behaviours and factors involved in the selection of, or attraction to, an oviposition site, and oviposition itself (Section 11.3), which is the actual deposition of eggs. Once the eggs have left the female (post-oviposition), they have many different strategies for ensuring adequate gas exchange, protection from environmental stresses, attachment to substrates, and defence from would-be attackers; these will be discussed in Section 11.4. The eggs of most aquatic insects develop without the benefit of parental care, but there are a few fascinating cases of parental care (Section 11.5).

11.2 Pre-oviposition

The oviposition site selected by the female is usually characteristic of the species and is important because her choice may influence the survival of the egg during embryonic development, as well as availability of food for the larva when it hatches. The female also has to survive to the end of the oviposition process, i.e. avoid predators and withstand any environmental stressors. Females may be particularly vulnerable at this stage so the choice of site may also reflect their ability to survive the experience. Thus oviposition site selection has important consequences for population dynamics and individual fitness in terms of minimizing pre-oviposition mortality and maximizing survival of the offspring. The range of oviposition sites exploited by aquatic insects is diverse, including both terrestrial and aquatic locations. Because the larval existence of most species is aquatic, however, terrestrial oviposition sites tend to be relatively close to water bodies, for example, in the littoral or riparian zone, on the banks or shores of water bodies, or in vegetation over-hanging water. The means by which different species select particular sites is also diverse. Selection of the oviposition site often involves a general attraction to a particular area, followed by a more specific reaction that determines the precise spot at which the eggs are laid within this area. Many different attractive and repellent cues may be involved in these processes.

11.2.1 Oviposition sites and strategies for oviposition

The oviposition sites and behaviours of aquatic insects are diverse and may be classified according to many different criteria. A list of the habitat characteristics of oviposition sites would be as long as the list of habitats in which aquatic insects occur, plus some terrestrial locations. Suffice it to say that

Aquatic Entomology. First Edition. Jill Lancaster & Barbara J. Downes.
© Jill Lancaster & Barbara J. Downes 2013. Published 2013 by Oxford University Press.

virtually every water body could potentially be an oviposition site for at least one aquatic insect species. Instead, it may be more insightful to consider oviposition sites according to the behaviours and traits of the females. Some important distinctions might include whether the female oviposits when in flight or when settled; the eggs are laid terrestrially, or in, or on the aquatic medium; eggs are deposited freely, attached to a hard substrate surface, inserted in soft substrates (e.g. mud), or inside plant tissues (endophytically); eggs are deposited singly or in coherent masses. There are many other possible criteria, and there will always be some species that do not fit neatly into any particular category. Closely related taxa will often have similar oviposition habits, but it is unwise to generalize without evidence because some species within the same genus may have quite different behaviours. For example, some caddisflies in the genus *Neophylax* (Uenoidea) oviposit above water (Beam and Wiggins 1987), yet some congeners oviposit underwater (Mendez and Resh 2008).

Females that oviposit when in flight may release eggs over solid substrates or over water, with eggs released singly or in batches, in a relatively unselective manner. In the case of solid substrates, this typically involves dropping eggs onto mud or damp ground, as is common among dragonflies of the family Libellulidae. Flying over water, females may release eggs whilst well above the water surface or as they dip the tip of the abdomen briefly into the water. Eggs may be released in one large batch, in a series of smaller batches or singly. Broadcasting eggs in large batches is common among stoneflies and some mayflies, although many stoneflies are poor fliers and may release eggs whilst running or skating over the water surface rather than being fully airborne. Female mayflies in the family Ephemerellidae typically extrude a mass or ball of eggs through the genital opening and that is attached ventrally and posteriorly to the female's abdomen. The egg ball is released over the water or by touching on the water's surface. Broadcasting eggs singly or in smaller batches is common among mayflies, but also among dragonflies and Diptera.

For females that oviposit when settled on the water surface or on an object that is out of the water,

the fate of their eggs depends on where exactly they settle. Many aquatic insects will lay eggs terrestrially, but close to the water, including many Megaloptera, Neuroptera and Trichoptera that settle and lay eggs attached to riparian vegetation or on stream banks. Some species, particularly Diptera, are able to settle on the water surface and lay floating eggs (e.g. some Culicidae) or squirt eggs freely into the water that then sink (e.g. some Chironomidae). Finally, many species will land on objects just above the water's surface (e.g. rocks, branches, macrophytes), and insert the tip of the abdomen into the water. Subsequently, they may squirt eggs freely into the water (e.g. some Trichoptera, Ephemeroptera), attach eggs to the object but underwater (e.g. some Hemiptera), or if the female selects a plant on which to settle, the plant tissues below the water may be cut and the eggs inserted (e.g. some Odonata).

The adults of many aquatic insects will enter the water to lay eggs. This is unremarkable in species in which the adult is aquatic (e.g. many Coleoptera and Hemiptera), but interesting for species that have terrestrial adults. These females may land on objects close to the water and then walk under the water, including some baetid mayflies (Eaton 1888) and psephenid beetles (Murvosh 1971), or may swim underwater, including some hydropsychid caddisflies (Deutsch 1984; Lancaster et al. 2010a). Females of some caddisflies have legs that are modified for swimming, with tibia and tarsi that are widened and often with a fringe of hair on the posterior edge (Deutsch 1985). As primarily terrestrial animals, gas exchange during the time a female is underwater may present a challenge. Adults of species that venture underwater are often covered with many hairs or setae that trap a bubble of air as the female enters the water, thus forming a compressible gas gill (Chapter 3) if the spiracles open into the air bubble. The life span of this gas gill presumably limits the time a female may remain submerged and laying eggs. Once underwater, eggs may be attached to substrates (many Trichoptera, Diptera, Plecoptera) or inserted into plant tissues (many Coleoptera, Zygoptera).

Species with larvae that are parasitic or egg predators, at least in the early instars, typically oviposit on the host and this also requires that females enter the water. The parasitic hymenopteran *Caraphractus cinctus* (Mymaridae), for example, swims underwater and lays eggs on the eggs of dytiscid beetles (Jackson 1961, 1966). Females of another parasitic wasp, *Agriotypus gracilis* (Ichneumonidae), land on rocks emergent from the water, crawl down the side of the rock into the water and then search for cased caddisflies, which are the hosts. Potential hosts are examined first by the wasp walking up and down the case to assess its size, and secondly by probing with the ovipositor, apparently to assess the hosts' development stage; they generally oviposit on prepupal or pupal hosts (Aoyagi and Ishii 1991). Similar search behaviour has been observed in a congener, *Agriotypus armatus* (Grenier 1970). Because water flow in the host caddisfly's pupal case is compromised once the host is killed (Bennett 2001), *Agriotypus* has a pupal respiratory filament that protrudes from the case. An obligate egg predator, gravid females of the dipteran *Acanthocnema* sp. (Scathophagidae) enter the water and lay their eggs directly onto the prey egg masses (Purcell et al. 2008).

Among the more unusual oviposition sites are those of the Belostomatinae (Hemiptera) where females lay their eggs on the backs of males (Section 11.5). Perhaps even more unusual are the marine Chathamiidae (Trichoptera) of Australia and New Zealand. The larvae of one chathamiid caddisfly, *Philanisus plebeius*, live and develop in intertidal rock pools. Females oviposit in the coelom of a starfish, *Patiriella* spp., embryonic development of the caddisflies occur in the starfish, and neonates then move away and into intertidal rock pools (Anderson and Lawson-Kerr 1977; Winterbourn and Anderson 1980).

It is apparent from the above discussion that many aquatic insects have quite specialized oviposition behaviours and may be very selective about where they lay eggs. One consequence of this selective behaviour is that eggs and egg masses of multiple individuals often are spatially aggregated, i.e. clustered close together. This behaviour is often called communal egg-laying, regardless of whether or not multiple females oviposit simultaneously, and it differs from the 'egg clustering' of many terrestrial Lepidoptera, in which a single female lays eggs in batches instead of one at a time (Stamp 1980; Courtney 1984). Communal egg-laying has been documented in diverse groups of aquatic insects, including Trichoptera (Hoffmann and Resh 2003; Reich and Downes 2003a), Ephemeroptera (Encalada and Peckarsky 2006; Lancaster et al. 2010b), and Diptera (Muirhead-Thomson 1956; Coupland 1991), and in groups with diverse oviposition behaviours, including caddisflies that oviposit terrestrially (Beam and Wiggins 1987), on the submerged underside of objects that protrude from the water (Hoffmann and Resh 2003; Reich and Downes 2003a), and on fully submerged objects (Lancaster et al. 2010a). How strongly egg masses are aggregated can vary widely among even closely related species (Beam and Wiggins 1987; Lancaster et al. 2003). How these aggregation arise is unclear; environmental characteristics of potential oviposition sites are important, but it is not necessarily the case that suitable sites are in short supply, suggesting that females interact with one another (Lancaster et al. 2003; Reich and Downes 2003b; Lancaster et al. 2010b; Reich et al. 2011).

11.2.2 Locating oviposition sites

Once a female has mated and is ready to lay eggs, she then needs to locate a suitable oviposition site. We have seen that the range of potential oviposition sites is very diverse; the range of behaviours and sensory cues used to locate sites is also diverse. If mating and oviposition occur at the same location and in a short space of time, there may be no or very little search behaviour. For example, some marine chironomids swarm, mate, and oviposit on the water's surface in the space of two hours or less (Cheng and Collins 1980; Soong et al. 1999); similarly, some mayflies emerge *en masse* from the water's surface, mate immediately, and deposit eggs whilst in flight (Edmunds Jr and McCafferty 1988). Females of other taxa may travel long distances before locating oviposition sites if oviposition occurs a long time after mating or far from the mating sites. For example, distances of up to 1 km may separate the mating and oviposition swarms in some populations of a short-lived chironomid, *Chironomus anthracinus* (Tokeshi and Reinhardt 1996). Even when mating takes place relatively close to the general area in which eggs are deposited, however, females may still need to locate a suitable oviposition site.

A diverse suite of long- and short-range cues may be used to locate oviposition sites, provided that females have the appropriate sensory receptors (Chapters 6, 7). In the first instance, females may use long-range cues, including visual and airborne olfactory cues. This may be particularly important in species with adults that are long-lived or capable of long-distance dispersal, because adults often may be far from water bodies (Chapter 9). For example, blood-feeding Diptera (e.g. Culicidae, Simuliidae, Tabanidae) may feed on hosts that are many kilometres away from water bodies. Short-range cues also may be visual and chemical, but the cues and sensory structures may be quite different (e.g. contact instead of olfactory chemical cues for short range site detection). Contact cues may be physical rather than chemical, including temperature, substrate texture, and moisture content (tactile cues). Cues can be attractants as well as repellents, stimulants, or deterrents, and some females may spend considerable time sampling the environment before egg laying. Among the aquatic insects, oviposition behaviours of mosquitoes are perhaps the best documented, given their disease-spreading habits, and there are several reviews (Bentley and Day 1989; Clements 1999; McCall 2002). The general behavioural patterns, however, are generic and apply to a diverse range of species. Although it is sometimes convenient to consider long- and short-range cues separately, and likewise the different kinds of cue (visual, chemical, mechanical), these are not isolated or independent processes and the female is exposed to a complex, interacting web of cues throughout the pre-oviposition process. Further, the oviposition mode of females does not necessarily reflect closely the kinds of cues used and, for example, females that typically deposit eggs whilst in flight may respond to airborne chemicals, but may also taste the water using contact chemoreceptors before deciding whether to oviposit.

Visually, many aquatic insects detect water from a distance by the horizontally polarized light reflected from the water surface (Horváth and Kriska 2008) (Chapter 6). Species differ in spectral sensitivity of the polarization-sensitive photoreceptors and different species generally detect polarization in the spectral region that is characteristic of the larval habitat (Schwind 1995). This phenomenon explains, at least in part, how some insects are able to select water bodies in which to oviposit on the basis of water colour and spectral properties. It also explains why females from many taxa mistakenly attempt to oviposit on oil spills (Horváth and Zeil 1996), plastic sheets (Csabai et al. 2006), asphalt roads (Kriska et al. 1998), and clean cars (Kriska et al. 2006). At smaller scales, visual cues originating in the different polarization patterns generated by smooth and turbulent water may be used by mayflies and caddisflies that preferentially oviposit on emergent rocks situated in fast-flowing areas of streams (Reich and Downes 2003b; Encalada and Peckarsky 2006). Objects rather than patterns of light reflection can also act as visual stimuli for oviposition site selection, as in some leptocerid caddisflies that are attracted to vascular plants that protrude above the water surface of ponds (Tozer et al. 1981).

Chemical cues used to locate oviposition sites can arise from a range of sources, including the environment (e.g. volatile organics), conspecific or congeneric individuals (eggs, larvae, pupae, or adults), and other organisms such as predators. Long-range chemical cues generally are airborne volatiles that originate from plants (phytochemicals) or decomposing organic matter, and often are detected using olfactory organs on the antennae. Contact chemoreception may involve sense organs on the antennae, tarsi, and mouthparts. Chemical cues from water bodies with a high organic content may be generic products of decomposition and attract a suite of species whose larvae typically occur in nutrient-rich water. For example, the volatile metabolites released by mats of cyanobacteria act as a habitat-finding cue for dispersing adult beetles of *Bembidion obtuscidens* (Carabidae), which inhabit the littoral margins of saline lakes (Evans 1982), and for ovipositing mosquitoes of *Anopheles albimanus* (Culicidae) (Rejmankova et al. 2000). Airborne volatiles specific to certain terrestrial plants are common attractants for ovipositing terrestrial insects whose larvae have particular food requirements (Bruce et al. 2005). Many aquatic plants have emergent parts (e.g. flowers, leaves) that may release species-specific compounds, so it is entirely plausible that ovipositing aquatic insects may be attracted to plants that are suitable food sources for larvae. Indeed, the North American milfoil weevil, *Euhrychiopsis lecontei*, is a specialist aquatic herbivore that feeds (larvae and adults), oviposits, and mates on the invasive freshwater macrophyte *Myriophyllum spicatum*, and recent work suggests that adult weevils are attracted to high concentrations of glycerol and uracil released by *M. spicatum* during rapid growth (Solarz and Newman 1996; Marko et al. 2005). Ovipositing females may use chemical cues also to oviposit preferentially in places that result in high larval growth and survivorship, as seen in some container mosquitoes that discriminate between water bodies based on the kind of leaf detritus (Trexler et al. 1998; Reiskind et al. 2009). Phytochemicals can also act as deterrents, as evidenced by the inhibitory effects of the water fern *Azolla imbricata* and duckweed *Lemna minor* on oviposition by some mosquitoes (McCall 2002).

Oviposition pheromones may be one mechanism for attracting gravid females to the same location and producing aggregations of egg masses. Oviposition pheromones that originate from eggs, larvae, and pupae have been identified in various mosquitoes and some blackflies (Bentley and Day 1989; Clements 1999; McCall 2002). In the mosquito *Aedes aegypti*, for example, volatile cuticular hydrocarbons in the larvae are released into the water and eventually become airborne. Gravid females detecting this odour plume via olfactory receptors on the antennae are attracted to oviposit in the same location (Mendki et al. 2000; Seenivasagan et al. 2009). These pheromones may be genus- rather than species-specific in culicids and simuliids, and a single pheromone may elicit similar responses from a range of species. Pheromones, or particular chemicals within a pheromone cocktail, may act as repellents as well as attractants (Ganesan et al. 2006); some female *Aedes* may be attracted to sites where pre-existing eggs of congenerics have been laid, whereas other females of other *Aedes* species avoid the same sites. The potential importance of oviposition pheromones in other, non-dipterous aquatic insects is less well understood.

It is now reasonably well established that chemical cues from aquatic predators and parasites can also deter or repel ovipositing insects (cues from 'enemies' are unlikely to be attractants!). Again, most of this evidence comes from mosquitoes and oviposition is deterred by cues from predatory fish (Angelon and Petranka 2002; Van Dam and Walton 2008), amphibians (Blaustein and Kotler 1993; Mokany and Shine 2003), and other aquatic insects such as notonectids (Chesson 1984; Blaustein 1998; Eitam and Blaustein 2004) and dragonflies (Stav et al. 1999). Mosquito oviposition also may be inhibited by cues from larvae of the same or other mosquito species that are parasitized by trematodes (Lowenberger and Rau 1994; Zahiri et al. 1997b; Zahiri et al. 1997a). While comparatively few studies demonstrate that non-culicid aquatic insects avoid ovipositing in waters with predators, such oviposition avoidance occurs in some aquatic beetles (Resetarits Jr 2001; Brodin et al. 2006) and phantom midges (Berendonk 1999), and this phenomenon may be more widespread.

11.2.3 Mate guarding during oviposition

Males that guard females after mating (Chapter 10) may abandon the female so she oviposits alone, or some may continue to guard the female during oviposition (contact or non-contact guarding). Females may incur certain costs from carrying males, including reduced locomotory ability and increased predation risk, but mate guarding also may be beneficial (e.g. by reducing harassment from other males), and sometimes the benefit may compensate for the costs of guarding. For example, females of the water strider *Aquarius paludum insularis* (Gerridae) carry males on their backs (in tandem) and the majority of individuals oviposit in tandem and underwater, after 18 hours of guarding (Amano et al. 2008). In this species, oviposition guarding is beneficial for females, because it enables pairs to dive, lay eggs in deeper water and in oviposition sites with a lower risk of egg parasitism by a scelionid wasp, *Tiphodytes gerriphagus* (Amano et al. 2008; Hirayama and Kasuya 2009), although there may be a trade-off in terms of egg survivorship (Hirayama and Kasuya 2010). Tandem pairs may dive deeper simply because two gerrids have more mass than one (and hence will sink more quickly), because 12 legs provide more propulsion than six, or because the combined air store (gas gill) of a gerrid pair is larger than two separate ones, thus allowing longer submergence. In another species of water strider, *Limnoporus* spp., the time between copulation and oviposition is short, eggs are laid on the margins of floating macrophytes, and females with guarding males generally lay more eggs than unguarded females because they are harassed less by other males (Spence and Wilcox 1986).

11.3 Oviposition

The posterior part of the abdomen and some abdominal appendages are modified to form an ovipositor in some adult female insects (Figure 11.1). True ovipositors are absent from many groups of aquatic insect (e.g. Ephemeroptera, Plecoptera), although secondarily derived structures may occur within these orders (Figure 11.2). Where special egg-laying structures are absent, eggs are released from the genital opening and often accompanied by various secretions. Ovipositors are typically withdrawn into the abdomen when not in use and only protruded during egg laying, although accessory structures such as cutting valves may be visible externally. Perhaps the simplest ovipositors are found in species that release eggs freely into the water or deposit

eggs on surfaces, but even then they may have multiple mechanoreceptors to provide sensory information and help coordinate the process. Special ovipositors enable females to insert eggs into special places, such as within plant tissues (endophytic oviposition, Section 11.3.2) or animal tissues, as seen in some endoparasites.

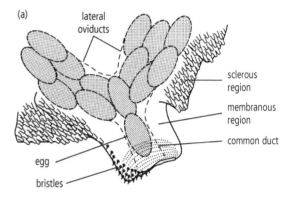

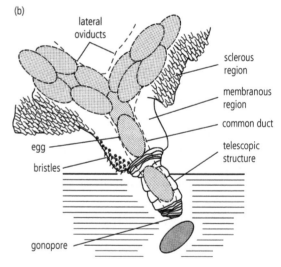

Figure 11.2 Schematic illustration of the secondarily derived ovipositor of the mayfly *Habrophlebia eldae* (Leptophlebiidae) in (a) its resting position, (b) during egg laying. The ovipositor, a tubular structure protruding from the seventh sternum, has a proximal sclerous region, a membranous distal region, and a telescopic structure, which is folded into the base of the ovipositor when at rest. Contact of the short, mechanosensory bristles on the ventral part of the membranous region with water triggers extrusion of the telescopic structure and egg laying. Eggs are laid one by one, and the telescopic structure returns to its resting position between each egg.

Source: From Gaino et al. (2009). Reproduced with permission from Taylor & Francis.

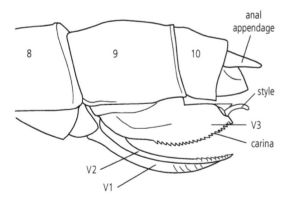

Figure 11.1 Lateral view of the terminal abdominal segments (numbered 8, 9, 10) and the ovipositor of a hypothetical damselfly that lays its eggs endophytically, based on *Lestes* sp. (Lestidae). The two pairs of cutting valves (V1, V2) are shown extended from the sheathing valves (V3). V1 and V2 move alternately, sliding along one another, to cut into the plant. The teeth on V3 form a carina, which the female leans against the substrate as an anchor whilst eggs are inserted into the cut plant material. The ovipositor itself is extended at the base of the valves only during egg laying.

11.3.1 Oviposition mechanisms and controls

The factors and processes that initiate and control the act of egg laying are diverse. Among the major taxa, the onset of oviposition may be associated with the time of copulation or mating. For many insects, mating initiates a readiness to oviposit, induced by substances transferred to the female during sperm transfer, and females that have not mated may be unable to oviposit. In other species that normally undergo fertilization, oviposition may occur without mating (i.e. parthenogenesis), but egg-hatching success may be less than for mated females. In these cases, mated females generally begin egg laying shortly after mating, whereas unmated females often retain their eggs longer. In yet other insects, mating may occur before mature oocytes are available, so the female must store sperm until maturation is complete and, in some cases (e.g. some Diptera), the act of mating may be required to stimulate the final maturation process.

There are many reports that peak oviposition is associated, in a species-specific manner, with environmental factors such as temperature, humidity, and wind, and with time of day. It is difficult to draw any generalities from these diverse observations, and environmental factors are certain to interact with one another. Many of these associations are likely to reflect the physiological constraints placed on adults (e.g. flight may be impossible in high winds, at low temperatures, or during heavy rain), the ability of species to locate oviposition sites in particular environmental conditions (e.g. diel patterns in light polarization and species' abilities to detect polarisation, Csabai et al. 2006), or simply a need to avoid periods when predators may be active.

Once a female is ready to oviposit and a suitable oviposition site has been located, the onset of egg laying may still require that the female receives appropriate stimulation from mechano- and chemoreceptors indicating that the immediate environment is suitable. Such receptors are usually located on the ovipositor, anal appendages or the terminal abdominal segments, and may be more numerous in species with elaborate ovipositors and complex egg-laying habits. The ovipositor is controlled by a set of muscles that have a particular sequence of activity, and the motor program controlling insect oviposition is complex. To perform complex movements, any mechanical system also needs sensory feedback. Damselfly ovipositors, for example, are well-equipped with sensory organs (Gorb 1994; Matushkina and Gorb 2002). These sensors may provide information about the relative positions and movements of the ovipositor and its component parts, as well as the mechanical character of oviposition substrates, which all help coordinate movements of the ovipositor.

11.3.2 Endophytic oviposition

Species that lay their eggs within plant tissues often have ovipositors comprising sharp-edged knife-like blades or fine-toothed saw-like blades for cutting into plants. Such structures are common among the Odonata and Coleoptera that lay eggs within plant stems. For example, the ovipositor of the beetle *Ilybius* (Dytiscidae) has two, saw-like blades (Jackson 1960). These are pushed into the surface of a suitable plant, then the blades are worked upwards in a rapid, saw-like action to cut away a flap of plant tissue. The egg is laid in the resulting hole and, once the ovipositor has been withdrawn, the flap of plant tissue covers the egg. In contrast, the endophytic ovipositor of damselflies consists of three pairs of valvulae (leaf-like appendages of the abdominal segments), two pairs for cutting, the third pair forms a protective sheath and may act as an anchor (Figure 11.1). Some species select particular plant species, or particular parts of plants, upon which to lay their eggs, and sensilla on the valves and styli may be involved in oviposition pant recognition (Matushkina and Gorb 2002; Matushkina and Lambret 2011). There is a positive correlation between the mechanical properties of the ovipositor and structural properties of the plant tissues (Matushkina and Gorb 2007). A mismatch between the ovipositor's cutting ability and a plant's resistance to cutting may result in prolonged oviposition duration and lower egg deposition rates, which presumably increases the female's vulnerability to predation and the risk of unsuccessful egg laying (Corbet 2004).

11.3.3 Unusual oviposition

Most aquatic insects are oviparous, as are most insects. This means that the egg is fertilized typically within the female, and the eggs are oviposited by the female in the environment. There are, however, some interesting deviations from this strategy, including ovoviviparous and viviparous insects. In ovoviviparous species, eggs are developed and fertilized normally, but they are retained inside the female's body until they hatch. Females of these species are often long-lived compared to closely related taxa. Adult females of an ovoviviparous species of *Callibaetis*, for example, may live for up to three weeks (Gibbs 1979), much longer than the one or two days of many other species in the same family (Baetidae). Embryos hatch within a few seconds (or less) of egg deposition, giving the appearance of live birth. The egg chorion of ovoviviparous species is often thin, and much thinner than that of oviparous species (Gaino and Rebora 2005). Among the aquatic insects, ovoviviparous species occur in the Ephemeroptera and Plecoptera, such as some members of the Capniidae (Hynes 1941). True live birth occurs in viviparous insects. Development occurs inside the females, but the egg does not develop a chorion and the female may provide additional nourishment to the developing embryo (i.e. the embryo is not dependent solely on the yolk supplied during oogenesis). Viviparity occurs in many terrestrial insects across many orders, such as some earwigs (Dermaptera), tsetse flies (Diptera), aphids (Hemiptera), and cockroaches (Dictyoptera), but we are unaware of any examples of viviparity in aquatic insects.

In larval or pupal paedogenesis, the adult and sometimes the pupal stage may be lost and oviposition occurs in a precocious stage. Oviposition of fertilized eggs is possible only if precocious stages are also able to mate, e.g. some mayfly females never become full adults, mating occurs between male imagos and female subimagos, and the female oviposits while still a subimago (Edmunds Jr and McCafferty 1988). For many insects that display paedogenesis, reproduction is almost exclusively by parthenogenesis, i.e. is no mating and eggs are unfertilized (Section 10.8). In a case of pupal paedogenesis, most females of the parthenogenetic blackfly *Prosimulium ursinum* fail to emerge from the pupa, eggs are laid within the pupal cuticle of the pharate adult, and the eggs are liberated when the pupa eventually disintegrates (Downes 1965). Such shortening of the life cycle and elimination of the adult stage has been suggested as an adaptation for life in arctic environments, where *P. ursinum* occurs. A particularly unusual example of larval paedogenesis and broadcast oviposition is found in the obligate, parthenogenetic mayfly *Eurylophella oviruptis*, Ephemerellidae (Funk et al. 2008). Some individuals oviposit in the normal ephemerellid way (the imago produces an egg ball, attached to the tip of the abdomen, which is released at the water's surface). For other individuals, however, the abdomen of the subimago bursts open, rupturing the oviducts and releasing eggs into the water. This abdominal bursting is suggested to result from over-inflation of the gut, which is no longer used for feeding (Section 14.3) and may be advantageous by allowing individuals to oviposit without the pre-oviposition mortality risks normally incurred by imagos (Funk et al. 2008). Presumably, there would be selection against this behaviour in sexual species because females typically mate as imagos, not subimagos, and a burst subimago would release unfertilized and non-viable eggs.

11.4 Post-oviposition eggs

The morphological characteristics of many insect eggs reflect the environment in which they are oviposited. These characteristics may be apparent before oviposition but, once outside the female's body, the eggs of many species undergo further morphological changes, including the appearance of adhesive or gelatinous coatings, silk wrappings, and attachment structures. Generally, these characteristics allow eggs and their developing embryos to survive in the external environment, e.g. exchange gases and protect the eggs from various threats and stressors during the development period.

11.4.1 Gas exchange in eggs

As discussed in Chapter 3, the movement of gases (O_2, CO_2) is closely associated with the movement of water and, typically, the challenge for many organisms is to take up O_2 without losing or taking up excess water (it is virtually impossible for a membrane to be permeable to oxygen and impermeable to water). As the outermost layer of the egg, the structure of the chorion reflects what kind of respiratory system an egg has (Hinton 1969). If O_2 enters and CO_2 leaves through the chorion, it follows logically that there must be holes large enough to allow exchange of these gases. Tiny holes in the chorion and their associated air channels are called aeropyles, which permit the movement of oxygen molecules between the environment and the developing embryo. Although smaller than micropyles, these aeropyles are formed in a similar manner to micropyles (microvilli of the follicle cells create channels during choriogenesis) (Section 10.2). The number and arrangement of aeropyles varies enormously among species. In addition, many aquatic insects lay their eggs in terrestrial environments, so we must consider the mechanisms for gas exchange in a range of conditions.

The majority of aquatic and semi-aquatic eggs lack a respiratory system, the chorion appears solid throughout, and aeropyles are absent. When no distinct respiratory system is present, chorionic respiration may be analogous to the cutaneous respiration of small-bodied larvae that lack a tracheal system. In some species, there may be no gas exchange at all and, presumably, the egg contains sufficient gases for development of the embryo. For one such species, *Nymphula nymphaeata* (Pyralidae, Lepidoptera), eggs are laid underwater but they lack aeropyles and embryos still develop normally, albeit more slowly, when gas exchange is prevented experimentally (Barbier and Chauvin 1974). Further, when the aeropyles of some insect eggs are plugged experimentally, oxygen uptake still occurs, albeit at a fraction of the normal rate, suggesting that the chorion may be permeable to gases in other ways, for at least some species. Many kinds of aquatic egg that lack a respiratory system are imbedded in a jelly-like substance, which serves various functions (Section 11.4.2), and may also limit the movement of gases, water, and other chemicals.

The majority of terrestrial eggs and the minority of aquatic eggs have an endochorion that is riddled with a meshwork of air spaces that holds a layer of gas. All species that have this intrachorionic gas

layer (also called a trabecular layer) also have aeropyles, but aeropyles may be present when air spaces are absent. The spaces between the struts of the intrachorionic air spaces are large enough to allow the movement of respiratory gases. When eggs are immersed in water, an air–water interface is established over the aeropyles, which are hydrofuge. This air layer functions as a plastron (Section 3.4), which allows gas exchange without any associated water gain or loss. As with the plastrons of other developmental stages, chorionic plastrons are largely restricted to species ovipositing in well-oxygenated waters, because plastrons are also efficient at extracting oxygen from the egg when oxygen pressure in the surrounding water is low. Chorionic plastrons are found in both aquatic and terrestrial eggs, because eggs laid terrestrially must be able to withstand soaking or potential submergence during heavy rain. Similarly, some eggs that are laid aquatically may need to tolerate periods of exposure to air. As long as there is a film of moisture over the egg surface, then the chorion can function as a plastron.

Because a lot of gas exchange between the water and the intrachorionic spaces occurs through the air–water interface at the aeropyles, the most effective plastrons will be those that have large aeropyles, many aeropyles, and/or an expansion of the aeropyles into a mesh network. Plastrons that consist of networks are typically distributed over most or all of the egg surface, as in some gomphid dragonflies (Trueman 1990), some stoneflies, some Belostomatidae (Hemiptera), and mosquitoes of the genera *Anopheles* and *Culex* (Hinton 1981). Stoneflies in the family Pteronarcidae have very well-developed plastrons (Miller 1939). The eggs of *Pteronarcys* have a flat base, which sits on the substrate, and a domed upper surface. The chorion has two layers (the endo- and exo-chorion), and on the domed upper surface these layers are separated by an interlamellar space (or intrachorionic layer) containing vertical columns, which branch out at the top forming tree-like structures that join one another, thus forming the lacunae of the plastron (Figure 11.3). The chorion of some Perlidae also appear to function as a sort of plastron, with an intrachorionic network of channels connecting the numerous aeropyles (Rościszewska 1991). Interestingly, the eggs of

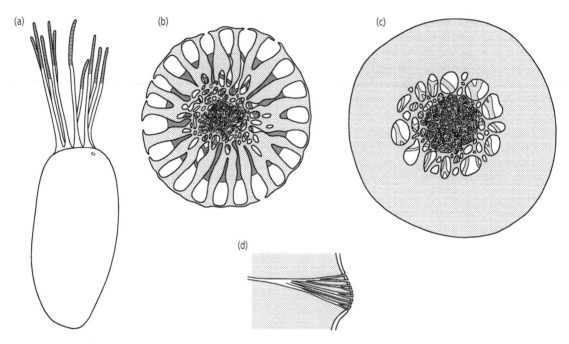

Figure 11.3 Eggs of an aquatic hemipteran, the water scorpion *Nepa cinerea* (Nepidae). (a) Outline of egg with respiratory horns. Striped area at tips of the horns indicates extent of the plastron-bearing region. (b) Cross-section through the apical, plastron-bearing section of a respiratory horn. (c) Cross-section through the basal region of a horn, below the plastron bearing region. (d) Branched aeropyle through the egg chorion.
Source: From Hinton (1961). Reproduced with permission from Elsevier.

these stonefly species are initially covered in a gelatinous sheet that fractures off within hours of oviposition, and only then are the aeropyles exposed to the water.

In a few insects, the chorion is extended to form one or more respiratory horns, the central part of which consists of some kind of air-filled meshwork. In some terrestrial insects the aeropyles are located at the tip of the horn, as in some Miridae (Hemiptera), and act as a kind of snorkel if the egg becomes submerged. Other species have plastron-bearing horns, in which the plastron is in the shape of a tube around the horn. Among the aquatic insects, plastron-bearing horns are characteristic of hemipterans in the family Nepidae (Figure 11.3) (Hinton 1961). The number of horns per egg varies among species (e.g. two in all species of *Ranatra* and *Cercotmetus*; 25 or 26 in *Borborophilus primitiva*), within some species (e.g. 6–10 horns per egg in *Nepa rubra*), and can also vary among eggs produced by a single female, e.g. *Laccotrephes* sp. (Hinton 1961). Typically, the plastron is confined to the apical parts of the horn and the basal part of the horn has a more solid chorion.

The air-filled mesh of the horn generally is continuous with mesh in the main body of the egg shell, which presumably facilitates gas exchange over the entire egg surface. Aeropyles, branched distally, also occur over the main body of the egg's chorion (Figure 11.3d). Most gas exchange, however, occurs through the respiratory horn and only when the entire horn is submerged will it function as a plastron; most of the time gas exchange is direct with the atmosphere.

The egg cocoons of hydrophilid beetles also function as plastrons, whether they are carried on the female (*Helochares, Spercheus*), float freely on the water surface (*Hydrochara, Hydrophilus*) or attached to vegetation. The cocoons of some hydrophilids (*Hydrophilus*) that oviposit on vegetation also have a stalk that opens at or above the water surface, providing a direct route for atmospheric oxygen.

11.4.2 Egg coatings and accessory glands

Several kinds of coatings are found on the eggs of aquatic insects: adhesives, gelatinous material, silk, and hydrophobic or hydrophilic materials. Some coatings may be multi-functional, e.g. gelatinous material may also act as an adhesive, but adhesives are not necessarily gelatinous, and separate materials may serve these different functions in a single species. These substances are often produced by the female accessory glands and the follicle cells of the ovarioles. If accessory glands are absent, as in all Ephemeroptera and some Plecoptera, these egg coatings are produced only by follicle cells and sometimes the oviducts. Female accessory glands producing these substances are often called colleterial glands (from the Greek *kolla* or *kollêtêrion* meaning 'glue'), but remember that the secretions of accessory glands also may serve other functions. The accessory glands are generally well-developed in insects whose eggs are glued to the substrate, embedded in a gelatinous material or packed into egg cocoons.

Gelatinous material, called spumaline (Hinton 1981), commonly forms a layer between and over the eggs, and a fine membrane often surrounds the entire mass of spumaline. Such gelatinous material is very widespread in the aquatic insects and may coat eggs that are laid singly (Coleoptera: Dytiscidae, Gyrinidae; Hemiptera: Corixidae) and also eggs laid in masses that form plaques attached to substrates (Trichoptera: Hydrobiosidae, Hydropsychidae, Polycentropodidae, Psychomyiidae, Philopotamidae; Plecoptera: Eustheniidae; Coleoptera: Psephenidae; Diptera: Dixidae), strings and loops (Trichoptera: Limnephilidae, Phryganeidae; Diptera: Chironomidae), or gelatinous spheres (Trichoptera: Limnephilidae, Leptoceridae). The shape of these egg masses is enormously variable between species, as is the arrangement of eggs within the mass (e.g. Reich 2004; Lancaster et al. 2010a). Eggs laid in spumaline often have a smooth, thin chorion—so thin that the eyespots of the developing embryo may be clearly visible—and egg size often increases as the embryo develops. Spumaline is initially fluid and is modified when it comes in contact with the outside air or water, often absorbing water and swelling markedly to several times its original mass. The amount of hydrating spumaline varies among species and, for example, among the Trichoptera that attach egg masses to submerged substrates, the Hydropsychidae typically have a thin

layer of spumaline whereas some Hydrobiosidae often have a thick, swollen layer. The chemical composition of spumaline is largely unknown, and available information suggests that the spumaline of some Trichoptera and Diptera comprises complex polysaccharides and glycoproteins, with variations in chemical composition among taxa (Endrass 1976; Hinton 1981). Its viscosity also varies among species, with oviposition location (terrestrial or aquatic), with age of the egg mass (often disintegrating as eggs hatch), and, for egg masses exposed to air, with humidity and water content of the surrounding medium. When spumaline-covered eggs or egg masses are deposited at the water's surface, the spumaline rapidly swells with water, yet the egg or egg mass is typically more dense than the water and it sinks, provided the spumaline surface is wettable. If the spumaline is non-wettable, eggs are normally attached to the substrate, otherwise they may float. The spumaline of egg masses is not necessarily a continuous, uniform mass of jelly but may have some morphological features that give the mass its structural integrity and provide protection for the eggs. The egg ropes of some chironomids, for example, often have fibrous threads running through them and the more jelly-like spumaline is attached to these fibres. Thus, the fibres anchor the mass to the substrate and give the mass its rope-like shape. In some egg masses, such as those of *Nemotaulius hostilis* (Limnephilidae), that are laid terrestrially and are thus prone to desiccation, the spumaline may have some sort of internal structure with channels running through the jelly and these appear to play a role in desiccation resistance (Berté and Prichard 1983).

Instead of a gelatinous coating, the eggs of most beetles in the families Hydrophilidae, Hydraenidae, and Spericheidae are arranged within a silken cocoon. This silk is produced by glands that are modified ovarioles and, possibly, also by oviduct glands (Hinton 1981). The female uses the two styli of the genitalia for spinning: the sticky, liquid silk flows over the styli, a drop is attached to the substrate or part of the cocoon, then pulled into a thread that dries and becomes elastic. Some species produce a fully enclosed cocoon that is submerged and attached to rocks or plants, or that floats freely (e.g. *Hydrophilus* and *Hydrous*). In *Sphercheus* and

Helochares (Hydrophilidae), the egg case is attached to the ventral side of the female's abdomen, which she carries around until the larvae hatch. Other members of this family, such as the South American *Anacaena lutescens*, do not form closed cocoons but may produce a compact group of eggs surrounded by a loosely woven silk net (Archangelsky and Fikáček 2004).

Broadcasting eggs may be one of the simplest forms of oviposition, but eggs then have the problem of breaking through the water's surface tension to enter the aquatic environment. Egg balls, typically 'explode' upon contact with the water, scattering eggs in many directions, suggesting that the egg coating has hydrophobic properties. Observations also suggest that a surfactant is released with the eggs, which breaks the surface tension and allows eggs to enter the water (Funk et al. 2006).

11.4.3 Attachment devices

Eggs that are deposited freely into the water, i.e. are not attached to substrates by the female, risk being moved by currents and, especially for eggs in running waters, being transported long distances. The spumaline-covered egg masses of some Chironomidae that are deposited freely into streams are often extremely sticky and this ensures that most drift only a short distance before becoming attached (Williams 1982). In some species, the individual eggs have a sticky coating and the eggs adhere to the first object (plants, rocks) that they encounter after oviposition, including some stoneflies in the family Leuctridae (Percival and Whitehead 1928). Alternatively, each egg may be equipped with one or more attachment devices, as is common in the Ephemeroptera, Plecoptera, and some Odonata.

Attachment structures on individual eggs are morphologically diverse. The attachment devices of mayflies have been examined in some detail and can be categorized initially into fibrous and non-fibrous structures (Koss and Edmunds Jr 1974). The non-fibrous structures include sucker-like discs or plates, adhesive projections, and adhesive layers. Adhesives, where present, are probably produced by the follicles as female Ephemeroptera lack accessory glands. Similarly, many Plecoptera also have sticky attachment structures. Commonly there are many small adhesive

knobs on the basal plate or bottom surface, as in some Pteronarcydae and Perlodidae (Percival and Whitehead 1928; Miller 1939). Alternatively, taxa (e.g. many Perlidae) have a short collar at one end of the egg and, in the middle of the collar is an 'anchor' comprising a stalk with a basal plate covered in adhesive knobs (Figure 11.4). Initially, the anchor is retracted into the collar and may be extended only upon contact with the substrate (Hinton 1981; Rościszewska 1991). The stalk may be very elastic, returning to its normal length after being stretching to four times its length, and these eggs may withstand very strong currents (Percival and Whitehead 1928).

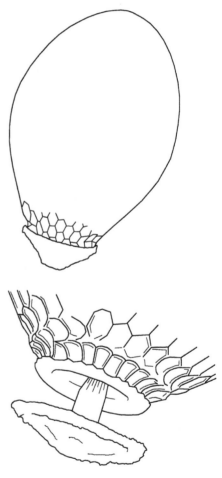

Figure 11.4 Egg of the stonefly *Oyamia gibba* (Perlidae), showing the entire egg (top), and close-up of the collar and extended anchor (bottom). The anchor is only extended upon contact with the substrate.
Source: Adapted from Isobe and Uchida (2009).

Fibrous attachment structures in the mayflies involve monofilamentous fibres that are arranged either in a loose collection or, most commonly, coiled tightly (Koss and Edmunds Jr 1974). Often the fibres terminate in knobs or granules. These fibrous attachment structures may be scattered across the egg surface (Figure 11.5), or aggregated at one or both poles of the egg to form polar caps or epithema (Figure 11.6), which may give the egg a rather obvious two- or three-part appearance. Both polar caps and lateral attachment devices may be present on a single

egg. The fibres may function as attachment devices by becoming entangled in vegetation, woody debris, or mineral substrates, or may be glued to the substrate if adhesives are present on the fibres or the terminal knobs (Figure 11.5d). Fibrous attachment structures are present also on the eggs of some Tipulidae (Candan et al. 2005) and some Odonata, particularly the Gomphidae. Analogous to the epithema of mayflies, the most elaborate attachment structures in odonates are polar aggregations of densely coiled filaments (Corbet 2004), or a single coiled

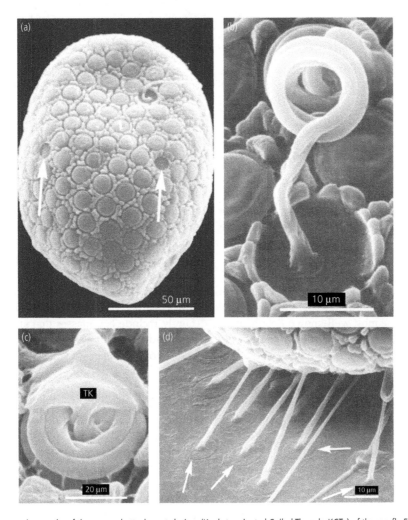

Figure 11.5 Electron micrographs of the egg and attachment devices (Knob-terminated Coiled Threads, KCTs) of the mayfly *Electrogena zebrata* (Heptageniidae). (a) Overview of egg with two micropyles indicated by arrows. The remaining round structures are the terminal knobs of coiled threads. (b) Close-up of a KCT showing a partially extended coiled thread underneath the terminal knob. (c) Close-up of a KTC showing the terminal knob (TK) and the coiled thread underneath. (d) Terminal knobs (arrows) of extended KCTs adhering to the substratum.

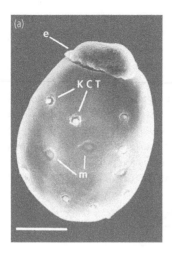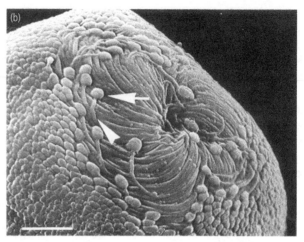

Figure 11.6 Electron micrograph of the egg and attachment devices of *Ephemerella ignita* (Ephemerellidae). (a) Whole egg showing the polar cap or epithema, e, two micropyles, m, and two lateral attachment devices, KCT, Knob-terminated Coil Thread. Scale bar represents 50 μm. (b) Close-up of epithema (apical part) showing coiled configuration of the threads (arrow head) and their terminal granules (arrow). Scale bar represents 10 μm.

Source: From Gaino & Bongiovanni (1992). Reproduced with permission from John Wiley & Sons.

filament (Gambles 1956). The filaments, often sticky, uncoil rapidly once the egg is wetted and are likely to become entangled in vegetation, thus anchoring the egg and preventing it drifting downstream or becoming buried in soft sediments.

11.4.4 Floating eggs (rafts)

Some eggs deposited on the water's surface float and remain floating throughout the developmental period. Perhaps the most well-known floating eggs are those of mosquitoes. In the subfamily Anopheline (e.g. *Anopheles*) eggs are laid singly and the chorion is typically extended laterally to form two distinct mid-lateral floats, or a broad continuous float surrounding the egg (Figure 11.7a). In the subfamily Culicines, eggs are either laid singly (e.g. *Aedes*) or in large batches forming rafts (e.g. *Culex*, *Coquillettidia*). Eggs of *Aedes* (Figure 11.7b) typically lack the lateral floats of Anophelines and are laid above the water line in sites that will eventually become submerged, and at which point the eggs float. Females of *Culex* deposit eggs in rafts above the water, with eggs attached to one another along the longitudinal sides via interlocking extrachorionic processes, such that individual eggs float vertically (Figure 11.7c). The exochorion of *Culex pipiens*, and

many other Culicines, is typically highly sculptured with a meshwork of tubercules and, in a raft, adjacent eggs are held together by interdigitation or interlocking of their tubercules along the longitudinal surfaces (Sahlen 1990). Should the raft become broken, the shape of the meniscus that forms around the eggs results in eggs moving towards one another to reform a raft, at least in part (Hinton 1968c). Not all mosquitoes, however, deposit floating eggs and other methods of oviposition within the culicid family include egg masses that are attached to the underside of floating macrophytes, as seen in some species of *Mansonia* and *Ficalbia*.

11.4.5 Egg enemies and defences

There is no guarantee that all the eggs laid will hatch successfully and egg mortality may result from desiccation, disturbance, predation, parasitism, and attacks from bacteria and fungi. These problems are avoided by ovoviviparous and viviparous species, but the overwhelming majority of aquatic insects are oviparous and minimizing egg mortality may be an important factor in maintaining population numbers. One simple strategy to minimize egg mortality may be to 'outrun the enemy' through rapid development and hatching of the larvae. The

(a)

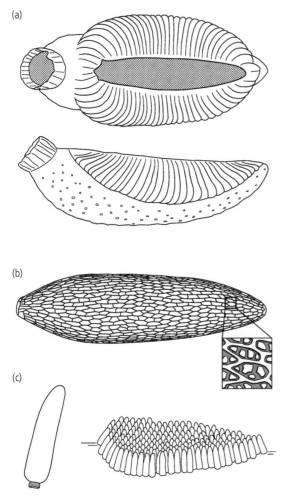

(b)

(c)

Figure 11.7 Illustrations of various forms of mosquito eggs (Culicidae). (a) *Anopheles rangeli* in dorsal (top) and lateral (bottom) views showing the lateral floats and a 'crown' at the anterior end (left) with the micropyle. (b) *Aedes dentatus*, which has an inconspicuous micropylar collar at the anterior end (left). The chorion is sculpted in a reticulated pattern with several tubercules in each cell of the reticulation (inset). (c) *Culex* sp. showing a single egg (left) and a floating raft of many eggs (right).

Sources: (a) Adapted from Linley and Lounibos (1993); (b) adapted from Linley and Turell (1993).

eggs of some species do develop very rapidly, but most will be vulnerable to some stressors or attackers and will have some form of defence. Consequently, many insects make choices about where to lay their eggs (within a range of possible locations), presumably selecting locations that are suitable for egg development. This decision-making process may

follow a hierarchical selection scheme using different cues at different spatial scales (Hoffmann and Resh 2003; Reich et al. 2011).

The problems of desiccation are often high for eggs that are laid in temporary water bodies and for eggs laid aquatically but near the water surface or in shallow waters, which risk exposure should water levels drop. Of course, eggs in terrestrial environments also risk desiccation, and they typically have desiccation-resistant structures and adaptations similar to those of terrestrial insects. For many species, the chorion may be sufficiently thick and dense to prevent water loss and, indeed, some mosquito eggs that are completely dry and exposed to high temperatures will remain viable for months (Roberts 2004). Eggs that have to cope with a potentially alternating aquatic and terrestrial existence perhaps face the biggest challenges. Selective oviposition in deep water may be one strategy to avoid desiccation from fluctuations in water level or, for species laying eggs terrestrially, oviposition high above the water level where the risk of being submerged is low. Spumaline may afford some protection to eggs simply because it loses water slowly to the air and this alone will increase egg survival. Species that inhabit temporary water bodies or that lay spumaline-covered egg masses terrestrially, appear to have a particular form of spumaline that is desiccation resistant, as illustrated by some caddisflies in the families Limnephilidae and Phryganeidae (Hinton 1981; Berté and Prichard 1983; Wiggins 2004). In these species, dried spumaline may be reduced to as little as 5 per cent of its initial weight, at which point a tough film or skin forms to cover the egg mass. This skin prevents further water loss and maintains moist spumaline underneath, in which embryos can develop and neonates can move. When water is added to dried egg masses, which may occur months after drying, the spumaline rapidly rehydrates and softens sufficiently for neonates to break out.

Hydrologic and hydraulic disturbances in running waters are often accompanied by substrate movement and this may result in egg mortality, especially for eggs that are attached to substrates. Eggs attached to rocks may be physically damaged if rocks tumble and roll, and, depending on the position and orientation of the rock when it comes to rest, eggs may be in unsuitable microhabitats (e.g. embedded in fine sediments) or exposed to the air and the attendant problems of desiccation. That versatile substance, spumaline, may provide some protection from mild physical abrasion, but it is unlikely to resist more vigorous scraping. Selective oviposition on large rocks may minimize the risks of substrate movement and, indeed, some species that oviposit on emergent rocks in rivers lay more eggs on large than small rocks (Lancaster et al. 2003; Reich and Downes 2003a, b). Given that rock size is often positively correlated with water velocity, this pattern may be driven primarily by females selecting rocks in fast flows (Encalada and Peckarsky 2006), but with the additional advantage that most eggs are laid on large emergent rocks that move only in the largest floods.

A great variety of predators and parasites can attack the eggs of aquatic insects, including mites (Proctor and Pritchard 1989) and a host of insects. Many reports of egg predation are largely anecdotal or isolated observations, and it is more difficult to determine whether such acts of predation are opportunistic feeding or an important source of nutrition for the predator (which may be obligate oophages), or whether egg mortality from predation has demographic consequences for the victim. Egg consumption is clearly important for species that are obligate egg predators throughout, or for most, of their aquatic life, such as larvae of the dipteran *Acanthocnema* sp. (Scathophagidae) (Purcell et al. 2008) and the cosmopolitan watermite *Hydrodroma despiciens* (Wiles 1982). Egg consumption may also be important in particular life stages, such as some hydroptilid caddisflies in the genus *Orthotrichia*, which appear to feed heavily on insect eggs only in the final larval instar. Periods of stress, such as food shortages, may coincide with opportunistic egg predation or cannibalism as a means to maintain reproductive status or simply survive (Pajunen and Pajunen 1991). Of course, eggs are also vulnerable to parasites and, for example, hymenopteran parasites have been identified from dytiscid eggs laid aquatically (Jackson 1961) and the terrestrial eggs of the dobsonfly *Sialis* (Sialidae) (Salt 1937; Pritchard and Leischner 1973).

Although egg predation obviously occurs and is widespread, there are many ways in which eggs may be defended against predators. As well as laying eggs in hard-to-access or cryptic locations (e.g. in crevices and cracks), eggs may be defended physically and chemically. Some eggs are camouflaged against visual predators in that they are dull in colour or may become covered in fine sediment or detritus, if an adhesive coating or sticky spumaline is present. Aposematic colouration is another form of camouflage, and although less common in aquatic than terrestrial insects, the egg chorion of some *Anopheles* (Culicidae) is patterned. The egg chorion and serosal cuticle may provide adequate physical protection to repel the foraging attempts of some potential predators. Spumaline and other secretions also may be an effective deterrent by creating a hard coating (Mangan 1992), or by glueing or clogging up the mouthparts or legs of the would-be attackers. Such defences do not work against all predators, however, and some are clearly able to penetrate the spumaline. For example, larvae of the dipteran *Acanthocnema* sp. (Scathophagidae), which prey upon trichopteran and dipteran eggs, are able to penetrate the spumaline and slash through the chorion to reach the egg contents (Purcell et al. 2008). Similarly, the watermite *Hydrodroma despiciens* has palps that are extremely effective at cutting through the spumaline of chironomid eggs (Wiles 1982). Further, late instar larvae of some caddisflies, *Orthotrichia*, plough furrows through the spumaline of hydrobiosid egg masses, apparently without difficulty. Chemical protection of eggs is well documented in terrestrial insects (Blum and Hilker 2002; Eisner et al. 2002), but relatively little information is available regarding aquatic insects. The exceptions are some mosquitoes where females lay their eggs in rafts and a drop of defensive fluid, of maternal origin, occurs on the upper (posterior) pole of each egg. The mosquito *Culex pipiens* occasionally oviposits in pools liable to rapid fluctuations in water level, and egg rafts that become stranded at pool margins may be vulnerable to terrestrial predators and scavengers, such as ants. However, the fluid on the upper side of the egg raft effectively deterred predation by some ants (Hinton 1968). So, are chemical defences of eggs largely absent in aquatic insects or is it simply that has no one has bothered to look?

Fungi and bacteria can attack the eggs of aquatic insects, as will be familiar to anyone who has attempted to hatch eggs in the laboratory. The extent to which this occurs in field situations is unclear, but it does occur and may be a significant cause of egg mortality (Russell et al. 2001).

Finally, in a world in which human beings routinely pollute every environment, the eggs of aquatic insects are often exposed to various waterborne toxins. As with other life stages, eggs are vulnerable to such toxins and developing embryos may be killed, deformed, experience delayed hatching or aberrant behaviour in later life stages. The egg chorion appears unable to prevent uptake of toxins, but spumaline may afford some protection, albeit not complete protection (Palmquist et al. 2008).

11.5 Parental care of eggs

Parental care is unusual among aquatic insects, although characteristic of the social and exclusively terrestrial insects, such as bees, ants, and termites. Among the few aquatic taxa that do display parental care, this care centres primarily around the eggs. In the Coleoptera, many Hydrophilidae deposit their eggs in discrete cocoons, which are often attached to vegetation or float freely at the water surface. In *Helochares* and *Spercheus*, the female carries the egg cocoon underneath the abdomen, and only drops the cocoon once the eggs hatch. Presumably, this protects the eggs from predators and the risks of being stranded in unsuitable locations.

Examples of parental care may be rare in the aquatic insects, but examples of paternal care (male care without input from females) are rare in any arthropod (Tallamy 2001). Curiously, some of the most extreme cases of paternal care occur in the giant water bugs, Belostomatidae. Females of the subfamily Belostomatinae (e.g. *Abedus*, *Belostoma*, *Diplonychus*) lay their eggs on the backs of males, and the males carry the eggs until they hatch. Females will occasionally carry eggs (Kruse and Leffler 1984; Estevez et al. 2006), potentially when males are in short supply (Kruse 1990), but paternal care is

most common. The eggs of these bugs are large (as are the adults) and gas exchange for the developing embryo is potentially problematic given the low surface area-to-volume ratio of the egg. To prevent the eggs drowning, the back-brooding males rise periodically to the water surface to aerate the eggs, and also increase gas exchange by increasing the flow of water over the eggs via brood pumping (doing push-ups), or by egg stroking (rhythmically brushing their hind legs over the eggs) (Smith 1976). There are costs to this behaviour as brood pumping and egg stroking require energy, and there are additional costs in terms of the male's reduced mobility (Kight et al. 1995) and foraging efficiency (Crowl and Alexander 1989). Failure to brood the eggs, however, results in reduced egg hatch (Munguía-Steyer et al. 2008) and the benefits presumably outweigh the costs.

Another unusual form of paternal care occurs in the belostomatid *Lethocerus*, where females crawl out of the water and deposit eggs on the stems of emergent vegetation. Although gas exchange in the terrestrial environment may not be problematic (Hinton 1981), the eggs are vulnerable to desiccation and to predation. Attendance and brooding behaviour of the male, however, ensures that the eggs are kept hydrated, free from predators and competing females (Ichikawa 1988, 1991; Smith and Larsen 1993), and that neonates also are protected from predators (Ohba et al. 2006). Female *Lethocerus* do not lay all their eggs in one batch. Instead, mating pairs alternate between bouts of copulation and oviposition, and this results in male–female pairs remaining close together until oviposition is complete. The female then departs and, until the eggs hatch, the male hydrates the eggs repeatedly, moving from the water to the eggs and standing in such a way that water draining off his body runs over the egg mass. As with the Belostmatinae, brood behaviour by male Lethocerinae may have associated costs, but failure to brood inevitably results in egg mortality (Ichikawa 1988).

CHAPTER 12

Development

12.1 Introduction

In discussing insect development, we focus on the processes that occur during each stage of the life cycle and the events that accompany transitions between life stages. There are no systematic differences between aquatic and terrestrial insects in their developmental patterns, but different developmental patterns can have profound ecological implications (e.g. for distribution patterns, population dynamics), thereby indicating this topic's great significance to any entomology text. Development can be divided into three main phases, and we will discuss each one in turn. The first is embryonic development or embryogenesis, which includes all events between fertilization of the egg and egg hatch (Section 12.2). Post-embryonic development incorporates larval development and the final transformation of larva into adult, also called the imago (Section 12.3). (Note: as discussed in Chapter 1, we refer to all juvenile insects as 'larvae' and do not use the term 'nymph', which is commonly reserved for the juveniles of hemimetabolous taxa.) The term 'eclosion' is used most often to describe the transition from larva to adult, but can also describe the transition from egg or embryo to larva. Metamorphosis, the marked morphological and physiological change from the final larval stage to the adult, is one of the most dramatic of biological phenomena (Sections 12.4, 12.5). Many people, including the philosophers of ancient Greece, have speculated about why many insects differ in the degree of change that occurs between egg hatching and sexual maturity, and the debate continues (e.g. Truman and Riddiford 2002; Konopová and Zrzavý 2005; Erezyilmaz 2006). For many aquatic insects, metamorphosis also accompanies a major habitat shift from aquatic to terrestrial environments, which can present particular challenges (Section 12.6).

There are three main patterns of development that differ in the extent of metamorphosis: ametaboly, hemimetaboly, or holometaboly. In the plesiomorphic (ancestral) pattern, ametaboly, a miniature version of the adult which is wingless, emerges from the egg. Subsequent stages increase in size, eventually develop reproductive organs, and continue to moult after sexual maturity. This group includes the wingless insect orders, such as the Thysanura or silverfish, none of which are considered aquatic. All the winged insects also go through multiple immature (larval) stages but, in contrast, undergo a marked change in morphology between the juvenile and winged adult stages. The winged insects can be subdivided into those that have hemimetabolous development (i.e. incomplete metamorphosis) or holometabolous development (complete metamorphosis), and both groups include aquatic insects. Hemimetabolous development is largely direct (as in the ametabolous taxa), with a gradual transformation through the larval instars and a marked change at the final moult from larva to adult. In contrast, the larval stages of holometabolous insects typically look completely different from the adult form, and pass through a pupal stage before entering the adult stage.

Although the main phases of development and various developmental pathways are relatively easy to identify, how long it takes an insect to move through each stage, and whether they manage to complete development, is very strongly influenced by environmental factors (Section 12.7). Accordingly, this results in different life-history patterns between species and potentially even between different populations within species. The study of the causes and consequence of different life histories is an enormous topic, and only a very brief discussion of the topic is possible (Section 12.8).

Before embarking on a discussion of development, it might be useful to consider a few definitions. In the broadest sense, development involves both growth and a more narrow definition of development. Growth literally refers to the increase in size or biomass of an organism through an increase in the numbers and sizes of cells, whereas development generally involves the formation of new cell types, tissues, and organs. In short, growth entails getting bigger whereas development involves the differentiation of tissues and, ultimately, the morphological and physiological progression toward reproductive maturity. Inevitably, growth and development are correlated in many ways, but some elements can be separated. Within particular life stages, for example, both growth and development occur in the embryos and larvae, pupae generally develop but do not grow, and adults typically do neither (other than sexual maturation in some taxa, Section 10.4). All larvae must feed in order to grow and, although some adults do feed, growth of somatic tissue is rare. Adult feeding appears to be mainly for the purposes of maintenance (long-lived adults) and maturation of reproductive organs and/or eggs, as in blood-feeders (Chapter 10). In some situations it may be useful to distinguish between growth and development *sensu stricto*, but the common practice is to refer to both processes collectively as development.

12.2 Embryogenesis to egg hatch

Embryogenesis (development of the embryo) begins once the egg has been fertilized and involves multiplication of cells (by mitosis) and their subsequent growth, movement, and differentiation into various tissues and organs. The stages of embryonic development are well-described for insects and there is a wealth of literature on the subject, so we refer the reader to that literature and to general entomology texts. Although all insects follow the same basic pattern of embryogenesis, there are, inevitably, variations on the theme. These variations, however, appear to follow evolutionary and phylogenetic lines, and the habitat (i.e. aquatic or terrestrial) is

less important. While the stages of embryogenesis are well known, the processes by which development comes about are less well understood and the subject of developmental biology. The duration of embryonic development varies enormously among and within species, in part due to environmental factors (e.g. temperature, Section 12.7.1), but can also vary among individuals within the same egg mass and that were laid contemporaneously by the same female, as seen in some stoneflies (Sandberg and Stewart 2004).

To escape from the egg, once embryonic development is complete, an insect must break through the various membranes that surround it, including the chorion and vitelline membrane (Section 10.2). Larvae may also have to make their way through various egg coatings, such as layers of gelatinous spumaline, before they are completely free. The general mechanism for hatching is primarily mechanical and begins with the insect swallowing fluid, and sometimes air, from a cavity within the egg (the amniotic cavity). When eggs are in the aquatic medium, the embryo typically swallows amniotic fluid, the resulting negative pressure within the amniotic cavity causes water to enter the egg through the micropyles, which, in turn, results in a net pressure increase on the chorion from inside the egg. The insect's abdomen may then contract to force haemolymph into the thorax and head, causing the latter to enlarge. At the end of embryonic development, the larvae of many taxa are curled or twisted inside the egg so straightening of the body, often combined with other movements and hydrostatic pressure, is often sufficient to rupture the chorion and egg membranes. Some insects, such as the Lepidoptera, simply eat their way out of the egg.

Rupturing of and escaping from the chorion may be facilitated in several ways. The chorion may have predetermined lines of weakness, typically running longitudinally or transversely, the latter separating an anterior egg cap from the posterior portion of the egg. The cap may remain attached to the rest of the chorion by a hinge as, for example, in some mosquitoes (some *Culex* and *Anopheles*), or may be completely detached, as in other mosquitoes (some *Aedes*). Larval movement and enlargement may be sufficient to break the chorion, but many insects also have egg

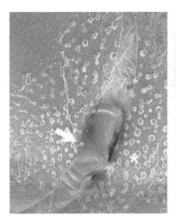 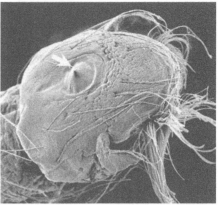

Figure 12.1 Electron micrograph of the egg burster of the mosquito *Aedes aegypti* (Culicidae) Left: the tip of the egg burster (arrow) is visible through a fracture in the chorion (magnification 700×). Right: Details of a larval head outside the eggshell with the egg buster clearly visible on the dorsal surface (arrow) (magnification 400×).

Source: Republished with permission of the Entomological Society of America, from Pereira et al. (2006).

bursters in the form of hard cuticular spines, cones, or plates on the body (usually the head) (Figure 12.1). Movements of the larva inside the egg bring the egg burster in contact with the chorion, forcing it to break. These movements may be rhythmic and repetitive, with the egg burster aimed at a line of weakness in the chorion, much like hammering a wedge into a fissure to fracture rocks. Continued movements free the insect from the chorion. Many aquatic Hemiptera (e.g. *Notonecta*, *Trichocorixa*, *Ranatra*) appear to make use of an inner egg membrane, sometimes called a 'blister' membrane (probably the vitelline envelope), to break the chorion during hatching (Davis 1964, 1965). The vitelline envelope swells as water enters the egg osmotically, forming a blister-like expansion which, via hydrostatic pressure, splits the chorion open at the anterior end of the egg into multiple flaps, through which the blister protrudes. By swallowing water, the embryo then swells and pushes into the blister, which eventually breaks from internal pressure and the larva escapes by further swelling and by active movements.

At egg hatch, the embryonic cuticles are shed and left within or just outside the egg. There are typically three cuticles secreted during embryogenesis and, except in some Diptera, the third is the cuticle of the first instar larva (Konopová and Zrzavý 2005). Some embryonic cuticles are flimsy and fragment readily, whereas others are more robust and may be clearly visible on or near the vacated chorion (Jackson 1957).

12.3 Larval development

Most insect growth occurs during the larval stages, but the rigidity of the exoskeleton puts a constraint on size of each instar, so it must be replaced periodically with a bigger exoskeleton—a process called moulting. Each time an insect moults, it enters a new stage or instar, and the duration of each instar (i.e. the period between successive moults) is often termed the stadium. Superficially, there may be no marked change in general body form between larval instars, with each successive instar being similar to the previous one, but the series of gradual changes typically results in distinct differences between first and final instars. Moreover, close examination of particular structures may reveal morphological (and functional) differences between instars, such as increases in the number of segments in antennae and cerci, and marked morphological changes in the gills and mouth parts of some mayflies (O'Donnell 2009). Even more extreme examples include species that display hypermetamorphosis (also called heterometamorphosis) in which different larval instars have distinctly different body forms, such as caddisflies in the family Hydroptilidae in which the first four larval instars are free-living and resemble beetle larvae, whereas the fifth, final instar larva builds a case and takes on a more typical caddisfly form (Figure 12.2). Moulting involves two distinct processes: apolysis, the separation of the existing cuticle from the underlying epidermis; and ecdysis, shedding of the old cuticle once the new one has formed. During the interval between apolysis and ecdysis, an insect is contained within the cuticle of the previous instar, a state described as being pharate (from the ancient Greek *pharos*, meaning 'cloak').

12.3.1 Types of larvae

As mentioned earlier, larvae of hemimetabolous insects largely resemble the adults, whereas larvae and adults of holometabolous insects are radically different. The most conspicuous difference between the larvae of these two groups is in the development of wings. In the hemimetabolous insects, the wings develop as external buds, which become larger at each moult, and finally appear, complete in the adult. Because wing buds are visible externally, hemimetabolous insects are often called exopterygotes. In contrast, the wings of holometabolous insects develop beneath the larval cuticle and

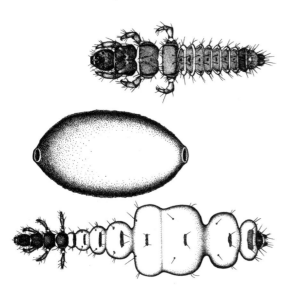

Figure 12.2 Example of hypermetamorphosis in larvae of the caddisfly *Leucotrichia* (Hydroptilidae). Showing a free-living early-instar larva (top), a final instar (bottom) and its retreat (middle). Note: figures are not to scale.

Source: Republished with permission, from Wiggins (2004).

are not visible externally, hence the name 'endopterygotes'. Holometabolous larvae also generally lack compound eyes (if present, larval eyes are morphologically different to adult compound eyes, Chapter 6) and the thoracic legs are often reduced or absent.

Among the holometabolous insects are many different types of larvae, most of which fall into one of a few basic forms (Figure 12.3). Oligopodous larvae have three pairs of jointed thoracic legs, a well-developed head capsule, and chewing mouthparts similar to the adults. There are two common forms of oligopod larvae: campodeiform larvae that are dorso-ventrally flattened, have a well-sclerotized thorax and abdomen, and are usually long-legged predators with a prognathous head; and scarabaeiform larvae that are round-bodied, with a weakly sclerotzed thorax and abdomen, and are usually short-legged. Within the oligopodous aquatic insects, campodeiform larvae are common (e.g. dytiscid beetles, Megaloptera, many caseless caddisflies, Figure 12.3a), scarabaeiform larvae are less common in the aquatic insects, but do occur in the chrysomelid beetles such as *Donacia*. Polypodous larvae (also called eruciform) likewise have three pairs of jointed thoracic legs, abdominal prolegs may be present also, and the body generally is poorly sclerotized (except the head capsule). Lepidoptera are the classic examples of polypodous larvae, but most cased caddisflies are also considered to be eruciform although abdominal prolegs are absent, except the hooks at the tip of the abdomen (Figure 12.3b, 12.3c). The third basic form is the apodous larva, which is very poorly sclerotized and lacks thoracic jointed legs. There are three different forms of apodous larva that differ in sclerotization of the head capsule and representatives of all three occur in the Diptera: eucephalous larvae have a well-sclerotized head capsule (e.g. most Nematocera, including Chironomidae, Simuliidae, Culicidae, Figure 12.3d); hemicephalous larvae have a reduced head capsule that can be retracted into the thorax (e.g. Orthorrhapha, Brachycera, such as Tabanidae, Empididae, Stratomyidae, Figure 12.3e); acephalous larvae have no head capsule (e.g. Cyclorrhapha, such as Syrphidae, Ephydridae, Muscidae, Figure 12.3f). These terms are often convenient to describe body forms, but larval form

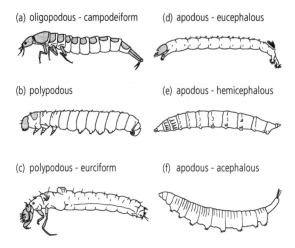

(a) oligopodous - campodeiform
(d) apodous - eucephalous
(b) polypodous
(e) apodous - hemicephalous
(c) polypodous - eurciform
(f) apodous - acephalous

Figure 12.3 Examples of different types of holometabolous larvae: (a–c) polypodous or oligopodous, with three pairs of true legs, (d–f) apodous, no true legs. (a) Coleoptera, Dytiscidae, (b) Lepidoptera, Pyralidae, (c) Trichoptera, Molannidae (case removed), (d) Diptera, Chironomidae, (e) Diptera, Tabanidae, (f) Diptera, Ephydridae.

can differ between apparently closely related species, and between instars of single species that display hypermetamorphosis.

12.3.2 Moulting

As a larva grows in volume, it gradually outgrows the external skeleton so it must be shed and a new, larger cuticle formed. Moulting insects are particularly vulnerable because they may be easy prey for predators and may be especially sensitive to physical and chemical stresses. Faults or unexpected stresses during the moulting process can result in significant deformities and potential fatalities. Muscles that move the limbs and body must be detached from the old cuticle and new muscle attachments to the new cuticle must be established quickly. New cuticle is soft and must harden sufficiently to resist the pull of muscles; muscular action on a soft cuticle can result in skeletal deformation and movement restrictions. New cuticle is also pale and field workers will be familiar with the sight of very pale larvae that have recently moulted (called 'teneral').

The moulting process is under endocrine and nervous control and is much the same for terrestrial and aquatic insects. Detailed descriptions of the physiological and biochemical processes are available in many entomological texts. It is, however, an amazing process and here we provide a brief description of the events during moulting and the formation of new cuticle. The mature cuticle before moulting (Figure 12.4a) consists of a basal layer of epidermal cells, on top of which lie sequential layers of endocuticle, exocuticle, and finally epicuticle. Apolysis, or the separation of the epidermal cells from the old cuticle, marks the beginning of a moult, and the insect within the loosened, but not yet shed, cuticle is the pharate next instar. In response to the hormone ecdysone, the epidermal cells separate from the old cuticle, forming an apolysial space (Figure 12.4b) and the number of epidermal cells increases through mitosis (Figure 12.4c). Moulting fluid (inactive) is released into the apolysial space (also called the ecdysial space), and this coincides with the formation of an ecdysial membrane on the underside of the old cuticle (Figure 12.4d). Soon after the apolysial space appears, the formation of new cuticle begins. Initially, the new epidermal cells develop an irregular surface they secrete a layer of cuticulin that covers the layer of epidermal cells (Figure 12.4e) and which becomes the epicuticle of the new exoskeleton. Once the new cuticulin layer is in place, the moulting fluid in the apolysial space is activated and the constituent proteinases and chitinases begin to digest some of the old endocuticle. The products of digestion are reabsorbed by the epidermal cells, through the cuticulin layer, and will be used in the production of new cuticle. The cuticulin layer is protected from digestion. New, unsclerotized procuticle is secreted below the cuticulin layer, pushing it up (Figure 12.4f). The fibres that attach muscles to the epicuticle are also resistant to moulting fluid. The pharate stage of larvae (the period between apolysis and ecdysis) is often quite long (hours to days), and movement during this period may be critical for feeding or predator avoidance so it is important that the muscles remain operative. Indeed, when pharate larvae with weakly sclerotized exoskeletons are viewed under a microscope, the new exoskeleton can sometimes be seen lying

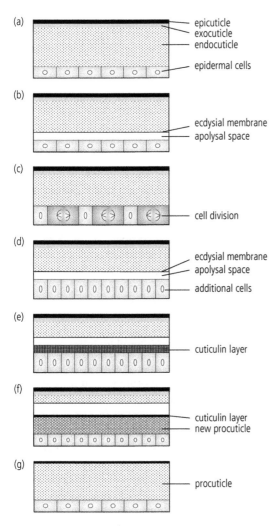

below the old one. Shortly before ecdysis (loss of the old cuticle), the moulting fluid disappears from the apolysial space. Most of this fluid appears to be reabsorbed by the insect, either through the epidermal cells or by flowing through the apolysial space to the anus or mouth, from where it enters the gut and potentially useful nutrients can be absorbed.

To facilitate ecdysis, some insects swallow water (or air for terrestrial species) to expand the gut and split the old cuticle along lines of weakness (ecdysial sutures). The muscular actions involved in ecdysis, such as wriggling and movements of the abdomen, are controlled by nervous motor programs, but the motor programs are initiated by hormonal action (at least in the well-studied species). Peristaltic waves of contraction pass anteriorly along the body and help to push the new instar out of the old cuticle (Figure 12.4g). The moulted cuticle, also called the exuviae, comprises primarily the exoskeleton and often looks like a hollow version of the insect—the ghosts of insects past. Some internal structures, such as the tracheae, gut linings and mouthparts, are also made largely of cuticle and often these too form part of the exuviae.

After ecdysis between larval instars, the water-filled gut that aided in splitting the cuticle expands the body size and stretches the new cuticle while it is still soft and pliable, thus ensuring that the new, hardened cuticle is spacious and provides room for further growth of internal tissues. If the cuticle is membranous, as in the larvae of many holometabolous species, body size can increase during a stadium, but otherwise size increases only between instars. Finally, the cuticle becomes hardened (sclerotized or tanned) and darkened or coloured. Most insects rest after moulting until the cuticle is hardened and able to withstand the pull of muscles, which rapidly develop attachments to the new epicuticle. Note that hardening and darkening are different processes and cuticle can become sclerotized without darkening, for example, over the eyes.

Moulted exuviae are often abandoned and, in aquatic insects, they may float or drift with wind and water currents. Exuviae are commonly found in invertebrate samples, caught in the surface film or accumulated in 'drifts' on the edge of water bodies. Some insects, however, often eat their moulted exuviae (provided that they have appropriate mouthparts)

Figure 12.4 Illustration of major steps in moulting. (a) Structure of mature cuticle before moulting, (b)–(f) steps during apolysis. Ecdysis occurs between (f) and (g). See text for description.

and this may be a means of recycling proteins or other nutrients (Mira 2000). Much of the research on consumption of exuviae has focused on terrestrial insects, such as cockroaches, but we have observed that some predatory caddisfly larvae consume their exuviae and the phenomenon may be widespread. Unfortunately, this means that using analysis of gut contents to estimate diet can lead to an over-estimation of cannibalism.

12.3.3 Numbers of instars

The numbers of larval instars in aquatic insects ranges from three in some Hemiptera to 50 in some Ephemeroptera (Ruffieux et al. 1996). Having many larval instars is probably the plesiomorphic condition. Species with the fewest larval instars and the least variation in instar numbers are generally holometabolous insects. In contrast, species with the most and indeterminate number of instars (i.e. variable within species) are typically hemimetabolous. There are, inevitably, exceptions to these general patterns, such as the Hemiptera, which are hemimetabolous and typically have few larval instars. Within species, variations in instar number may be related to environmental factors (Section 12.7) or sexual dimorphisms. In some Plecoptera, for example, females are larger than males and have more larval instars (Frutiger 1987).

12.4 Metamorphosis of hemimetabolous insects

At metamorphosis, the final instar larvae of hemimetabolous insects are transformed directly into adults with functional genitalia and wings (except those that are wingless, Section 9.6). The process of adult emergence is much the same as moults between larval instars. The pharate adult increases in volume by swallowing air and/or contracting the abdominal muscles and forcing the haemolymph into the thorax and head. The larval cuticle splits longitudinally along the dorsal surface of the head and thorax, and the adult crawls out. Although the adults of hemimetabolous insects generally resemble their larvae in appearance, the degree of resemblance varies among taxa and differences are perhaps greatest when larvae and adults occupy

different environments (i.e. aquatic versus terrestrial). Some larval structures are lost at metamorphosis whereas others may be retained in the adult even though they are apparently non-functional. For example, the external gills of larvae are typically lost but remnant gills are found on the adults of some Plecoptera, Odonata, and Ephemeroptera (Štys and Soldán 1980; Zwick 2000).

Progressive development of some adult structures occurs in the larvae, such as the wing buds, which are visible and enlarge between instars, but functional, membranous wings appear only at eclosion. In general, the wings develop such that the lateral margins of the wing buds become the costal margins of the adult wing (i.e. the anterior edge of an outstretched wing). The exceptions are the Odonata in which the wing buds develop in an upright position, and the margin nearest the midline of the body ultimately becomes the costal wing margin. The genitalia typically develop at the very end of the juvenile stage and are rarely visible until the final moult to the adult. Most internal organ systems of the larvae are smaller and/or less well-developed versions of those found in the adults, growing and developing gradually during larval life. Flight muscle rudiments, for example, are present in larvae and are attached to the integument at points that correspond to the locations of future sclerites. Some such muscles may be involved in larval leg movements as well as adult flight (bifunctional muscles), whereas others remain non-functional until the adult stage.

12.4.1 Mayfly subimago and imago

Uniquely among insects, Ephemeroptera undergo hemimetabolous metamorphosis and have two life stages that are winged. The winged subimago emerges from the final larval stage, and soon moults to the adult imago stage. The transition from larva to subimago is usually referred to as emergence. The subimago is usually duller in colour than the imago, as is most evident in the subimagine wings, which appear dull or translucent compared to the shiny, transparent wings of the imago. The subimago often appears hairy, with a fringe of hairs on the outer and hind margins of the wings, and the wings and abdomen are typically clothed with microtrichia and microspines. Because of these structures, the subimago is hydrofuge (Chapter 3), a quality that makes it possible to transform from larva to subimago whilst underwater and also to emerge into the terrestrial environment without becoming entrapped by the water surface (Section 5.2). These hairs, microtrichia, microspines, etc., are shed with the submaginal cuticle, except in species in the subfamily Oligoneuriinae, which shed the subimaginal cuticle from the body but not the wings. Although subimagos are winged and superficially resemble the adult much more than the larvae, they are usually sexually immature and the transition from subimago to imago often involves reduction or loss of remnant larval features, such as mouthparts and parts of the alimentary system (Section 14.3) (Harker 1999), and maturation or expansion of adult features, such as the genitalia, caudal filaments, forelegs, and eyes, especially in males (Figure 12.5) (Edmunds Jr and McCafferty 1988). In a few exceptions, the subimago is the reproductive stage and an imago stage may be absent, such as female Polymitarcydiae and Palingeniidae in which the female subimagos mate with male imagos.

Bizarrely, limbs may be non-functional, truncated, or lost entirely inbetween the subimago and imago of some species. In *Dolania americana* (Behningiidae), for example, forelegs of adult males are intact, but the meso- and metathoracic legs are often missing or truncated (Peters and Peters 1986). Even more extreme are adults of the genera *Campsurus* and *Tortopus* (Polymitarcyidae), which have highly atrophied legs and are effectively leg-less (McCafferty and Bloodgood 1989). In females of the 'stump-legged' *Campsurus segnis*, all three pairs of legs are aborted stumps, each consisting of weakly defined segments (Morgan 1929). Similarly, middle and hind legs of the males are too short to be any use, whereas they have longer forelegs that are able to grasp females during mating. Adult activities of these short-lived species may consist of a single flight, during which mating and egg-laying occur.

The duration of the subimago and imago stages varies among species, but there are two general patterns. Most species conform to a 'long-lived' pattern, the plesiomorphic condition, with a subimaginal period of eight hours to two days, and an adult period of one day to two weeks. In the shorter lon-

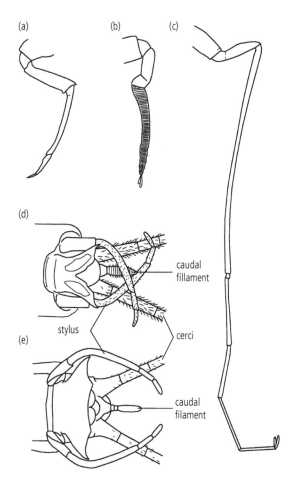

Figure 12.5 Body parts of the imago and subimago of the mayfly *Ephoron leukon* (Polymitarcyidae). Foreleg of (a) female imago, (b) male subimago, and (c) male imago. Tip of abdomen of male (d) subimago and (e) imago. Parts drawn to the same scale, to illustrate changes in length and proportion.
Source: Adapted from Ide (1937).

gevity pattern, male subimagos last only a few minutes and adults a few hours at most; female longevity is of similar duration, except in taxa where the female adult is absent and duration of the female subimago is roughly equivalent to the male imago.

12.5 Metamorphosis and emergence of holometabolous insects

A distinct pupal stage, between larval and adult stages, occurs only in the holometabolous insects. The larvae and adults of holometabolous insects are morphologically very different and it is during the pupal stage that larval tissues transform into adult tissues. The pupa is often a non-feeding, relatively quiescent stage, although many pupae can move. The pupal cuticle typically serves as a mould in which adult features develop and, consequently, the pupa resembles the adult more closely than it does the larva. The fate of the larval tissues and the development of adult features vary among taxa in accordance with how much the larva resembles the adult. In some groups, the same tissue contributes to the larval, pupal, and adult forms, as in some Coleoptera. At the other end of the spectrum, larval tissues of some taxa are degraded completely through histolysis, as in the Diptera with apodous larvae. New adult tissues are built from specific groups of undifferentiated cells called imaginal tissues or discs, which may be recognizable as early as embryogenesis, or created de novo at metamorphosis. The hormonal milieu of the larva generally suppresses differentiation of imaginal tissues, and the hormonal changes that accompany metamorphosis are an important trigger for tissue differentiation. Energy is required for metamorphosis to occur and this energy may be produced (at least in part) from fats and/or glycogen stored in the larval stages. For most insects, there is little change in net protein content during pupation, although there may be qualitative changes during the transition from larval to adult tissues. The developmental sequence during metamorphosis, biochemical changes, endocrine and genetic control have been studied intensively. Although terrestrial rather than aquatic taxa have been the focus of this work, the processes are probably much the same.

12.5.1 Types of pupae, pupal cells, and pupation habits

There are several morphological types of pupae depending, primarily, on the arrangement of limbs and mandibles. At the moult from larva to pupa, the wings and some other adult features that have been developing internally are everted and become visible externally, although not fully expanded. There are two main types of pupa. In exarate pupae, the appendages (e.g. wings and legs) extend freely from the body; in obtect pupae they are glued down by a

secretion produced at the larval-pupal moult, and obtect pupae typically are more heavily sclerotized than exarate pupae (Figure 12.6). Some pupae are capable of movement, and these may be exarate or obtect. The obtect pupae of many mosquitoes, for example, normally rest at the water surface where thoracic respiratory horns penetrate the water surface for gas exchange with the atmosphere. When disturbed or at risk of predation, however, they dive away from the surface. A further differentiation among pupae is based on whether the mandibles are articulated (movable)—movable mandibles are typically only used to escape the pupal cocoon and pupae generally do not feed. Pupae with articulated mandibles are called dectitious and are always exarate (e.g. Megaloptera, Trichoptera). Mandibles are immobile in adectitious pupae, and they may be exarate (most Coleoptera) or obtect. Whether pupal mandibles can move is functionally important as they can be used to break out of the pupal cocoon.

There is considerable variation among the aquatic insects in their pupation habits, with some larvae leaving the water to pupate on land (e.g. some Coleoptera, Megaloptera, Neuroptera) and others pupating in the aquatic medium. Some pupae are largely sedentary and typically buried in mud (some Coleoptera) or attached to the substrate (Trichoptera), whereas others are free-living and active (most Culicidae, Chironomidae, and Ceratopogonidae). Sedentary pupae are particularly vulnerable and unable to escape from predators or move in response

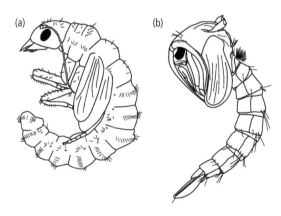

Figure 12.6 Examples of different types of pupae. (a) Exarate pupa of a megalopteran, Sialidae. (b) Obtect pupa of a dipteran, Culicidae.

to abiotic stresses and, hence, most pupate in a cell or cocoon that provides some protection. Note, however, that sedentary pupae may also be capable of movement within the pupal cell and during adult emergence.

When pupation occurs on land, a common strategy is to burrow into the ground, although some pupal cocoons are above ground. Taxa that construct pupal cells in the mud along the margins of water bodies (typically above the water line) include many dytiscid beetles and Megaloptera. To build a pupal cell, a simple behavioural sequence is followed by some beetles (Formanowicz and Brodie 1981) in which the larva first digs a hole by excavating mud pellets with the legs and head. The larva then crawls into the hole and enlarges it by wriggling and by ejecting more mud pellets grasped in their mandibles. Finally the hole is sealed from within by using the mandibles to place pieces of mud in the opening. Similarly, megalopteran larvae also dig a pupal chamber into damp, loose soil. No cocoon is spun and spines on the abdomen help keep the pupa free from direct contact with the soil (Elliott 1996). In contrast, larval spongillaflies (Sisyridae) that are ready to pupate leave the sponges they feed on, swim to the water's edge, climb out of

the water onto plants, trees, or other objects (including sides of bridges) and finally spin a silk cocoon in which to pupate. Some water penny beetles (Psephenidae) in the genus *Psephenus* pupate in the splash zone of rocks and simply attach the pupa to the rock surface (Murvosh 1971).

Among the most conspicuous, underwater, attached pupae are those of the Trichoptera. Taxa that inhabit cases as larvae simply convert the final instar larval retreat into a pupal case, thereby retaining the energy and silk protein already invested. Front and back openings to the tube or case then may be closed with a capping stone, or a perforate silken membrane, sometimes with fragments of rock, sand, or detritus (Figure 12.7), which keeps out predators yet allows circulation of water through the case for gas exchange. Non-case building caddisflies also pupate in cocoons, often with rock fragments on the outside forming a pupal cell. Within the Trichoptera there are two main types of pupal enclosure (Wiggins and Wichard 1989). In the first, pupating larvae construct a closed cocoon of parchment-like silk that is typically rigid, often brownish in colour and semi-transparent. The metamorphosing pupa is often visible through the cocoon wall. Water flows over the external surface of the cocoon, which may

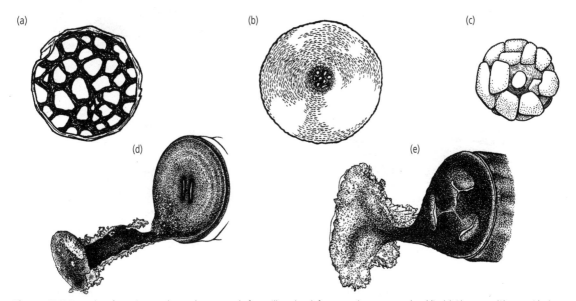

(a)

(b)

(c)

(d)

(e)

Figure 12.7 Examples of anterior pupal case closures made from silk and rock fragments by some cased caddis. (a) *Phryganea* (Phryganeidae), (b) *Eobrachycentrus* (Brachycentridae), (c) *Setodes* (Leptoceridae), (d) *Neothremma* (Uenoidae), (e) *Sericostriata* (Uenoidae).

Source: Republished with permission, from Wiggins (2004).

be semi-permeable to O_2, CO_2, H_2O, and various ions, although there is some debate about the physico-chemical properties of the cocoon (Wiggins and Wichard 1989; Weaver III 1992). The cocoon may be attached directly to the substrate, as in the Hydroptilidae (Figure 12.8a), or enclosed within a pupal cell of rock and detritus fragments glued together with silk. Spaces between the fragments allow water to enter and exit the cell (Figures 12.8b–12.8d). Within the pupal cell, the closed cocoon may be distinct from the cell walls, or attached by various devices

such as silken skirts and loops (Figures 12.8c, 12.8d). In the second main type of pupal enclosure, the cocoon is open and typically attached to the wall of the pupal cell. Open meshes or holes at each end of the cell permit water to flow into the cocoon and directly over the surface of the pupa, i.e. the pupal cuticle (Figures 12.8e, f).

Some of the more amazing feats of pupation are found in the torrenticolous taxa, such as Diptera in the families Simuliidae and Belphariceridae, where larvae and pupae occur in very fast-flowing water

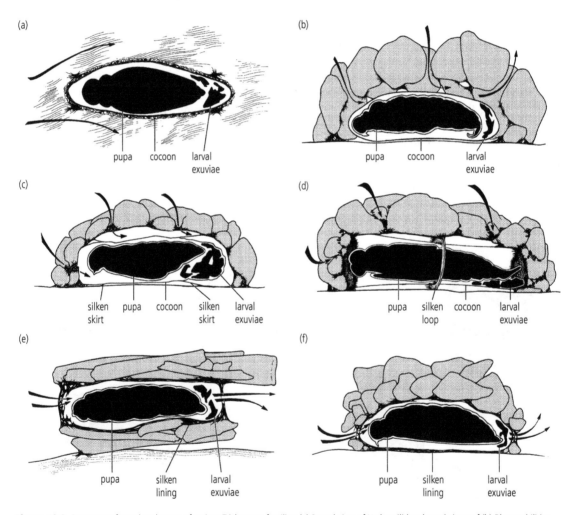

Figure 12.8 Structures of pupal enclosures of various Trichoptera families. (a) Dorsal view of Hydroptilidae; lateral views of (b) Rhyacophilidae, Hydrobiosidae, and Glossosomatidae, (c) Philopotamidae, (d) Stenopsychidae, (e) most cased caddisflies, (f) Hydropsychidae and Polycentropodidae. Arrows indicate direction of water flow.

Figure 12.9 Lateral view of a simuliid pupa inside the slipper-shaped pupal case. The pupa is oriented so that water flows from left to right, with the spiracular gills directed downstream.

(Figure 12.9). The larvae of many such species spin a pupal cocoon using silk produced by modified salivary glands, and that is firmly attached to the substrate. In the simuliids, the pointed end of the streamlined, slipper-shaped cocoon faces upstream and largely covers the pupa, but the cocoon is open at the downstream end with the anterior end of the pupa and gill filaments typically protruding into the flowing water. A complex sequence of species-specific behaviours is required to create a cocoon without the larva and subsequent pupa being washed away (Stuart and Hunter 1995; Stuart 2002). Consequently, simuliid species also differ in the shape and texture of the cocoon (proportions and patterns of coarse versus fine silk strands), and these are often useful characters for identification.

12.5.2 Prepupation and events prior to pupation

The last larval stage is often quiescent for some time (days or weeks) before ecdysis to a pupa, and the insect may be a pharate pupa for part of this time. During this period of quiescence, insects are often referred to as prepupae, although this does not usually represent a distinct morphological stage separated by a moult from the previous larval instar. The spatial distribution of final instar larvae may change if pupation sites differ from the places larvae normally occur. This may entail movements out of the water, as in the spongillaflies and many beetles, or simply a redistribution of individuals within a water body. Such behaviour can result in aggregations of pupae, as in many Trichoptera. Pupal aggregations can arise if individuals simply respond to the same environmental cues and coincidentally end up in the same locations. Some late instar caddisflies, however, congregate (gather in response to one another) without physical or chemical stimuli

(Gotceitas and Clifford 1983; Gotceitas 1985). True, congregative behaviour is rare or rarely reported in larval aquatic insects, so the significance of this behaviour is unclear.

For many insects, pupation proper begins when the larval cuticle is shed via moulting to reveal the pupa. If this occurs within the confines of a cocoon or case, then the larval exuviae typically is squashed into the posterior end of the enclosure (Figure 12.8) or, in some taxa, may be discharged. If pupae are free-living, then the moult from larva to pupa is straightforward and bodily movements allow the pupa to break free of the larval cuticle. Apolysis but not ecdysis sometimes occurs in the final instar moult such that the larval cuticle is not shed but forms the protective 'shell' within which the pupa develops (e.g. the puparium of some cyclorrhaphous Diptera).

12.5.3 Abiotic stresses, gas exchange, and pupal enemies

There is a great deal of metabolic activity during the pupal stage as tissues are transformed, and therefore pupae must be able to exchange gasses (e.g. take up O_2, expel waste CO_2) throughout this developmental period. This may be straightforward for taxa that pupate on land (provided they are not buried in mud), although desiccation or immersion may be problematic. The psephenid beetle *Psephenus herricki* pupates on the splash zone of rocks close to the water's surface and cannot tolerate drying out or complete submergence (Murvosh 1971). Similarly, a drop in water levels after pupation could be a significant source of mortality for taxa that are attached to substrates and pupate aquatically, as observed for the caddis *Agapetus monticolus* (Marchant and Hehir 1999).

Gas exchange may present challenges for taxa that pupate underwater and all the modes of gas exchange used by larvae and aquatic adults (Chapter 3) may be exploited by pupae. Indeed, many different gas exchange strategies may occur within a single family, such as the Chironomidae (Langton 1995). More rarely, some pupae have open tracheal systems, such as many Diptera that exchange gases directly with the atmosphere via prothoracic respiratory horns when at the water surface (e.g. many

Culicidae, Chironomidae, and Ceratopogonidae), or which have respiratory horns embedded in the tissues of emergent macrophytes and obtain oxygen from the aerenchyma (e.g. some other Culicidae and Ephydridae). The respiratory horns of other dipteran pupae do not reach the air–water interface and gas exchange is via a plastron on the horn and/or cuticular diffusion. Gas stores are exploited by the lepidopteran *Acentria* (Crambidae), which pupates in a two-chambered silken cocoon. The air-filled chamber in which the insect lies acts as a gas store and gases are exchanged by diffusion between the two chambers (Chapter 3). In contrast, pupae of many taxa have closed tracheal systems, specialized gas exchange structures are absent, and diffusion through the cuticle is likely to be adequate for metamorphosis.

Prepupae and pupae are vulnerable to predation and parasitism from a suite of taxa, including other aquatic insects. Chironomid predation of caddisfly pupae is common, for example, with infestation rates that can vary markedly between different populations of the same species and can be as high as 82 per cent (Rutherford and Mackay 1986; Mendez and Resh 2008).

12.5.4 Eclosion: appearance of the adult

Strictly speaking, a holometabolous insect becomes an adult immediately after apolysis of the pupal cuticle and formation of the adult cuticle. Until the pupal cuticle is shed (ecdysis), this individual is pharate, i.e. fully developed as an adult but remaining within its pupal cuticle. This distinction is important when considering the so-called pupal movements that are actually movements of the pharate adult trying to escape from the pupal cocoon and/or cell. The well-developed, articulated mandibles of dectitious pupae are moved by the pharate adult's muscles and used to cut its way out of the pupal cocoon, and the pupa moves away from its cocoon before emerging as an adult. The adult mouthparts of many Trichoptera are non-functional and the sole purpose of the adult mandibular muscles is to move the pupal mandibles at emergence, and these muscles subsequently degenerate. Similarly the pharate adults of some adectitious pupae frequently make use of special spines on the pupal

cuticle (cocoon cutters) to escape from the pupal cocoon. Pupae that are buried in underground chambers often have backward-facing spines that enable the pharate adult to wriggle out of the pupal cell and through the substrate. Pupal cells often have inbuilt lines of weakness at which the shell typically breaks. In cased caddisflies, for example, the cap used to seal the larval case at the onset of pupation (Figure 12.10) is also the weak point at which the pupal case breaks.

Only after the pharate adult has escaped the pupal cocoon and/or cell does ecdysis of the pupal cuticle and emergence of the adult typically occur. This transition is accomplished in a similar manner to moults between larval instars, with the pharate adult increasing its volume by swallowing air and contracting the abdominal muscles to force the haemolymph anteriorly. The increased pressure that results from this activity splits the pupal cuticle longitudinally down the thorax and/or the head (the ecdysial line). The pupal mouth is sealed over in obtect pupae, but the pharate adult is able to swallow air that enters the pupal case via the pupal tracheal system. The adult climbs out of the exuviae and extends its wings, limbs, antennae, etc.

When pupation occurs underwater, as in many Trichoptera, the pharate adult breaks out of the pupal cocoon and swims to the water surface, in what may be a positive phototaxic response (Morgan 1956). Pupal legs may be flattened or have a

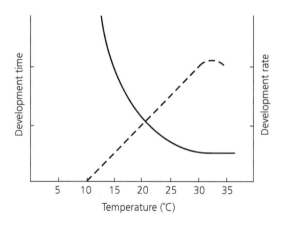

Figure 12.10 Schematic illustration of the general relationships of temperature with time to development (solid line) and with developmental rate (dashed line).

fringe of hairs that aid swimming. Eclosion occurs once the insect reaches the water surface and sometimes at the margin of an emergent object.

12.6 Habitat transition

Metamorphosis in aquatic insects typically coincides with a habitat transition from aquatic to terrestrial environments, as well as the transformation from immature to adult insect. This is a critical period during which the insect is vulnerable to predators in and out of the water. Consequently, eclosion is usually rapid (seconds to minutes) and often occurs at night, especially at dusk and dawn. This coincides with times of day that some predators are inactive (e.g. those that rely on visual cues to locate prey), although predation from bats (nocturnal foragers) and net-spinning spiders (sit-and-wait predators) may be high. It is more likely that emergence is timed primarily to optimize conditions of humidity and temperature. The cues or stimuli that trigger emergence are diverse and include abiotic variables such as temperature, photoperiod, humidity, and also endogenous circadian and seasonal clocks that control the timing of hormonal, physiological, and behavioural events over daily and seasonal cycles (Myers 2003; Danks 2005; Saunders 2010).

Newly emerged adults (teneral adults) have a soft cuticle that can be expanded by swallowing air, taking air into the tracheal system and increasing haemolymph pressure through muscular activity. This is particularly important for expanding the wings, which are typically folded and soft immediately after metamorphosis. Until the new cuticle is fully hardened and flight is possible, teneral adults commonly shelter quietly at or near the emergence site. Emergence may be highly synchronized (all individuals emerge in a very short period of time), especially if adults are short-lived and mating must occur soon after emergence, as in many Ephemeroptera. Synchronized emergence may be advantageous because it increases the probability of mating success and predator satiation (Corbet 1964; Sweeney and Vannote 1982). In many other species, males and females do not emerge at the same time: a phenomenon called protandry if males emerge first, protogyny if females emerge first. Asynchro-

nous emergence by sex may serve to maximize mating success, at least in species with short-lived adults (Butler 1984).

Metamorphosis and adult emergence occur at one of three general locations: at the air–water interface, underwater with the adult immediately exiting the water, or on land after the aquatic stage has left the water. In the latter situation, the pharate adults of some species may travel long distances terrestrially and emerge far from water, such as some stoneflies that emerge in forests and high up in trees (Hynes 1976; Kuusela and Huusko 1996). Emergence at the air–water interface is straightforward if pupae are buoyant and typically occur at the water surface, as in many aquatic Diptera such as Culicidae. If pharate adults must swim to the water surface, then a gaseous layer may form between the pharate adult and the larval or pupal cuticle. This provides buoyancy and allows the insect to float at the water surface long enough to eclose. Complete underwater emergence, as in some mayflies, requires that the adult (or subimago) is hydrofuge, rapidly swimming to the water surface and taking flight.

12.7 Environmental influences on development

Many environmental factors can directly or indirectly influence growth and development, and temperature is perhaps the most important. Other important factors include photoperiod, humidity (for terrestrial life stages), water chemistry, food resources (quality and quantity), and various biotic processes such as competition and predation. These factors are often correlated (e.g. temperature and photoperiod), so it is often difficult to discriminate between causal and covarying factors. Environmental factors will vary markedly in place and time, with the net result that insect life histories will vary between species, but also spatially and temporally within species.

As well as affecting development over the long term, environmental factors can act as cues or stimuli to initiate or terminate particular developmental events, such as egg hatch, onset of pupation, or adult emergence. Temperature is a common stimulus because there are upper and lower limits beyond which no development occurs, but other factors can also function as cues. For example, oxygen concentration can promote or prevent egg hatch: eggs of some *Aedes* mosquitoes hatch only when immersed in deoxygenated water (Clements 1992), whereas the eggs of some dytiscid beetles will hatch only in oxygenated water (Jackson 1958). Water can be an important cue in temporary water bodies, and mosquito eggs that lie dormant in dry sediments may be stimulated to hatch by wetting and/or the vibrations associated with rainfall (Roberts 2001). Not all mosquito eggs will necessarily hatch on the first wetting; multiple cycles of wetting and drying may be required for some mosquito eggs to hatch (Irving-Bell et al. 1991; Vitek and Livdahl 2009), and the hatching of one egg may stimulate the hatching of others (Edgerly and Marvier 1992).

12.7.1 Temperature

Insects are poikilothermic ectotherms (body temperature is closely correlated with environmental temperature), and heat plays a major role in their growth and development (effects of extreme temperatures are discussed in Chapter 4). For each species there are upper and lower temperature limits beyond which no development occurs. Between the limits, a rise in temperature generally will speed up metabolic processes, thereby increasing developmental rate and reducing the time required to complete development (Figure 12.10), but typically with a trade-off of smaller adult body size. Development time decreases in a hyperbolic manner. In contrast, development rate (the reciprocal of development time) is directly proportional to temperature and increases linearly within the optimal range, but begins to decrease as temperature approaches the upper lethal temperature. The development rate is zero where the line intersects the abscissa (*x*-axis) (Figure 12.10), and this is the lower temperature threshold below which there is no development. It is not especially useful to report development time in terms of calendar time (minutes, hours, days), because development is temperature-dependent. Instead, a more meaningful concept might be physiological time, or the total amount of heat required for an insect or life stage to complete development. In practice, physiological time is expressed as

degree-days (or day-degrees) above the threshold required for development, assuming unlimited food and no other physiological stresses. The degree-days required for development of a particular species and particular life stage is a constant amount, also called the thermal constant. For example, with a threshold temperature of 3 °C, if eggs completed their development in 20 days at a constant temperature of 6 °C (3 degrees above threshold × 20 days = 60 degree-days), only 15 days would be required at 7 °C (4 degrees above threshold × 15 days = 60 degree-days). The effects of temperature on development has been studied in many aquatic insects, especially during embryogenesis, with particular emphasis on hemimetabolous taxa, such as mayflies, stoneflies, and dragonflies, and some Diptera (e.g. Jackson and Sweeney 1995; Pritchard et al. 1996; Gillooly and Dodson 2000). The ecological consequences of temperature-dependent development, however, are quite complex and exceptions to hypothesized patterns are common (McKie et al. 2004).

12.7.2 Voltinism

The profound effects of temperature on development mean that an individual's lifespan and the duration of a generation may vary spatially and temporally with the climate. Many aquatic insects take a year to develop and are considered univoltine (i.e. one generation per year). If temperatures are high relative to degree-day requirements, then there may be two generations per year (bivoltine), or more (multivoltine). In contrast, generation times may exceed one year (semivoltine) and semivoltinism may be common among species inhabiting cold environments where temperatures infrequently exceed the temperature threshold required for development, in species with nutritionally poor diets, or where food is scarce. Voltinism may be indeterminate or flexible in many species, with uni-, bi-, multi-, or semivoltine populations occurring along environmental gradients such as temperature, altitude, and latitude (Flenner et al. 2010). Even within populations there may be variations in voltinism called cohort splitting. Such splitting can occur when a portion of the cohort reaches the adult stage within one or even two seasons, but the remaining proportion fails to complete development and consequently adds an additional year (or many months) to the life cycle. Cohort splitting is common and has been reported for diverse taxa, including Diptera (Carter 1980), Ephemeroptera (Svensson 1977; Lee et al. 2008), Odonata (Corbet 1957; Flenner et al. 2010), and Plecoptera (Schultheis et al. 2002; Nesterovitch and Zwick 2003). Within a generation, development may be homodynamic (continuous), with individuals progressing from egg to adult without obvious pause and, in this situation, one generation follows immediately after the other. Heterodynamic species are those which undergo one or more periods of diapause (Section 12.7.3), and this is common in uni- and semivoltine species in habitats subject to extreme environmental fluctuations.

Semivoltine life cycles are particularly interesting because these result in the simultaneous existence of multiple cohorts or generations that may not interbreed and therefore may be genetically different from one another. For example, in a nine-year study of a semivoltine mayfly, *Ephemera danica* (Ephemeridae), very poor weather during the mating and oviposition period in one year may have been responsible for the small year-class observed in alternate cohorts over the next six years (Wright et al. 1981). Cohort splitting ensures some genetic mix between cohorts (Schultheis et al. 2002) and may ensure that populations are less vulnerable to extinction from stochastic environmental events.

12.7.3 Dormancy, diapause, and quiescence

Insects have evolved a variety of adaptive mechanisms to cope with the environment when it becomes unfavourable for activity or development, including dormancy, quiescence, and diapause (Danks 1987). Following the terminology of Košťál (2006), dormancy is the most general of these terms and describes any state of suppressed development that is adaptive (i.e. ecologically or evolutionarily meaningful) and is accompanied usually by suppressed metabolism. In this broad sense, dormancy encompasses both quiescence and diapause. Quiescence is an immediate response to a decline in any limiting environmental variable (e.g. extremes of temperature, water, oxygen, food) during which the

insect typically reduces activity and physiological responses are slow. As soon as the environment becomes more favourable, quiescence ends and the insect resumes normal activity. In contrast, diapause is a genetically regulated process in which the developmental programme is interrupted and directed away from direct morphogenesis and into an alternative diapause programme with its own particular succession of physiological events (i.e. the insect may continue to change during the phases of diapause). Diapause is typically associated with periods of adverse environmental conditions, but diapause is initiated usually before the adverse conditions begin, and termination does not necessarily coincide with a return to more favourable environmental conditions. Diapause typically lasts for several months and, in a few cases, for a year or more. The resistance of insects to harsh conditions is usually increased during diapause, including cold-hardiness and desiccation-resistance (Chapter 4). Because metabolism is typically depressed, diapausing insects use less energy and nutrient reserves are conserved (Hahn and Denlinger 2007).

Diapause can happen at any life stage (embryo, larva, pupa, adult) and usually occurs in only one stage, although repeated periods of diapause may occur during the larval stage of long-lived species. Metabolic suppression may be highest in the non-feeding, immobile stages (e.g. egg, pupa), and is usually less deep in diapausing, free-living larvae and adults that are mobile.

Diapause is obligatory for some insects, such that every generation enters diapause at the same life-cycle stage and individuals terminate diapause contemporaneously. Diapause may also be facultative with some individuals completing their life cycle without any dormancy whereas others enter diapause one or more times. Facultative diapause within populations can result in cohort splitting (Section 12.7.2), for example, some individuals in the population do not enter diapause and are univoltine, whereas some others enter diapause and are semivoltine. Cohort splitting potentially increases the likelihood that at least some individuals will encounter environmental conditions suitable for survival. Similarly, not all individuals within a diapausing cohort necessarily terminate diapause simultaneously, which leads to asynchronous development and, in some cases, cohort splitting. In a life-history context, such asynchronous development can be a form of risk-spreading or bet-hedging strategies that can have profound impacts on population dynamics (Hopper 1999; Evans and Dennehy 2005) and may be common in temporally variable environments. Facultative embryonic (egg) diapause is widespread in the Plecoptera and it has been suggested that the dormant eggs, in combination with variable embryonic development rates, represent a kind of seed bank from which larvae can be recruited to the population over prolonged periods of time (Zwick 1996; Sandberg and Stewart 2004).

12.7.4 Phases and control of diapause

The processes of diapause can be divided into three main phases: prediapause, diapause, and diapause termination. Prediapause typically begins well before environmental conditions become unfavourable. Entry into diapause is a slow process, with gradual changes, not a suddenly triggered event. The environmental cues that induce diapause are diverse, but the most common, especially in temperate and polar regions, are seasonal alternations of photoperiod (day/night length) and decreasing temperature. Because diapause is often long and the insect does not feed during this time, the accumulation of energy reserves before diapause is important to the survival and fitness of the individual once diapause is terminated. The initiation of diapause itself may be difficult to determine and is often defined as the point at which direct development ceases. This often coincides with a decrease in metabolic rates. Diapause termination is also a gradual process (like diapause initiation) and usually depends on the reception of environmental stimuli. The physiology of termination often begins long before the insect shows behavioural signs that diapause is over.

The phases of diapause are controlled proximally by hormones, and ultimately by genetics. Much of the relevant research work comes under the heading of 'seasonal biological clocks' and, although insects are commonly used as study organisms, aquatic insects feature rather infrequently. The hormonal changes associated with insect diapause have

been investigated by many researchers and there is an extensive literature on the subject. Powerful modern techniques in analytical biochemistry and molecular genetics have provided the tools for examining the genetic regulation of diapause (Denlinger 2002; Saunders et al. 2004) and this is an active research area but, again, aquatic insects have yet to figure prominently.

12.8 Life histories

As discussed above, virtually all aspects of insects' reproduction and development can be influenced by the abiotic and biotic environment. The net effect of all these influences determines life-history patterns. The terms 'life history' and 'life cycle' often are used interchangeably, which is unfortunate because a distinction between the two is useful. A life cycle is a qualitative description of the sequence of morphological stages and physiological events that link one generation with the next, and the components of the life cycle are the same for all members of a species. In contrast, a life history describes the quantitative and qualitative details of the variable events associated with the life cycle and, therefore, life histories are likely to vary between individuals or populations of a single species. So, following Butler (1984), the life cycle of a mayfly includes egg, multiple larval stages, subimago and adult stage, and incomplete metamorphosis. In contrast, the life history of a particular mayfly population may include egg diapause, asynchronous egg hatching, rapid development, bivoltinism, facultative parthenogenetic reproduction, and high fecundity, among many other possible traits. The quantitative elements of life histories make them much more difficult and time consuming to describe than a simple life cycle.

There is an extraordinary diversity in the life histories of aquatic insects, many descriptive studies of species-specific life histories and also various attempts to synthesize and identify patterns (i.e. sets of species that share common sets of traits). Such syntheses typically focus on taxonomic groups or particular habitat types, and the major categories of life-history traits include development and mortality rates of various life stages, phenology (timing of egg-laying, pupation, and eclosion), diapause, voltinism, adult size, and fecundity. All these traits are affected by environmental variables such as temperature, photoperiod, quality and quantity of food resources, and predation pressure. The general motivation behind research into life-history patterns and life-history theory is to understand why such patterns arise (e.g. evolutionary and physiological processes), and also the consequences of particular patterns for population dynamics, geographic distribution patterns, etc. Information on aquatic insect life histories is abundant, although not exhaustive (Resh and Rosenberg 2010). However, species are often very flexible in their life histories and attempts to identify patterns are often fraught with difficulty, not least because the more a species is studied, the more flexible it usually appears (Clifford 1982). In contrast, this very flexibility makes aquatic insects ideal model systems for testing ideas in ecology and evolution regarding trade-offs in life-history traits (Córdoba-Aguilar 2008).

PART 5

Trophic Relationships

In Part 5, we consider how aquatic insects locate, consume, and digest their food. Food underpins many types of investigations in both basic and applied ecological research, covering aspects ranging from understanding what drives the distribution of individual species to unravelling how interactions play out between species, whether and how trophic cascades occur in food webs, and variation in the production of plant and animal biomass across ecosystems. For basic ecology then, an understanding of diet and nutrition is needed for questions that span very different levels of investigations (individuals to ecosystems). Likewise, aspects of food and diet can be highly pertinent to a range of questions in applied ecological science. Good examples include situations where insects are used in biological control—that is, where species are used to control the abundance of another (typically, an invasive, alien species). The success of such programs is hugely dependent on very detailed information about food acquisition and behaviour. Another example is where we need to understand the ramifications of human impacts that inadvertently shift the basal resources of food webs, causing effects that ripple across ecosystems and may cause quite unexpected shifts in abundances of species, such as top predators. For example, excess nutrients are added to freshwaters from run-off from fertilized paddocks and can cause algal blooms, with direct ramifications for herbivorous species that feed on algae. Shifts in the abundances of herbivores can propagate up food chains by altering the amount of food available to predatory species. Similar effects can be caused by the removal of vegeta-

tion, particularly trees, from catchment areas surrounding rivers, lakes, and wetlands. Again, such removal can cause a significant drop in the amount of terrestrial plant material falling into freshwater systems, with concurrent impacts on the species that rely on those food sources.

Investigating the diet and nutrition of insects is not as easy as it may sound. Collectively, aquatic insects eat a wide range of foods, and many individual species have broad diets that may span both animal and plant material. Diets may also shift ontogenetically, that is, as the insect develops. Even when insects have a broad diet, not all potential foods in their environment are equally available. An insect's ability to move about and detect potentially acceptable food in its environment will have an impact on its diet. Insects may switch between foodstuffs depending on availability. For example, predators may switch between prey types or even shift to eating plant material under some circumstances. Some predators actively hunt their prey, whereas others sit and wait for unsuspecting prey to pass by, with some species even using toxins or constructing nets to catch them in a manner akin to terrestrial spiders. Such behavioural differences will result in the acquisition of different species of prey. Another critical aspect is the morphology of insect mouthparts, which varies greatly between species and restricts the types of food that insects can grasp, manipulate, process, and move into their mouths. Mouthparts can vary even between species that appear to have similar ways of gathering food (e.g. roving predators) and even between species in the same genus. Such variability means lumping species into broad categories (e.g. carnivores) and taxonomic generalizations about diet are likely to be of only limited, practical utility. It is also important to recognize that some insects are non-feeding at some life-cycle stages—pupal stages are an obvious example—but the adults of many aquatic insects do not have functional mouthparts and do not feed at all. In these cases, it is the energy acquired and stored during the larval stages that will be critical to survival.

Dietary differences between species are also reflected in the morphology of the gut and how digestion proceeds. While overall gut structure is broadly similar across the insects, there are differences between species in whether they are able to cut up or grind their food prior to chemical digestion. Insects also vary in where and how enzymes are deployed to break down food, with some insects using extra-oral digestion (i.e. enzymes deployed onto food outside the mouth) and others able to begin digestion in the preoral cavity with salivary enzymes. Insects that consume wood have particular problems with digestion, because wood contains high amounts of cellulose and lignin, which cannot be digested by most animals. Wood-eating insects may have special pouches that open into the gut and that contain microorganisms (e.g. fungi, bacteria, protozoans) capable of producing the enzymes that can break down cellulose and lignin. The interactions between such microorganisms and insects lie at the foundations of understanding how food is acquired and digested when woody products are commonly consumed in the ecosystem. Finally, for all species, the ability of insects to absorb nutrients from their food while protecting themselves from toxins or an excess of substances (such as water) has ramifications for growth and development. Exposure to poor-quality food or food that is chemically protected (e.g. many plants) can slow development and reduce the fecundity of adults.

The above summary should serve to demonstrate that understanding diet and nutrition relies on knowing many basic aspects of insect morphology, physiology, and behaviour, and we introduce these in Part 5.

CHAPTER 13

Feeding devices and foraging strategies

13.1 Introduction

The food items consumed by aquatic insects are enormously diverse, and so are the various mechanisms employed to acquire food. Many ecological studies classify aquatic insects into 'functional feeding' groups, based primarily upon similarities in the way in which food is acquired and secondarily on the nature of the food consumed. Any given feeding mode can result in the consumption of several different food types and, therefore, most aquatic insects are polyphagous to some extent. In many situations, the functional feeding group classifications may be more advantageous than grouping species along taxonomic lines, because even closely related species may have very different ecologies or behaviours. As with all classification schemes, however, there are difficulties as many species do not fit neatly into any one category and diet or feeding behaviour may change with development.

The form and function of insect mouthparts and associated head structures are often a good reflection of their diet and feeding mechanisms (see Chapter 1 for description of mouthparts). The feeding strategies of insects are as diverse as the foods they consume. There are literally thousands of papers published on insect mouthparts and approximately 34 classes of general mouthpart arrangements (Labandeira 1997). Insects can be opportunistic in their diet, but only within the limits of their feeding apparatuses so any discussion of feeding strategies inevitably centres on mouthpart structure and function. Feeding usually involves the movements of multiple mouthparts, coordinated in a manner that results in food being acquired and then transported into the mouth. Modifications and variations in mouthpart morphology are nearly infinite, but some mouthparts are best suited for particular functions. For example, the heavily sclerotized and muscular mandibles are often modified for crushing or cutting; in contrast, the jointed maxillae, maxillary, and labial palps can potentially swing and rotate in multiple directions and are often important in manipulating and transporting food into the mouth.

This chapter deals mainly with aquatic life stages. The food and feeding of terrestrial stages (i.e. many adult aquatic insects) are much the same as for terrestrial insects, at least in groups with adults that actually feed. We begin by discussing, generally, the kinds of food resources potentially available to aquatic insects (Section 13.2). The rest of the chapter covers various functional feeding groups, including predators that are primarily carnivorous (Section 13.3); parasites (Section 13.4); shredders, chewers, and xylophages (Section 13.5); algal piercers/bursters (Section 13.6); grazers (Section 13.7); collector-gatherers (Section 13.8); filter feeders (Section 13.9). Much of this discussion focuses on the feeding strategies, mechanisms, structures, and apparatuses used to somehow capture food and transfer it into the mouth (e.g. mouthparts). What happens to food once it enters the alimentary system is discussed in Chapter 14.

13.2 Food of aquatic insects

The foods consumed by aquatic insects span a wide range of types. The primary distinctions are often between animal prey; 'large' plants (macrophytes, mosses); plankton and attached biofilms (often dominated by photosynthetic algae); and detritus, although insect species are not necessarily restricted to consuming food from any one class.

Aquatic Entomology. First Edition. Jill Lancaster & Barbara J. Downes.
© Jill Lancaster & Barbara J. Downes 2013. Published 2013 by Oxford University Press.

Many aquatic insects consume the tissues of other animals. Mostly they attack live prey and, compared to terrestrial insects, comparatively few aquatic insects are specialized carrion feeders that feed exclusively on the corpses of other animals. Many aquatic insects will scavenge on carcasses, such as fish (Kline et al. 1997; Chaloner et al. 2002; Walter et al. 2006) or fish eggs (Nicola 1968), but they are typically generalist feeders and scavenging bouts may be opportunistic, not the norm. Consumption of eggs (oophagey) occurs in many groups and predators may consume eggs of other insects (Section 11.4.5) or unrelated groups, such as amphibian eggs (Richter 2000). Oophagey may be opportunistic or obligate, and obligate oophages tend to have life cycles that are shorter than the egg development period of their prey, presumably to ensure that the life cycle can be completed before food runs out (e.g. Purcell et al. 2008). The typical prey of aquatic insects are other invertebrates, generally smaller than the predators, but some will consume tissues of larger prey and vertebrates such as fish fry and amphibian larvae (Caldwell et al. 1980; Formanowicz 1982; Le Louarn and Cloarec 1997), or even small snakes and turtles (Ohba 2011). Foraging cannibalism is common among aquatic insects and, although cannibalism may have important consequences for communities and population dynamics (Johansson and Crowley 2008) and provide a means for survival in suboptimal habitats (Wissinger et al. 2006), these are primarily opportunistic feeding events, not selection of conspecifics. Sexual and filial cannibalism are unrecorded in aquatic insects and sibling cannibalism appears to be suppressed in some predatory species where neonates are strongly aggregated (Hildrew and Wagner 1992; Ohba et al. 2006).

There is a long-standing view that the living tissues of macrophytes (mosses, vascular plants) are rarely used as a food source by aquatic insects. While it is true that there are proportionally far more phytophagous insects in terrestrial environments, there are aquatic insects that consume and digest tissues of vascular plants and bryophytes, and representatives can be found in multiple orders (Newman 1991). The toxic terpenoids and phenolics that are common in bryophytes are often presumed to deter would-be consumers, but moss fragments appear in the guts of many aquatic invertebrates

(Suren and Winterbourn 1991) and some species, such as caddisflies in the genus *Scelotrichia* (Hydroptilidae), appear to be moss specialists (Cairns and Wells 2008). The impact of plant feeders can be so significant that they are used as biological control agents of aquatic plants that have become pests/weeds in areas outside their normal geographical range (Cuda et al. 2008). Biological control of weeds by phytophagous insects is often unsuccessful, but introduction of the floating fern *Salvinia* in many tropical and subtropical areas has been controlled by *Cyrtobagous salviniae* (Curculionidae). The adults of this weevil feed on the plant buds, whereas the larvae feed externally on roots and tunnel through buds and rhizomes (Room 1990). Furthermore, at least one caddisfly, *Desmona bethula* (Limnephilidae), repeatedly leaves the water to feed on the aerial portions (leaves, flowers, fruits) of living aquatic and semi-aquatic plants (Erman 1981).

Phytoplankton (e.g. planktonic algae, bacteria) in the water column are the primary food resources for rather few insects and various Crustacea (e.g. copepods, cladocerans, ostracods) are probably the dominant consumers of phytoplankton in most freshwaters. In contrast, the filamentous and attached algae growing on various substrates are consumed by many different aquatic insects. The biofilms that form on surfaces are often referred to collectively as aufwuchs (or periphyton for biofilms on plant surfaces; epilithon for biofilms on stones) and include attached algae, detritus, and a suite of microorganisms (e.g. bacteria, fungi, protozoa), often enmeshed within an extracellular matrix. Algal tissues, particularly diatoms, are often the critical nutritional component of this resource in environments with plenty of sunlight. Bacteria, including methanotrophs that fix methane, may be important food sources where light and oxygen are in short supply, such as the profundal zone of lakes (Grey et al. 2004).

Detritus generally refers to the dead parts of terrestrial plants, including leaves and wood. Some of the earliest true insects probably fed on decaying plant material, and this food source is exploited widely by terrestrial and aquatic insects. Although there are many wood-boring insects in terrestrial environments, fewer aquatic insects are xylophagous, but they do exist (Section 13.5). Detritus is further divided into large and small fractions: particles

> 1 mm (maximum dimension) are classed as CPOM (coarse particulate organic matter); particles < 1 mm as FPOM (fine particulate organic matter), respectively. Other or more numerous size categories are, of course, possible and may be desirable in some circumstances (Boling Jr et al. 1975). The size of an insect and its feeding structures will ultimately determine whether particles are considered large or small by the insect. Nevertheless, distinguishing between different-sized particles is useful because insects consume large versus small particles differently: large particles must be reduced to smaller pieces that will fit into the mouth; smaller particles need to be collected in large numbers and manipulated into the mouth. Detritus is in fact a very complex food source because dead plant tissues are associated with other organic particles, such as insect faeces, and a diverse microbial community of bacteria, fungi, and protozoa. Many of these microbes are involved in decomposing the detritus, whereas others play different functional roles. Insects that consume detritus inevitably consume this microbial community also, which can increase the nutritional value of the food quite significantly (Chapter 14).

13.3 Predators

Animals that attack and consume other animal prey, when the prey are normally alive when first attacked, are typically called carnivores or predators. It is important to note that some other authors also include herbivory as a form of predation and, although this may seem peculiar, it makes perfect sense in the context of many ecological theories about consumer–resource dynamics. To confuse the issue further, many aquatic insects are not strictly carnivorous, but have somewhat omnivorous diets that may include significant amounts of algae or detritus (Lancaster et al. 2005). In this book, the word 'predator' means an insect that is entirely or predominantly carnivorous.

The feeding habits of predators can be divided into those that swallow their prey whole or in solid lumps (engulfers) and those that suck body fluids and/or the liquid products of extraoral digestion (piercers). In terms of feeding strategies, predators may be active foragers that hunt prey, or ambush

(sit-and-wait) predators that remain immobile and wait for unsuspecting prey to approach. Both strategies are used by engulfers and piercers alike, and some species may use both active and sit-and-wait strategies (Lancaster et al. 1988). Most predatory aquatic insects are polyphagous and consume a wide range of prey types, limited primarily by their size and ease of capture, although some specialization does occur. Among some larval dytiscid beetles, for example, northern European *Dytiscus latissimus* and *D. semisulcatus* prey mainly on cased caddisflies; *Acilius* spp. on microcrustacea; and *Cybister lateralimarginalis* on odonate larvae (Nilsson 1996). Piercing predators are generally carnivorous and consume only animal prey (except some semi-aquatic Hemiptera which also scavenge, and the Corixidae that consume some plant/detrital material), whereas engulfers may be strictly carnivorous or true omnivores, consuming and digesting both plant material and animal prey (Malmqvist et al. 1991; Lancaster et al. 2005).

13.3.1 Engulfers

Most aquatic insect orders have representatives that are engulfing predators, and their mandibles are typically modified for grasping and tearing prey, whereas other mouthparts, such as the maxillae, help transport the prey into the mouth. Small prey are typically swallowed whole. Larger prey are torn or bitten into smaller pieces, or seized by the rear end and engulfed whole as far as the thoracic segments or the head capsule. The head, and sometimes the thorax, is bitten off and rejected, presumably because these more heavily sclerotized body parts are more difficult to digest or process. Such partial prey consumption, or tissue-selective feeding behaviour, is likely to be widespread in aquatic insects (Winterbourn 1974; Malmqvist and Sjöström 1980) and even closely related species may differ in which tissues are consumed (Martin and Mackay 1982, 1983).

The mandibles of engulfing predators are generally elongated and sickle-shaped, often with sharp apical teeth or serrated inner margins used to grasp prey (Figure 13.1a, 13.1b). Symmetry in mouthparts is common, e.g. left and right mandibles of almost identical size and shape (Figure 13.1a), but asym-

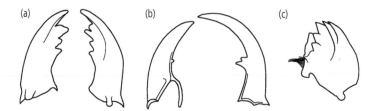

Figure 13.1 Mandibles of various larvae with different feeding habits. (a) Ventral view of mandibles of a carnivorous dobsonfly, *Archichauliodes* sp. (Corydalidae). (b) Dorsal view of asymmetric mandibles of the larval beetle *Hydrophilus triangularis* (Hydrophilidae), a specialist consumer of snails. (c) Ventral view of the left mandible of a detritivorous stonefly, *Klapopteryx kuscheli* (Austroperlidae), with stout cutting teeth and a well-developed molar region.

Source: (b) Republished from Wilson (1923), accessed from the NOAA Central Library Data Imaging Project. (c) Adapted from Albariño (2001).

metry is also common and may be pronounced in predators that specialize on certain kinds of prey. For example, some hydrophilid beetles that feed on snails have asymmetric mandibles (Figure 13.1b) (Wilson 1923; Archangelsky and Durand 1992) and snail shell chirality (left or right coiled) may determine which snail species are most likely to be consumed by which beetles (Inoda et al. 2003). Beetle larvae of the family Gyrinidae are somewhat unusual in that they capture prey with finely channelled mandibles, through which poison is injected to paralyse or kill the prey. The prey is then torn into smaller pieces and ingested. Many adult aquatic beetles are partly predators and partly scavengers, and use their mouthparts to tear at carrion. Indeed, tinned cat food is often used as bait to catch aquatic beetles—beetles detect the bait with olfactory cues (Chapter 7) so fresh, smelly bait is most effective. In most engulfing predators, the mouthparts generally do not chew the prey, and further grinding and crushing of the food into smaller pieces is typically carried out in the gut (by the proventriculus, Section 14.2.2).

Mandibles often play a dominant role in prey capture for many taxa, but other mouthparts may also be crucial. In predatory stoneflies, the galea of the maxilla have long, sharp spines at their tip that appear to be adapted for stabbing and holding prey, along with the adjacent teeth on the lacinea (Figure 13.2) (Sephton and Hynes 1983; Stewart and Stark 1988). The mouthparts of larval Odonata are unique and prey are captured using a 'mask', an elongated labium with a pair of hinged, hook-like labial palps, and various setae, spines, and teeth for skewering prey (Figure 13.3). At rest, the mask is folded at the

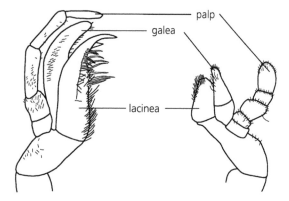

Figure 13.2 Examples of maxillae of larval stoneflies that are (left) predatory, *Stenoperla australis* (Eustheniidae) and (right) detritivorous, *Nebiossoperla alpina* (Gripopterygida).

Source: From Sephton and Hynes (1983). Reproduced with permission from CSIRO Publishing.

junction of the post- and prementum and held between the bases of the legs. It is extended very rapidly, by means of an increase in the pressure of the body fluid produced by the abdominal muscles, and prey are grasped by the labial palps.

An example of unusual predators are most species of cased caddisflies in the genus *Ceraclea* that feed exclusively on freshwater sponges (Resh 1976; Resh et al. 1976). Because sponges are sessile filter feeders, these predators do not require sharp spines or pointed mouthparts to immobilize their prey, but they do have modified mandibles that are well-suited to collect sponge fragments. Some species, such as *C. fulva*, also use the siliceous spicules of the sponge to build their cases (Corallini and Gaino 2003) and cannot complete their life cycle in the absence of the sponge.

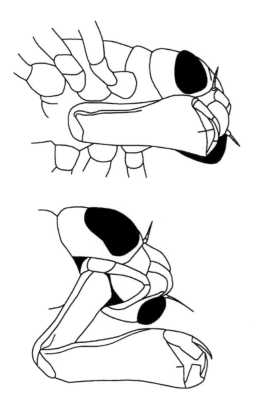

Figure 13.3 Ventral-lateral view of the labial mask of a dragonfly larva retracted (above) and partially extended (below).

Source: Adapted from Wesenberg-Lund (1943).

Among the net-spinning caddisflies are several predatory species, typically in the families Ecnomidae and Polycentropodidae, that use their nets to capture prey before engulfing them. The nets of many such species are usually built in still or slowly flowing water, or in slow-flow microhabitats around rocks or other objects, where they are not exposed to very strong currents that could damage the net (Townsend and Hildrew 1979; Edington and Hildrew 1995; Lancaster et al. 2009). These are ambush predators that construct silken nets with a tube or retreat in which the larva rests, and one or two catching surfaces that operate as snares in which prey become entangled as they move around the environment. Like web-spinning spiders, these predatory caddisflies are alerted to the presence of prey by movements of the silken threads (Tachet 1977), and then they rush out to engulf the prey. Unlike spiders' silk, the nets of caddisflies are not sticky and rely on prey getting entangled rather than glued

to the net. Some other polycentropodids, such as *Neureclipsis bimaculata* are predatory filter feeders that typically spin trumpet-shaped nets in lake outflows and catch drifting zooplankton in their nets, as well as other stream invertebrates (Petersen et al. 1984). They too tend to spin nets in relatively slowly flowing water and the size of the net opening is inversely related to velocity to minimize drag and increase filtering efficiency. Like web-spinning spiders, *Neureclipsis* larvae can attach excess prey to the silken walls of their retreats, so prey can be stored and consumed at a later time (Petersen et al. 1984).

13.3.2 Piercers

Predators with piercing mouthparts are found primarily in the Neuroptera, Coleoptera, and Hemiptera. In the Neuroptera, larvae of the Osmylidae stab prey with their long slender jaws, a salivary secretion is injected into the prey to paralyse it, and then the contents of the prey are slowly sucked out. Larval spongillaflies (Sisyridae) feed on freshwater sponges and use their long, piercing mouthparts to penetrate the sponge and suck out fluids. Larval dytiscid beetles have modified suctorial mandibles with a groove or channel, which typically lies along the inside edge. For most types of piercing predator, when prey are seized between the mandibles, digestive enzymes are injected into them and the fluids that result from extraoral digestion are sucked into the alimentary canal.

Virtually all the aquatic Hemiptera are predators with opisthognathous, suctorial mouthparts (stylets enclosed in a rostrum) and extraoral digestion. The exceptions are the family Corixidae, which lack a distinct rostrum, but possess stylets capable of piercing and sucking. Feeding habits of corixids are likely to be different from those of other water bugs, and also likely to be diverse within the family Corixidae. They do pierce and suck when feeding on prey, but they also ingest solid food, which is masticated by buccopharyngeal teeth (Savage 1989). All other aquatic Hemiptera have broadly similar, piercing-sucking mouthparts (Figure 13.4) (Cranston and Sprague 1961; Stewart and Felgenhauer 2003). The proboscis is a long, narrow cylinder (a modified labium), composed of usually three or four jointed segments, and it is folded underneath

the head when at rest (extending ventrally and posteriorly), but projects anteriorly when feeding.
Inside the proboscis are needle-like feeding stylets
that are modified mandibles and maxillae (Figure
13.4b). The maxillary stylet forms the food canal
and a smaller salivary canal. The mandibular stylets
are heavier and have a series of thick, backward-
pointing barbs at the tip whereas the maxillary
stylet has slender barbs. The mandibular barbs hold
the stylet bundle in place in the prey body while the
tip of the maxillary stylet rasps the tissues. When
feeding, the rostrum rests on the surface of the prey
and the stylets are pushed into the prey. Secretions
of the salivary glands are injected into the prey and
the products of extra-oral digestion are then sucked
into the alimentary system by a food pump.

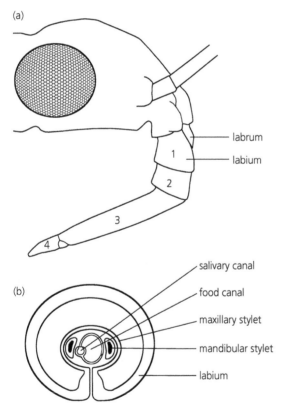

Figure 13.4 Feeding rostrum of *Gerris remigis* (Gerridae) (a) Side
view of gerrid head showing a four-segmented rostrum. (b)
Cross-section through the fourth labial segment, showing the
maxillary stylet, which forms the food and salivary canals, and the
mandibular stylets.

Source: From Cranston and Sprague (1961). Reproduced with permission from
John Wiley & Sons.

13.3.3 Raptorial or prehensile appendages

The simplest predators primarily use their mouthparts to seize prey, but some have additional raptorial or prehensile appendages that are used for prey
capture. The phantom midge *Chaoborus* is an ambush
predator of the water column of ponds and lakes.
Largely transparent and able to maintain perfect
neutral buoyancy (Section 5.3.1), it hangs almost
invisible in the water column. Prey that come too
close are seized by the prehensile antennae (Figure
13.5) and manipulated into the mouth. By being
nearly transparent, chaoborid larvae are inconspicuous to their prey and potential predators (Johnsen
2003). Unfortunately, successful feeding bouts and
the presence of prey in the gut increases visibility
and the risk of mortality from other predators
(Giguère and Northcote 1987).

Many predatory aquatic insects have forelegs
(and occasionally mid legs) that are modified to
capture, hold, and manipulate prey, and these modified legs may play no role in locomotion. Caddisfly
larvae of the family Hydrobiosidae typically have
chelate forelegs that are held parallel to the head
(Figure 13.6). Leg segments may be modified in several ways to form chelae. In some hydrobiosids,
such as *Taschorema*, the femur is produced to form
an elongate spur, often with spines on the dorsal
surface of the spur, which forms a chelate appendage with a reduced (and often fused) tibia and tarsus, and the tarsal claw (Figure 13.6d). In others,
such as *Apsilochorema*, there is no pronounced femo-
ral spur and the femur bears spines on the ventral
surface, which forms a pincer with the elongate and
curved tarsal claw (Figure 13.6a).

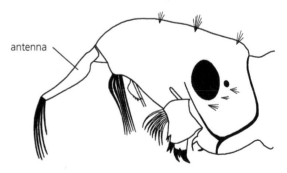

Figure 13.5 Prehensile antenna of a phantom midge larva,
Chaoborus (Chaoboridae).

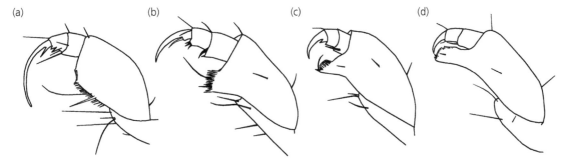

Figure 13.6 Larval forelegs of four genera in the caddisfly family Hydrobiosidae, forming a gradient from the plesiomorphic condition exhibited by (a) *Apsilochorema obliquum* to the strongly chelate foreleg of (d) *Ethochorema trubidum*, with intermediate forms of (b) *Ulmerochorema rubiconum*, and (c) *Psylobetina cumberlandica*. Morphological variations include a gradient from a field of spines on the ventral margin of the femur (the largest segment illustrated) in *Apsilochorema*, that are fused to varying degrees to form a chela at the ventral apex of the femur with a serrate dorsal margin in *Ethochorema*; and changes in length and of curvature of the tarsal claw.

Using forelegs to feed is perhaps most widespread in the aquatic Hemiptera and some species have well-developed, raptorial forelegs with a femur–tibia type of claw similar to that of praying mantises. The piercing mouthparts of the Hemiptera are less than ideal for capturing mobile prey, so the use of the forelegs to handle prey is important for this group. Predatory forelegs are functionally and morphologically different from the mid- and hind legs, and modifications commonly involve elongation of the coxa and other leg segments, alterations in the degree of rotation and junction of the leg joints, and alterations in the muscle arrangement of the prothorax (Gorb 1995). Such modifications allow the dexterous movements necessary for manipulating prey by the forelegs, whilst the mid- and hind legs are adapted primarily for locomotion and often rapid swimming (Section 8.3.1). With raptorial forelegs also comes the potential to capture more than one prey item at a time. Some water scorpions (Nepidae) can accommodate up to three prey items simultaneously: one skewered on the stylets during feeding, and one being held in the claw of each foreleg (Bailey 1985; Cloarec 1991). Such behaviour may be advantageous when prey are aggregated and the time between prey encounters may be long. Finally, most predatory aquatic insects attack and consume prey that are smaller than themselves, but a few are able to handle very large prey. Perhaps the most spectacular example is the belostomatid *Kirkaldyia* (= *Lethocerus*) *deyrolli* that specializes in eating vertebrates, such as tadpoles, frogs, fish, and occasion-

ally snakes, lizards, and turtles (Ohba and Nakasuji 2006; Ohba et al. 2008b; Ohba et al. 2008a; Ohba 2011). These bugs rely on using their raptorial forelegs and claws to hook into the prey, and also use the remaining legs to pin it in place.

13.4 Parasites

Both ectoparasites and endoparasites can be identified among the aquatic insects, and both are primarily fluid feeders. The feeding strategies of endoparasites are very similar, regardless of whether they are aquatic or terrestrial species; they live inside host tissues and take up bodily fluids. Most aquatic Hymenoptera are endoparasitic in the larval life stage. Ectoparasitic aquatic insects occur in various orders, including Hymenoptera (Bennett 2008), Trichoptera (Wells 2005), Coleoptera (Peck 2006; Whitaker Jr 2006), and Diptera (Tokeshi 1995). The hosts of these parasites are also diverse and comprise mainly other aquatic insects and invertebrates, although some vertebrates may be hosts. Identifying true ectoparasites is often problematic as some aquatic insects also live in phoretic, mutualistic, or commensal relationships (Tokeshi 1993) that may superficially appear to be parasitic. Ectoparasites generally have some means of clinging to the host (e.g. chironomids may attach their silken tubes to the host, and beaver-beetles use their forelegs to cling to the host's hairs) and those that parasitize cased caddisflies benefit, like the host, from the protection of the case. The mouthparts of ectoparasites are typically modified for piercing the host's skin and tapping into body tissues. Parasites of hosts with heavily sclerotized exoskeletons usually attach their mouthparts to the thin cuticle of gills or intersegmental membranes, which connect the harder sclerites (Giberson et al. 1996).

13.5 Shredders, chewers, and xylophages

Aquatic insects that chew, mine, or gouge plant tissues (living or dead) are generically termed shredders. The distinction between detritus and living plant tissues is important in terms of the nutritional value of the food (Chapter 14), but less important in terms of the morphological adaptations and feeding strategies employed. Mouthpart morphology of shredders is, however, diverse and can influence which plant tissues particular species are able to consume and the nutritional benefits gained (Clissold 2008). The mouthparts of shredders are often suited also for scavenging, and species that might normally be classified as shredders are often found feeding on large animal carcasses, such as some caddisflies (Kline et al. 1997).

Shredding the living tissues of macrophytes is common in the aquatic Lepidoptera and Coleoptera (Curculionidae, Donaciinae). Larvae of aquatic Lepidoptera are obligate herbivores and have stout mandibles with cutting teeth (Habeck and Balciunas 2005) to chew the tissues of macrophytes in much the same way as the caterpillars of terrestrial Lepidoptera chew terrestrial plants. The smaller particles created by chewing are then transported into the mouth by the coordinated movements of the other mouthparts. Adults and larvae of aquatic weevils are strictly phytophagous, but may differ in the way that they utilize plant food. For example, larvae and adults of *Hydrotimetes natans* feed underwater on the submerged macrophyte cacomba (*Cacomba caroliniana*), but larvae mine the stems and adults chew on the leaves and stems (Cabrera-Walsh et al. 2011); adults of *Cyrtobagous salviniae* feed on the buds of *Salvinia* whereas larvae feed on roots and rhizomes (Room 1990). Similarly, adult and larval Donaciinae (Chrysomelidae) are phytophagous with chewing mouthparts, but feed in different ways. Adult Donaciinae are primarily terrestrial but associate with the host plants and make a

characteristic pattern of holes in the leaves, whereas larvae are aquatic and generally chew roots and rhizomes.

Detritus shredders typically feed on large pieces of detritus (> 1 mm, CPOM), including wood, which they chew, mine, or gouge. The microbial assemblage associated with detritus (e.g. decomposing fungi and bacteria) are consumed at the same time. The mouthparts of shredders are generally modified for grinding. Mandibles tend to be short and stout, often with cutting teeth and a well-developed molar region for grinding (Figure 13.1c). The maxillae and labia are also often short. In shredding stoneflies, the mandibles often have well-developed prostheca (a movable process at the base of the mandible) (Stewart and Stark 1988), which transports food to the mouth. The lacinea of the maxilla is often scoop-shaped with combs of teeth, hairs, or spines (Figure 13.2), presumably for manipulating food toward the mouth, and the galea is also variously adapted for combing and raking.

Obligate xylophagy (consuming only immersed wood) is well documented in the Diptera (especially Chironomidae and Tipulidae), Trichoptera, Coleoptera, and some Plecoptera, and an even more diverse group are facultative xylophages (McKie and Cranston 1998; Hoffman and Hering 2000). The mouthparts of facultative species tend to fit the generic model for most shredders. Obligate xylophages, however, often have further adaptations for mining wood, and virtually all have strong, heavily sclerotized mandibles. Larvae of the caddisfly *Lype phaeope* (Psychomyiidae) scrape off layers of wood with their mandibles, and collect the fragments on the forelegs (Spänhoff et al. 2003). Comb-like bristles on the tarsi of the forelegs, bristles on the front edge of the labrum and hairs on the galea then pass the wood fragments into the mouth. Robust, scoop-shaped mandibles suitable for gouging occur in the xylophagous larvae of *Lara avara* (Elmidae) and are used to slice off thin pieces of wood (Steedman and Anderson 1985). Similarly, wood-eating Tipulidae, such as *Tipula* and *Lipsothrix* spp., often have broad mandibles with few teeth and a sharp ventral cutting surface that they use for cutting pieces of wood (Dudley and Anderson 1987).

13.6 Algal piercers/bursters

Some herbivorous insects pierce individual algal cells, typically those of filamentous algae, and consume the cell contents. This behaviour is seen in some adult and larval beetles in the family Haliplidae and some larval caddisflies in the family Hydroptilidae. Neither of these groups have stylet-like mouthparts that are often associated with piercing. They are small-bodied insects and cutting mandibles may be sufficient to burst the cell membrane and release the contents of the algal cell, which must then be moved into the mouth. Perhaps 'algal bursters' is a more appropriate description of this feeding strategy.

Beetle larvae of the haliplids *Peltodytes* and *Haliplus immaculicollis* have chelate forelegs with which they grasp an algal filament, such as *Spirogyra*, and pass it backwards in a hand-over-hand fashion until they reach the end (Hickman 1931). They then push the filament forward, at the same time puncturing each cell and sucking the contents up grooves in the mandibles. In contrast, *Haliplus cribarius* and *H. triopsis* do not have chelate forelegs and feed in a very different manner, often on attached *Nitella* and *Chara*. Feeding may commence on any part of the filament, larvae scrape off the outside layer of the alga with their mandibles and loosened material is sucked up through the mandibles.

Within the hydroptilid caddisflies, most specialized macroalgal eaters occur in the tribe Hydroptilini. Like some haliplid larvae, some hydroptilids also have chelate forelegs (Figure 13.7), but it is unclear whether these are adaptations for manipulating algal filaments whilst feeding, for gripping algae in swift currents, or both (Wells 1985). Many algal-bursting hydroptilids have asymmetric mandibles, with one mandible pointed or bearing one or more sharp cutting blades, and the other mandible simple or with a serrated inner edge (Figure 13.7). These are used simultaneously to grasp and puncture the cell membrane in order to release the cell protoplasm, which is then consumed. Larvae of some Hydroptilini (*Hydroptila consimilis* and *Oxyethira* sp.) move along the filaments until they encounter cells small enough to puncture (Keiper and Foote 2000; Keiper 2002). A cell is clutched between the mandibles and bitten, with the left man-

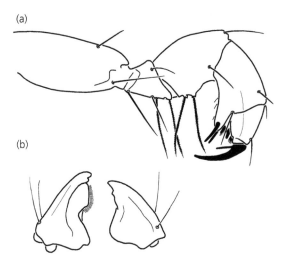

Figure 13.7 Algal-bursting caddisfly larva *Hellyethira simplex* (Hydroptilidae) showing (a) chelate foreleg and (b) asymmetric mandibles. The mandible shown on the left is equipped with cutting blades; the one on the right is more simple in form.

Source: From Wells (1985). Reproduced with permission from CSIRO Publishing.

dible maintaining a stable grip and the right mandible moving, aided by rocking the head to apply pressure on the mandible. In young larvae, multiple bites (e.g. 10–20) may be required before the pointed mandible bursts the cell and the protoplasm is released. Similar feeding behaviour occurs in *Ochrotrichia wojcickyi*, but these larvae are more generalist feeders that scrape periphyton as well as pierce algal cells, and their mandibles are less specialized than those of *H. consimilis*. Not all Hydroptilidae are algal cell bursters, however, and some other phytophagous species feed on bryophytes (Cairns and Wells 2008). Members of the genus *Orthotrichia* have rather different mouthparts, with more slender mandibles and a labrum with a tooth-shaped process, and rather different feeding habits with some species that appear to consume insect eggs (Wells 1985), and still others that are parasitic (Wells 1992, 2005).

13.7 Grazers

Grazers remove attached biofilms (i.e. aufwuchs) by pressing various brushes, rakes, combs, brooms, gouges, and excavators on the mouthparts against the substrate surfaces, and then moving these structures back and forth. Mouthpart morphology of grazers influences which algae they can consume and hence their impact on the taxonomic structure and physiognomy of algal assemblages (Steinman et al. 1987; Wellnitz and Ward 1998). Many grazing insects consume a mixture of algal filaments, attached and loose single-celled algae, and fine detritus. Fine material may be consumed coincidentally when associated with filaments or attached algae, or brushing behaviours may be aimed specifically at sweeping up fine, loose material. Inevitably, abrasion of the mouthparts on substrates results in wear and tear of feeding structures (Arens 1990), and this must reduce feeding efficiency. Repair of mouthparts is impossible, but new ones are generated with each moult. It is perhaps convenient that at least some grazers go through many larval instars before reaching maturity (e.g. many mayflies). Mouthpart wear is not peculiar to grazers, but probably occurs in all or most aquatic and terrestrial insects, and may impair feeding capacity (Chapman 1964; Massey and Hartley 2009).

Various brushing structures are common features on the mouthparts of many grazers. In a remarkably detailed study, Arens (1989) described the comparative functional morphology of the mouthparts and brush-like structures of grazing insects (Ephemeroptera, Trichoptera, Coleoptera, Diptera), in addition to some other invertebrates and fish. Many studies have focused on grazing mechanisms and it appears to be a surprisingly complex process involving three distinctly different operations: scraping-collecting, transporting, and crushing of the food. Specialized devices are required for each step and mouthparts must move in the correct sequence for efficient feeding. First, attached algae must be scraped from the substrate and placed within reach of the transport apparatus. These loosened algae and other loose particles must be collected together and transported to the mouth. Finally, algae must be crushed mechanically and separated from excess water before entering the mouth. Species in running waters have the additional problem that the mouth, or preoral cavity, must be shielded against the current to prevent loosened algae drifting away rather than entering the mouth. In grazing insects, the large headshields typical of many dorso-ventrally flattened mayflies (e.g. the Heptageniidae), hair fringes on the labrum,

or an enlarged labrum can all provide screening from the current.

In the first step, where algae are scraped or swept from the substrate, mandibles with a sharp or toothed cutting tip often serve as gouges and dredgers (common in grazing caddisflies and beetles), or as cutting tools to bite algal filaments. Other scraping apparatuses include brushes and rakes, often on the maxillae or labial palps. A brush is composed of many slender, bristle-like fine structures, usually with two or more kinds of bristle within a brush (Figure 13.8). Individually, a single bristle is unlikely to scrape much algae but, arranged in clusters or rows, they can be effective tools. A rake is a single row of fine structures serving as prongs (Figure 13.9). Loose algae may also be collected or swept up by brushes on the maxillae and on the labial palps. Once algae has been removed or swept with a scraping tool, the second step is to collect the particles and transport them to the mouth. A common solution is to trap these fine particles in fields of hairs, combs, or other tools, then clean these hairy traps and move the algae towards the mouth. Scraping brushes are often supported by brooms with long, sweeping bristles that immediately catch algae dislodged by the brushes (Figure 13.8). Combs and brushes then sweep up the algae from these collecting areas and transport it into the mouth. The hypopharynx and labium are commonly central collecting places for detached algae. In *Ameletus*, for example, the hypopharynx is a pilose pad upon which the maxillary scraping rakes (Figure 13.9b) unload algae. Third, the epilithic algae collected by the combs and structures is mixed with a lot of water and needs to be concentrated if the insect is to avoid swallowing excess water.

Getting food into the mouth is obviously an important stage of feeding, but digestion of the harvested algae is not possible until siliceous cells of diatoms have been cracked open and the cellulose walls of other algae have been torn open. Many diatoms can pass through the gut unharmed (Peterson et al. 1998) and, presumably, these have escaped crushing. Some mechanical crushing may be carried out within the gut by the proventriculus (Figure 14.3), but many grazing insects lack a grinding proventriculus. Instead, both these functions of food concentration and cell-cracking may be carried

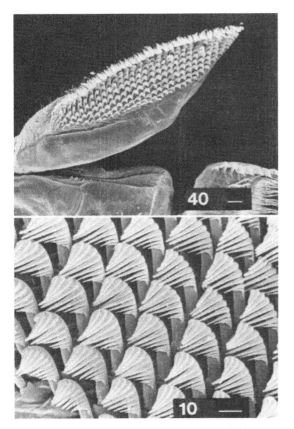

Figure 13.8 Maxilla of the larval mayfly *Rhithrogena* (Heptageniidae) showing some brushing structures on the ventral side. Top: Scraping brush on the maxillary palp made of rows of comb bristles and a fringe of pinnate hairs forming the broom on the upper edge. Bottom: Close-up of comb bristles of the scraping brush. Scale bars are in μm.

Source: Reproduced with permission, from Arens, W. (1990). Wear and tear of mouthparts: a critical problem in stream animals feeding on epilithic algae. *Canadian Journal of Zoology*, 68(9), 1896–1914. © 2008 Canadian Science Publishing or its licensors.

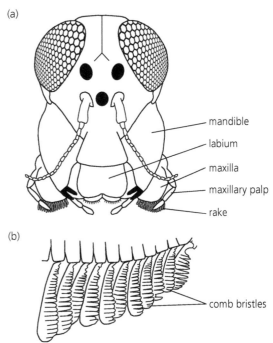

Figure 13.9 Mouthparts of mayfly *Ameletus inopinatus* (Siphlonuridae). (a) Frontal view of head showing mouthparts and rake on maxilla, (b) close-up of part of rake, made of comb bristles, on the left maxilla.

Source: Adapted from Arens (1989).

out by the ridged, molar surfaces of the mandibles, which act as a filter press to strain the algae and remove water and a masticator to break algal cells. Crushing of the algae is often aided by small grains of sand that often occur in biofilms, become mixed with the food and act as grist to the mill.

Although many inferences can be made about mouthpart function and movement sequence from morphological investigations, visual, video, or film analysis of feeding behaviour by live animals can often be the most illuminating. In some mayflies, the initial scraping–collecting phase appears to have two stages and precedes particles being

transported to the mouth for processing. For example, *Rhithrogena pellucidula* (Heptageniidae) first has a labial brushing cycle in which the labial palps sweep up loose material and this is followed by a maxillary scraping cycle in which brushes on the maxillary palps scrape attached material (McShaffrey and McCafferty 1988). Various combs then remove particles from the brushes and transport them to the mouth. In contrast, *Ephemerella needhami* (Ephemerellidae) first has a maxillary brushing cycle in which loose material (often on the surface of filamentous algae) is swept up by the galinea–lacinia of the maxillae and deposited on the hypopharynx and labium. This cycle is followed by a mandibular biting cycle in which the mandibles bite off sections of algal filament, which are then transported to the mouth (Figure 13.10) (McShaffrey and McCafferty 1990). Any one species, however, may display more than one feeding cycle depending on the food source available. For example, when feeding on fine detritus the mayfly

Cloeon dipterum uses primarily the labial palps and laciniae in food collection, and rarely the mandibles. For the same species, however, the mandibles are used much more when feeding on filamentous algae (Brown 1961).

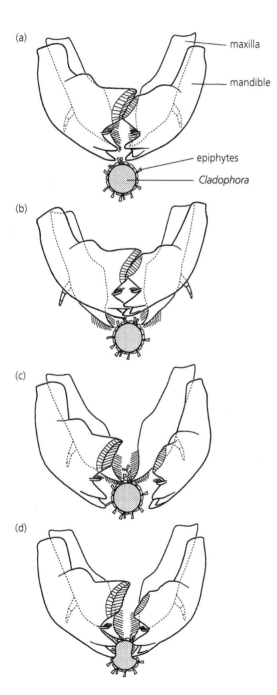

(a)
— maxilla
— mandible
— epiphytes
— *Cladophora*

(b)

(c)

(d)

13.8 Collector-gatherers

Collector-gatherers are also referred to as deposit feeders and they typically feed on fine organic particles (< 1 mm, FPOM), algae and their associated microorganisms, that are sedimented or deposited on various surfaces. The brushing mouthparts typical of some grazers are often well-suited to feeding on deposited particles and, inevitably, there is some overlap in the diet and feeding strategies of different species. For example, the mayfly *Stenacron interpunctatum* (Heptageniidae) appears to do little scraping and feeds predominantly by brushing up loose material with brushes on the labium and maxillae, or by using the brushes to filter fine particles entrained in the current (McShaffrey and McCafferty 1986). Small-bodied collector-gatherers, such as many larval Chironomidae and small-bodied stoneflies such as the Leuctridae, pick up individual particles or clusters of particles. Indeed, most chironomids are collector-gatherers at some point in their larval development (Berg 1995). Like grazers, the food of collector-gatherers is often mixed with a lot of water when it reaches the mouth and the mandibles may act as a filter press and masticator to concentrate and mechanically crush the food. The tube-dwelling chironomid larvae of *Microtendipes pedellus* have an unusual but effective collecting behaviour (Walshe 1951). The larva extends out of

Figure 13.10 Schematic illustration of the maxillae and mandibles of the grazing mayfly *Ephemerella needhami* (Ephemerellidae) at various stages during brushing and biting cycles, as the mayfly feeds on a strand of filamentous alga (*Cladophora*) coated in epiphytes (e.g. diatoms). In this illustration, the maxillae lie behind the mandibles. (a) Resting position of the mandibles and maxillae in a non-feeding mayfly. The algal filament is aligned parallel to the long axis of the body and may have been manipulated into position by the labial palps. (b) The head has moved forward and the maxillae have opened to begin brushing. (c) The maxillae brush epiphytes off the algal filament with spines and setae on the tip of the maxillae, while the mandibles open. Food particles are deposited on the lingula and superlingulae. (d) Start of the mandibular biting cycle. The mandibles bite the algal filament, which will eventually be forced into the mouth. Each cycle lasts 0.3–0.5 s at room temperature.

Source: Springer-Verlag (McShaffrey, D. and McCafferty, W.P. (1990). Feeding behavior and related functional morphology of the mayfly *Ephemerella needhami* (Ephemeroptera: Ephemerellidae). *Journal of Insect Behavior*, 3, 673–688.). With kind permission from Springer Science and Business Media.

its case, spreading a salivary secretion with its anterior proleg and then withdraws into the tube dragging the secretion and adhering particles with it.

13.9 Filter feeders

Filter feeders are analogous to collector-gatherers in that they feed on FPOM, but they collect particles that are suspended in the water column rather than those lying on substrates (Wallace and Merritt 1980). In running waters, filter feeders tend to be somewhat sedentary, passive feeders that allow the current to sweep particles into some device that intercepts the flow. In contrast, filter-feeding insects inhabiting standing waters tend to be either mobile in the water column and actively sweep particles into the mouth (e.g. mosquito larvae) or sedentary, tube-dwellers that filter particles from the currents created through their tubes (e.g. some Chironomidae). Some filter-feeding tube-dwellers also venture partially or completely out of the tube and collect particles from the surrounding substrate and, therefore, may also be classified as collector-gatherers (Berg 1995).

Like collectors, filter feeders are often somewhat indiscriminate in the materials they consume, provided that the particles are small (i.e. well below 1 mm). Anything small enough to get caught on the filtering device is likely to be consumed and, thus, they typically consume a mixture of algae, fine detritus, meiofauna, bacteria, and associated microorganisms. Consequently, the composition of their diet is mixed and often dictated by the availability of materials rather than any dietary choices.

13.9.1 Filtering with body parts

Among the aquatic insects of standing waters, some mosquito larvae are often described as filter feeders (also called suspension feeders or plankton feeders) (Merritt et al. 1992a). The mouthparts are covered in setae arranged into various combs and sweepers called mouth brushes. Movements of these brushes, often accompanied by pharyngeal contractions, create currents and eddies around the mouth and small particles are carried into the mouth. Some argue that this is not true filter feeding, as the brushes act as paddles and do not necessarily sieve particles out

of the water (Merritt et al. 1992b), see also Section 5.5.2.

In running waters, the simplest method of intercepting drifting particles is with fringes of hairs, often equipped with microtrichia or setae, and that are held up into the current. Periodically, particles are cleaned off the hairs by passing them over the mouthparts, which often also have brushing devices. These insects require some means of avoiding dislodgement and maintaining their position in the face of flow forces that can be considerable (Chapter 5). Filtering fringes may be on the mouthparts, as in mayflies of the genera *Tricorythus* and *Murphyella*, or on the forelegs, as in mayflies of the genera *Oligoneuriella, Isonychia,* and *Coloburiscoides* (Wallace and O'Hop 1979; Braimah 1987a; Eplers and Tomka 1995), with anchoring provided by the mid- and hind legs. The long hairs on legs used for filter feeding may superficially resemble swimming hairs, but thrust for swimming is usually provided by the hind legs (Chapter 8), whereas filtering hairs are typically on the fore or mid-legs. As with grazing insects, feeding involves a set of coordinated movements of mouthparts and filtering devices. In *Oligoneuriella rhenana* (Oligoneuriidae), for example, the forelegs with filtering structures are spread away from the head to capture particles, then moved under the headshield where brushes on the labial and maxillary palps comb particles off the legs and into the reach of the galea–laciniae, which pass food particles into the mouth (Eplers and Tomka 1995). A similar filtering system is used by some caddisflies, such as *Oligoplectrum* and *Brachycentrus*, which anchor their cases to the tops of stones, spread hair-fringed legs (mid- and hind legs) into the current to catch particles, which are then scraped off by the forelegs and eaten. Filter feeding is well studied in many larval Simuliidae (Currie and Craig 1987), which are attached to the substrate via a circlet of posterior hooks anchored in a sticky, silk pad, and intercept particles using a pair of cephalic or labral fans (modified mouthparts) (Figure 13.11). Particles retained on the fans are removed by mandibular brushes and setae on the labrum when fans are retracted into the cibarial cavity. The maxillae then clean the mouth parts and pass particles into the gut. (Note: only some larval simuliids are filter feeders; others lack fans and are scrapers or predators.)

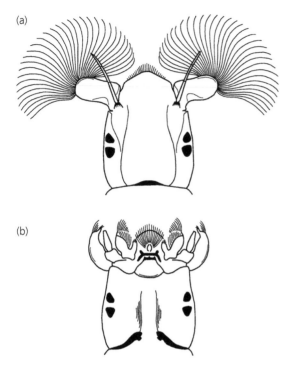

Figure 13.11 Head of filter-feeding simuliid, showing (a) dorsal view with extended labral fans and (b) ventral view (labral fans removed).

Source: Adapted from Miall (1903).

13.9.2 Filtering with tubes, burrows, and nets

Insects that live in tubes or burrows can also filter-feed, as seen in many Chironomidae (Oliver 1971; Berg 1995). Using salivary silk secretions, chironomid larvae construct catchnets across their tubes, which often project above the substrate surface. In running waters, the current carries food into the net (e.g. *Rheotantytarsus*); in standing waters, larvae use body undulations to create a current through the tube, from which they filter particles, often by having particles stick to the sides of the tube (e.g. *Chironomus*). Larvae ingest all or part of the net and its contents and then rebuild the net. Mayflies in the family Polymitarcyidae live in burrows well below the sediment surface (e.g. *Ephoron*, *Tortopus*) or in wood (e.g. *Povilla*). Their burrows are ventilated by currents created by beating abdominal gills, particles are filtered from the water on forward-facing filtering brushes on the forelegs and mandibles, and transferred to the mouth by the maxillary palps (Hartland-Rowe 1953).

The most elaborate filter-feeding structures are built by caddisflies. Perhaps the simplest are those of the family Philopotamidae, whose sausage-shaped nets are composed of many superimposed layers of mesh, forming a dense felt that catches tiny particles (Wallace and Malas 1976; Edington and Hildrew 1995). Food particles (mainly diatoms and detritus) that collect on the inner wall of the net are gathered by downward sweeping movements of the brush-like labrum.

The Hydropsychidae are perhaps the master net-spinners among freshwater insects, typically spinning their nets in fast-flowing water, and they are among the most frequently studied filter-feeders. Net design varies in relation to the immediate microhabitat, but typically consists of a tube, where the larva sits, and a large vestibule, one side of which is roughly perpendicular to the flow, supported by pieces of stone, twigs, etc., and which filters particles (Hynes 1970; Edington and Hildrew 1995). The larva holds itself in position using the anal claws, the mid-, and hind legs. The forelegs are held close to the head, which moves back and forth across the catch net. Particles are picked off with the mandibles and forelegs, edible particles are passed into the mouth and inedible particles rejected, suggesting some capacity for dietary selection. The hydropsychid catching net is made of a regular mesh, which varies between species and instars, and with various environmental variables such as temperature and current velocity. Within species, some variations or imperfections in net meshes occur naturally, but the frequency of these anomalies often increases with pollution (Petersen and Petersen 1983). Perhaps the most unusual hydropsychid net is that constructed by *Macronema*, which is embedded in the substrate and functions much like a pitot tube to draw water through the net (Hynes 1970).

13.9.3 Feeding using filters

The fluid mechanics of how insects filter feed has been discussed in Section 5.5.3 and here we will consider only the implications for feeding and diet. Geometry and dimensions of the filtering structure are important in determining the particle size range that can be intercepted, and many fans or hairs are branched. The sizes of pores or spaces between filaments is clearly important in determining the sizes of particles that can be caught. The filtering devices of most insects, however, do not act as simple mechanical sieves and they commonly capture particles much smaller than the pore spaces (Wotton 1976; Braimah 1987a; Brown et al. 2005). The fact that a single strand of hair placed in a river will accumulate fine particles is evidence that particle capture is not just a case of sieving. Once a particle reaches the surface of a filter, whether it is retained (i.e. captured) depends on several physical processes (Rubenstein and Koehl 1977). One hypothesis, for example, is that attachment efficiency could be increased if the filter surface is sticky. Some simuliids and culicids produce a mucus-like substance in glands in the labrum and at the base of the maxillae, but it does not appear to aid in capture of food particles (Fry and Craig 1995; Fry and Craig 1996). The function of this substance is unclear, but it may facilitate ecdysis or the formation of mouthparts during pharate stages.

Because flow has such a strong impact on filtering efficiency, flow patterns can affect locations of individuals over a few millimetres on single substrate particles (Whetmore et al. 1990; Lacoursière 1992; Hart et al. 1996), and the presence of other filter feeders can decrease or even increase feeding efficiency (Cardinale et al. 2001). Further, because filtering devices differ among species, we typically see species-specific distribution patterns within streams (Alstad 1982; Ross and Merritt 1987; Tachet et al. 1992; Zhang and Malmqvist 1996; Malmqvist et al. 1999). However, the small-scale relationships between flow and feeding efficiency or density of individuals is often weak because, in part, there is considerable phenotypic plasticity in the structure of hydropsychid filtering nets (Loudon and Alstad 1992; Plague and McArthur 2003) and simuliid labral fans (Zhang and Malmqvist 1997; Lucas and Hunter 1999), and even their behaviour and body posture during feeding (Hart and Latta 1986; Hart et al. 1991). Thus, individuals can modify their feeding apparatuses and behaviours to accommodate different flows, yet maintain feeding efficiency.

CHAPTER 14

Diet, digestion, and defecation

14.1 Introduction

The foods consumed by aquatic insects are enormously varied, and Chapter 13 illustrates how the diverse range of mouthparts and food capturing strategies is associated with diet. The structure and function of the gut system also reflect the mechanical properties and nutritional composition of the food eaten. The most important aspects influencing gut structure are whether the food is solid or liquid, and plant or animal in origin (Dow 1986). The diets of some species are simple to categorize according to these criteria, yet others may be generalist feeders such as, for example, some caddisfly larvae that consume both plant and animal material (Lancaster et al. 2005), and some Corixidae that consume both solid and liquid foods (Savage 1989). Some insects, particularly the holometabolous taxa, will change diet at different life stages, e.g. many larval Trichoptera are detritivorous, but switch to plant pollen and nectar as adults. There are many examples among the terrestrial insects of fluid-feeding on both plant fluids (sap-suckers) and animal fluids, but relatively few aquatic insects feed on plant liquids (but see algal bursters in Section 13.6). It is noteworthy that the gut system of most aquatic grazers and shredders is more similar to the relatively simple gut found in generalist terrestrial insects (e.g. cockroaches), than the gut of terrestrial insects that consume solid plant tissues (e.g. caterpillars of Lepidoptera). This is attributable, at least in part, to the structural and nutritional differences between freshwater algae and terrestrial plants, and to the added nutrition from the microfauna and flora that is often associated with detritus in aquatic systems.

This chapter begins with a general description of the structure and functions of the main parts of the insect alimentary system (Section 14.2), an arrangement that is common to most aquatic and terrestrial species. Larvae and adults of the same species often have very different diets and, accordingly, their guts differ morphologically and functionally. These differences are perhaps most extreme where adults are short-lived and feed little or not at all, such as the mayflies (Section 14.3). Whether the guts served any function in non-feeding adults is the subject of some debate. Not all the food consumed by an insect is digested or absorbed and the undigested material is excreted, along with wastes of various metabolic processes (Section 14.4). Lastly, we consider the digestive processes and how these differ with the nutritional quality of food, the potential roles of microorganisms in digestion, and the potential impacts of secondary plant compounds (Section 14.5).

14.2 Structure of the alimentary system

The gut is normally a continuous tube between the mouth and anus, and peristaltic muscle contractions move food through the gut. The rate at which food moves through the gut varies with many factors, including the insect's physiological state, age, kind of food, and temperature. There are three main gut regions (Figure 14.1): the foregut (stomodeum), midgut (mesenteron), and hindgut (proctodeum), with sphincters controlling the movement of material between adjacent regions. The anterior end of the foregut connects with the preoral cavity. The different regions of the gut are modified to serve particular functions: the foregut is the site of ingestion, storage, and grinding; the midgut is the site where most digestion and absorption occurs; the hindgut is where water, salts, and various other molecules

Aquatic Entomology. First Edition. Jill Lancaster & Barbara J. Downes.
© Jill Lancaster & Barbara J. Downes 2013. Published 2013 by Oxford University Press.

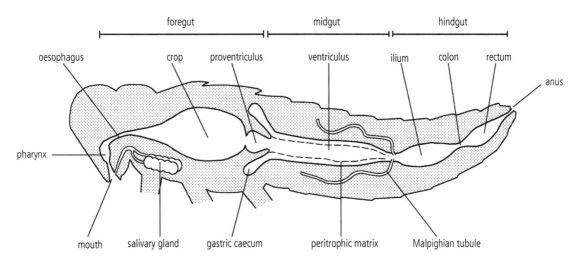

Figure 14.1 Schematic illustration of the overall structure of the alimentary tract.

are absorbed before faeces and associated wastes are ejected through the anus. The gut epithelium is one cell-layer thick and rests on a basement membrane, which is surrounded by a variably developed muscle layer. The fore- and hindgut are both lined with cuticle, which is shed at each moult.

Gut morphology is closely related to diet, particularly with respect to whether the food is solid or liquid, and plant or animal in origin. Insects that consume solid foods typically have wide, straight, short guts with strong musculature and a gut lining that is modified to provide protection from abrasion. Insects that consume liquids typically have long, narrow, convoluted guts to allow maximum contact of the food with the gut walls, whereas protection from abrasion is less important. Liquid feeders often have special adaptations to deal with the excess water. If food is low in nutritional value, it is also often very abundant (e.g. terrestrial plant tissues), but insects must process large amounts of material. Their guts are often short, have relatively little storage area (foregut), and a relatively large midgut to maximize exposure to digestive enzymes and nutrient uptake. In contrast, a nutritionally rich diet (e.g. animal tissues) may be available only sporadically (e.g. predators and blood-suckers often have large but infrequent meals), and the foreguts of these insects typically have large storage capacity relative to the midgut (Figure 14.2).

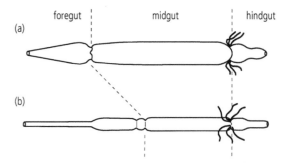

Figure 14.2 Relative sizes of the fore-, mid-, and hindgut regions of late instar larvae of two caddisflies: (a) *Halesus radiatius* (Limnephilidae) which is primarily detritivorous, (b) *Polycentropus kingi* (Polycentropodidae) which is carnivorous.

Source: Springer Verlag (Sangpradub, N. and Giller, P.S. (1994). Gut morphology, feeding rate and gut clearance in five species of caddis larvae. *Hydrobiologia*, 287, 215–223). With kind permission from Springer Science and Business Media.

14.2.1 Preoral cavity

Food first enters the alimentary system via the mouth, which opens into the preoral cavity (sometimes called the buccal cavity) and it is into this cavity that the salivary glands release their products. The preoral cavity is bounded by the mouthparts and is often divided into an upper cibarium and a lower salivarium (Figure 14.3). Fluid-feeding insects often have a cibarial pump in which there are prominent dilator muscles attached to the cibarium.

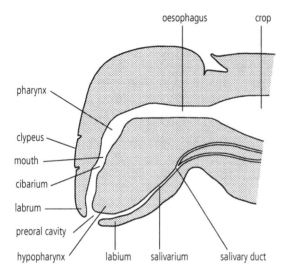

Figure 14.3 Schematic illustration of the preoral cavity, indicating different regions of the alimentary tract.

Paired salivary or labial glands are present in most insects, but they vary enormously in form and function. Typically, saliva is a watery, enzyme-containing fluid that dilutes and lubricates the food, and adjusts its pH and ionic content. The most common salivary enzymes are for digesting carbohydrates, but some predators have protein- and/or fat-digesting enzymes. In blood-feeding insects (e.g. adults of many aquatic Diptera), the saliva typically has no enzymes, but may contain anticoagulants and blood-thinning agents. Saliva and digestive enzymes may be exported to the food while it is outside of the body (extra-oral digestion), and then the food is sucked up in a liquid form, as in the predatory Hemiptera and larvae of some dytiscid beetles. Extra-oral digestion is common among the insects (Cohen 1995) and increases digestive efficiency by reducing the intake of undigestible material. In some species, the salivary glands have taken on functions unrelated to digestion. Among the aquatic insects, perhaps the most well-known example is the larval caddisflies that produce a proteinaceous silk in the salivary glands, which in some species may be so big that they occupy much of the body cavity (Glasgow 1936; Engster 1976b). Note: some other aquatic insects can also produce silk, but not necessarily in modified salivary glands. For example, larvae of some mayflies in the family Polymictarcyidae produce silk using the Malpighian tubules (Hartland-Rowe 1958; Molineri and Emmerich 2010).

14.2.2 Foregut

The lining (epithelium and associated cuticle) of the foregut is often folded longitudinally, and these folds stretch when food is pushed down the foregut. The foregut is typically divided into the pharynx, the oesophagus and the crop (a food storage area), and there is sometimes a grinding organ called the proventriculus, or gizzard, in insects that ingest solid food. Dilator muscles attached to the pharynx are especially well-developed in sucking insects and form the pharyngeal pump. Many insects, however, have some sort of pharyngeal pump, typically used for drinking by adult aquatic insects. The oesophagus is usually narrow, but dilated posteriorly to form the crop where food is stored. Some digestion may occur during storage in insects whose saliva contains enzymes or that regurgitate digestive fluid from the midgut, but there is no absorption across the foregut lining as the cuticle lining makes it impermeable. The proventriculus may serve various functions, including acting as a valve to regulate the rate at which food enters the midgut, a filter for separating liquid and solid components of the food, or as a grinder to break up solid material. Consequently, the structure of the proventriculus is very varied (Gibbs 1967). When used as a grinder, the proventriculus may contain heavily sclerotized teeth, ridges, and spines (Figure 14.4). Solid foods may be fragmented but identifiable whilst in the foregut, and subsequently ground to a pulp by the proventriculus. Thus, many ecological studies which examine gut contents to describe insect diet, often examine only the contents of the foregut if a grinding proventriculus is present.

14.2.3 Midgut

In most insects, the midgut is the principal site for secretion of digestive enzymes, for digestion and absorption, and the midgut lining is highly permeable. Typically, the foregut projects somewhat into the midgut and, in many insects, blind-ending, lateral diverticula called 'gastric caeca' arise at or near

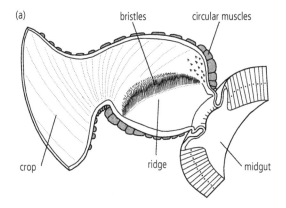

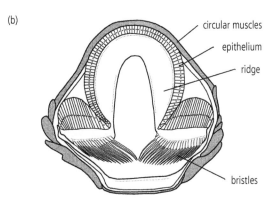

Figure 14.4 The larval proventriculous of the caddisfly *Plectrocnemia conspersa* (Polycentropodidae), showing (a) a longitudinal section, and (b) transverse section through the middle of the proventriculous. The proventriculous is a small, roughly spherical chamber displaced dorsally and surrounded by very large, circular muscles. The epithelium has two folds with greatly thickened cuticle forming ridges (oriented ventro-lateral and dorso-lateral) that are covered in stiff bristles.

Source: From Gibbs (1967). Reproduced with permission from John Wiley & Sons.

the anterior end of the midgut (Figure 14.1). These may be sites where digestive enzymes are produced or sites of significant absorption of digestive products. A thin sheath, the peritrophic matrix (often called the peritrophic membrane, even though it is not cellular in structure) provides spatial separation of digestive processes; it generally separates the midgut epithelium from the food and protects it from abrasion or pathogen attack (Hegedus et al. 2009). The peritrophic matrix consists of a network of chitin fibrils in a matrix of protein and glycoprotein, but has many pores that provide passage to all but the biggest molecules. The matrix is produced more or less continuously;

it often extends into the hind gut and sloughed-off fragments may be associated with faecal pallets as they are excreted. Sponge-eating caddisfly larvae of the genus *Ceraclea* appear to have an extra thick peritrophic matrix, perhaps to prevent the sharp, siliceous spicules of the sponge from damaging the underlying epithelium (Corallini and Gaino 2001). The peritrophic matrix may play a role in sequestering and possibly detoxifying ingested toxins such as DDT, as illustrated by the observation that copious amount of the matrix, laden with the insecticide, were excreted by mosquito larvae that were resistant to DDT (Abedi and Brown 1961). On the other hand, ingested heavy metals may prevent normal formation of the peritrophic matrix in some aquatic larvae (Abedi and Brown 1961). The main tubular part of the midgut, the ventriculum, lacks any structural differentiation, but functional differentiation often occurs and different regions may have cells that are specialized for digestion or absorption of particular materials. In larval *Aedes* mosquitoes, for example, the anterior midgut is the site of fat absorption and storage, and the posterior region absorbs carbohydrates.

14.2.4 Hindgut

The entry of the Malpighian tubules into the gut usually denote the start of the hindgut (Figure 14.1). The Malpighian tubules are effectively the insect kidney; they filter the haemolymph and produce urine, which is released into the hindgut. The pylorus lies at the most anterior end of the midgut and its well-developed muscle layer (pyloric sphincter) regulates the movement of materials from the mid- to the hindgut. The ileum is typically a narrow tube that directs undigested food to the rectum for final processing. In many xylophagous insects, the ileum is dilated to form a fermentation pouch housing endosymbiotic bacteria or protozoa capable of digesting the cellulose and lignin found in wood. Pouches off the hindgut occur in some xylophagous aquatic insects (Figure 14.5), such as *Tipula* and *Lipsothrix* (Tipulidae) (Sinsabaugh et al. 1985; Dudley and Anderson 1987), but not in others, such as *Lara avara* (Elmidae) (Steedman and Anderson 1985). The products of digestion, when liberated by the microorganisms, may be absorbed across the

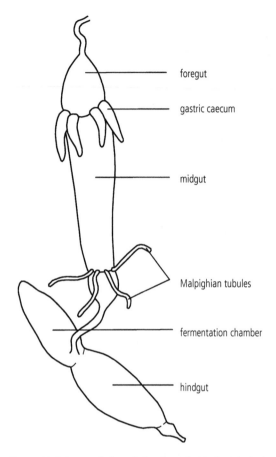

Figure 14.5 Gut morphology of a larval cranefly, *Tipula abdominalis* (Tipulidae) showing a rectal lobe on the hind gut, described as a fermentation chamber.

Source: Republished with permission of The Ecological Society of America, from Sinsabaugh et al. (1985).

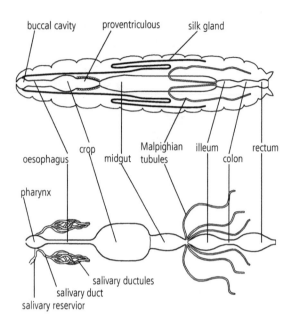

Figure 14.6 Dorsal of the gut morphology of a larval (top) and adult (bottom) caddisfly in the genus *Cheumatopsyche* (Hydropsychidae). The larvae are omnivorous filter-feeders; adults are likely to consume mainly pollen and nectar. The adult crop is often dilated with air. In the larva, the very prominent silk glands (modified salivary glands) are shown lateral to the alimentary canal for clarity; in the thoracic region they normally lie ventrally (cf. Figure 14.1). The larval silk ducts enter the labial spinneret ventral to the buccal cavity. In the adult, the two salivary ducts unite below the pharynx and pass into a small salivary reservoir, which opens on the underside of the hypopharynx.

Source: Adapted from Korboot (1964).

ileum wall. The rectum lining is mainly thin, but includes thick-walled rectal pads that absorb water, ions, and small organic molecules. The contents of the hindgut are generally fluid when they enter the rectum, but many terrestrial insects produce dry faeces, which illustrates the effectiveness of this absorption capacity.

14.3 Gut structure and function of non-feeding insects

The gut structure described in the previous sections refers generally to larvae (and some adults) that feed. Within a single species, however, larvae and adults often have very different diets (e.g. detritivo-rous larvae and blood-sucking or nectar-feeding adults) and have correspondingly different gut morphologies (Figure 14.6). The adults of some aquatic insects are renowned for being short-lived and feeding very little or not at all.

The mayflies are perhaps the most extreme example, and complex changes occur in gut structure and function during the transition between life-cycle stages (Pickles 1931; Harker 1999). Functional mouthparts are lost at the end of the larval stage and gut structure undergoes extensive changes after each moult to subimago and to imago. The structure of the imaginal gut, however, suggests that it performs some active function, if not in feeding. The midgut of the imago is gas-filled and undergoes rhythmical waves of contraction, which is an energetically expensive activity for a non-feeding animal. During the transition from larva to adult, the midgut becomes inflated with air and enlarged to occupy a large part of the thorax and abdomen (Figure 14.7a). A consequence of this expanding midgut is a compression of the hindgut (Figure 14.7b). For example, the hindgut of *Cloeon dipterum* occupies abdominal segments 7 to 10 in the larva, but is compressed and pushed into segments 9 and 10 in the imago (Harker 1999). Air enters the midgut through the mouth and perhaps indirectly from the tracheal system. The peritrophic matrix of the midgut is lost and the epithelial cells change form, but appear to remain functional in some way. Once larvae cease feeding, food remnants are not cleared from the gut system but, in association with remnants of the peritrophic matrix, they form a plug at the junction of the mid- and hindgut, which prevents any air escaping in a posterior direction. The mouth is open and lacks any closing mechanism, but a one-way valve at the foregut–midgut junction allows air to enter the midgut but not to escape in an anterior direction.

The function of this air-filled midgut in adult mayflies is unclear. It has been suggested to have an aerostatic function as well as playing a role in sperm ejaculation by males. Further suggestions are that males can change their specific gravity by altering the degree of gut dilation and that this leads to the rise and fall patterns seen in the dancing flight of mating swarms (Pickles 1931). Evidence to support these hypotheses, however, is weak. Alternative, plausible explanations include (1) the suggestion that the inflated gut reduces the volume of the haemocoel and channels the haemolymph close to body tissues, and the gut contractions move the haemolymph and metabolites around the body (Harker 1999). As adults cannot feed, fat reserves are the main source of energy for flight (Sartori et al. 1992), but most fat reserves of mayflies are in the abdomen not the thorax where the flight muscles are located. Gut contractions force the haemolymph to circulate and thus

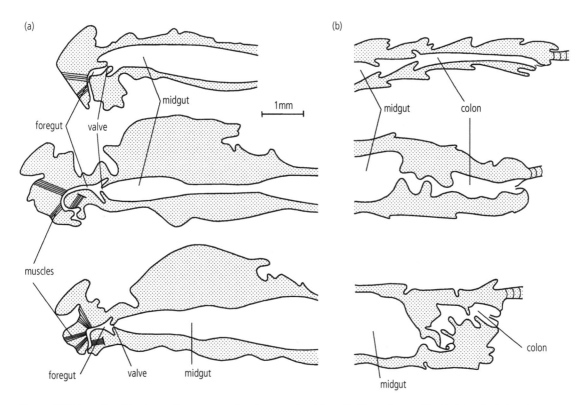

Figure 14.7 Illustration of the changes in gut structure from larva to male adult mayfly, *Cloeon dipterum* (Baetidae). Longitudinal sections through the (a) anterior and (b) posterior region of the gut of the final larval instar (top), subimago (middle), and imago (bottom).

Source: From Harker (1999). Reproduced with permission from John Wiley & Sons.

increase the rate at which fuel is delivered from the abdomen to the thoracic flight muscles, and this may be particularly important for species where males fly more than females. (2) An inflated gut ensures full abdominal distention (a bit like the rigidity found in a fully inflated balloon) and may facilitate flying. Mayflies that were prevented experimentally from inflating their guts had shriveled abdomens (a bit like the limpness of a deflated balloon) and had difficulty flying (Funk et al. 2008). (3) Intriguingly, an air-filled gut may assist in oviposition in some parthenogenetic mayfly species (Funk et al. 2008) (Section 11.3.3).

14.4 Excretion and defecation

Waste materials that must be removed from the insect body include undigested food and the waste products of metabolism. The distinction here is between food materials that pass directly from the mouth to the anus (sometimes essentially unchanged structurally or chemically) and excretory wastes that arise from metabolic processes that occur in cells of the body. In practice, insect faeces, in liquid or pellet form (frass), contain both undigested food and metabolic excretions. All insects must conserve certain ions, particularly sodium, potassium, and chloride, that may be metabolically important but limited in the insects' food resources or easily lost through diffusion into water. These ions enter the urine passively and must be actively reabsorbed if they are to be retained. Two processes, therefore, are involved in the production of waste products, excretion and osmoregulation (maintenance of body fluid composition). Osmoregulation is dealt with in detail in Chapter 4, where it is particularly important in relation to the chemical composition of the water in which aquatic insects live. Here we focus primarily on the elimination of wastes.

14.4.1 Excretory organs and nitrogenous wastes

The Malpighian tubules, in concert with the rectum and/or ileum of the hindgut, are the main organs of excretion and osmoregulation, as discussed in Chapter 4. These are long, thin tubules that float free in the haemocoel and join the gut at the mid–hindgut junction (Figure 14.1). The Malpighian

tubules produce the primary urine, filtered from the haemolymph, which is composed of nitrogenous wastes, water, sugars, salts, and amino acids. Once in the hindgut, reabsorption of water and certain metabolites and solutes (especially potassium, K^+) may occur. Waste products and excesses of useful substances are voided with the faeces. There is, of course, enormous variation among species in the structure and function of Malpighian tubules and the hindgut, variations that may be associated with differences in diet. Some species have simple tubules that are undifferentiated along their lengths, other may have complex tubules with distinct regions that differ in structure and function.

Although some nitrogen is stored in the fat body or the haemolymph, most insects excrete nitrogenous metabolic wastes in at least some life stages. Many insects, such as predators or blood-suckers, will consume far more nitrogen than they need, and their nitrogenous excretions may be quite significant. Many aquatic insects excrete nitrogen primarily in the form of ammonia (Staddon 1955, 1959; Gardner et al. 1983), which is the simplest form of waste but it is relatively toxic. Ammonia, however, is readily diluted in water and water conservation is unimportant for most aquatic insects. For terrestrial insects that need to conserve water, wastes are chemically more complex and usually consist of uric acid or urates (salts of uric acid), which are non-toxic and can be excreted essentially dry.

14.4.2 Defecation

An enormous amount of literature considers the way that insects acquire and digest food, but much less attention is given to the process of defecation (Weiss 2006). Insects that are fluid feeders generally eliminate the excess fluid through the anus and there are no solid wastes. Insects that consume large amounts of solid food, especially plant material that is nutritionally poor, eliminate a steady supply of frass, which consists primarily of undigested material. Engulfing predators may also produce large amounts of frass if their prey have sclerotized exoskeletons or if large amounts of plant material are consumed at the same time. If an insect's diet changes ontogenetically, then so will the characteristics of its waste.

Among the aquatic insects that produce solid faecal material, there is some variation in the shape and nature of the faeces, and the texture varies with the kind of food consumed (Ladle and Griffiths 1980), presumably also in relation to the grinding potential of the mouthparts and proventriculus. In the well-formed, cohesive pellets of some species, such as the Odonata and simuliid Diptera, faeces are enclosed in a sac composed of the peritrophic matrix, which is periodically delaminated from the midgut. Other species, such as the detritivorous larvae of some caddisflies, produce pellets that are much more friable and likely to disintegrate quickly.

There is some discussion in the ecological literature about faecal pellets as a potential food source for aquatic insects in general (Wotton 1980; Wotton et al. 2003) and some evidence that coprophagy may be important for particular species (Petersen and Wiegert 1982). In contrast to terrestrial insects (Weiss 2006), however, there is less information on whether aquatic insects have particular behavioural or morphological adaptations related to waste disposal or whether any make use of their own faeces for defense or other functions. Some larval chironomids, however, may use their own faecal pellets during tube construction (Ladle and Griffiths 1980). In larval dragonflies, the faecal pellet is voided promptly and vigorously in a form of ballistic ejection. Prompt ejection may be beneficial to avoid interfering with respiration, given that oxygen uptake in dragonflies is via rectal gills (Chapter 3). Pellets of the libellulid *Erythemis simplicicollis* (Libellulidae) may be projected as far as 19 cm vertically and 60 cm horizontally (Tennessen and Painter 1994), and those of *Anax imperator* (Aeshnidae) may travel at least 30 cm out of the water (Corbet 2004), but the possible function of this ballistic behaviour is unclear. Shelter-dwelling or site-faithful caterpillars of many Lepidoptera also distance themselves from their frass by various means, such as ballistic ejection, 'turd hurling', and 'butt flicking', apparently to remove any cues that frass may provide to predators (Weiss 2006). Given that larval dragonflies are often sedentary, ballistic frass ejection may also reduce the risk of predation, but this hypothesis requires testing.

14.5 Nutrition, digestion, and absorption

The nutritional quality of putative food resources for aquatic insects depends as much upon the physico-chemical characteristics of the food as on the capacity of the insect to digest and absorb the products of digestion. Estimates of assimilation efficiency (*AE*) or approximate digestibility is often used to describe the balance between food quality and digestive capability and, in its simplest form, assimilation efficiency is the mass difference between the food ingested (ingesta) and the faecal material (egesta), often expressed as a proportion or percent of the ingesta:

$$AE = (I - F) / F$$

where I = dry mass of food consumed, F = dry mass of faeces produced. Other nutritional indices (Waldbauer 1968) include growth efficiency (*GE*, or efficiency of conversion of ingested food), and metabolic efficiency (*ME*, or efficiency of conversion of digested food):

$$GE = B / I \text{ and } ME = B / (I - F)$$

where B = dry mass gain of the insect. Performance is expressed in terms of the relative rates (gram per gram) of consumption and growth. The utility of these nutritional indices depends, of course, on being able to quantify ingesta and egesta. This is problematic for fluid feeders and for species that produce watery or friable faeces, and assessing their nutritional efficiency is much more difficult.

14.5.1 Food quality

Important characteristics used to assess the nutritional quality of putative food resources include carbon-to-nitrogen ratios (C:N ratios), chemical components (lipid, protein, carbohydrate), structural chemicals (e.g. cellulose, cuticle), and toxins (allelochemicals or secondary plant compounds). These factors may influence the choices that insects make when selecting food, and also the ability of the insect to digest and assimilate the food consumed. Some insects do self-select dietary components from a range of food options (Waldbauer and Friedman 1991) and experiments on various terrestrial insects have shown that different species select different ratios of protein:carbohydrate, which appear to correspond to species-specific patterns of growth and development. There is ample evidence that aquatic insects select and reject certain food items, but whether this self-selection results in particular nutrient ratios is unclear.

14.5.2 Enzymes and food absorption

As mentioned earlier, digestion may be initiated by enzymes in the saliva, but most digestion is dependent on enzymes secreted by the midgut epithelium. Most digestion goes on in the midgut, but some insects may regurgitate substances (including enzymes) from the mid- to the foregut and, in these situations, there may be digestion in the foregut. Many different digestive enzymes are produced by insects and the enzymes are usually classified by the substrates upon which they act, e.g. carbohydrases, proteinases, and cellulases. Within these groups, however, enzymes may be specific to particular molecules or parts of molecules. For example, the proteinases produced in the midgut may be endopeptidases, which split proteins into polypeptides; or exopeptidaeses, which split individual amino acids off the polypeptides. The enzymes produced by various insect species reflect qualitatively and quantitatively the normal constituents of the diet. For example, carnivores produce mainly lipases and proteinases; omnivores produce carbohydrases, lipases, and proteinases. If diet changes with life stage, then the digestive enzymes produced will also vary. Enzyme activity, and hence digestion, is strongly influenced by pH, buffering capacity, and redox potential of the gut. These may differ markedly between different parts of the gut and between species, and there appear to be no general patterns with respect to diet.

The majority of food absorption occurs in the midgut and to a lesser extent in the hindgut. The hindgut is primarily the region of water and ion resorption in connection with osmoregulation (Chapter 4), but may be important for absorption of small organic molecules in insects with endosymbionts in the hindgut. Most organic molecules move passively across the gut wall and into the haemolymph, i.e. along a concentration gradient, although special carriers may facilitate rapid movement of some materials. Once in the haemolymph, molecules can reach most organs and tissues, where they again cross cell walls to be used in metabolic processes.

For generalist feeders and insects that consume a wide range of potential food items, it can be difficult to determine which food items are actually digested and absorbed (i.e. which ones contribute to the insect's nutrition), and which pass through the gut undigested. Techniques such as natural abundance stable isotope analyses can be useful in estimating the relative contribution of some major food types to insect diet, such as animal prey versus algae in omnivorous predators (Lancaster et al. 2005) and terrestrial detritus versus algal resources in generalist feeders (Junger and Planas 1994). Stable isotope analyses are, however, much less able to distinguish between food resources of a similar type (e.g. between prey species that are all algal grazers) and assessing the diet composition of their consumers requires different, more sensitive approaches. Molecular detection techniques (Symondson 2002; Sheppard and Hardwood 2005; King et al. 2008) provide a range of alternatives, including enzyme electrophoresis (Giller 1986), immunological methods using mono- and polyclonal antibodies (Reynolds and Scudder 1987; Reilly and McCarthy 1990), and more recently developed DNA-based techniques (Zaidi et al. 1999; Morales et al. 2003).

14.5.3 Microorganisms and digestion

A diverse array of microorganisms (bacteria, fungi, protozoa) may be present within insect bodies, often in symbiotic relationships (Gibson and Hunter 2010; Ferrari and Vavre 2011). Some occur within cells and can be vertically transmitted from mother to offspring, whereas others are extracellular and are more typically transmitted horizontally (between insect hosts) or acquired by the insect from the environment. The potential roles these microorganisms can play in an insect's biology are also fantastically diverse (e.g. the bacterium *Wollbachia* is well-known as a reproductive parasite) and only some aid digestion. At time of writing, this is a very active research area (energized at least in part by new molecular techniques), but aquatic insects are rarely subjects for investigation, except for some disease vectors.

Microorganisms that are in the gut (usually extracellular) often form symbiotic or mutualistic interactions with the insect host and can play a role in food digestion, producing essential vitamins for the host, and keeping out potentially harmful microbes (Dillon and Dillon 2004). Those that aid in digestion or nutrition, are common in insects that have nutritionally suboptimal diets (Douglas 2009), and seem to be absent from the guts of predators. In some species, gut-inhabiting microorganisms may be permanent residents (microbial symbionts) or ingested and egested with feeding activities. One important role that microorganisms can play is in providing chemicals that the insect cannot synthesize, but that are essential to metabolism, such as sterols (required for moulting hormone) and carotenoids (used in visual pigments), and thus they must be provided in the diet or by microorganisms. A second role for gut microorganisms is in digesting lignocellulose in many plant- and wood-eating insects, which typically lack cellulases (enzymes to break down cellulose). Microorganisms may be present but play no apparent role in other species. For example, the harpellates are obligate, symbiotic trichomycete fungi that colonize the digestive tracts of arthropods, including larval black flies (Simuliidae) (Lichtwardt 1986; McCreadie et al. 2005), but the 'host' rarely seems to benefit.

For insects with diets that contain a lot of detritus from terrestrial plants or woody material in general (i.e. cellulose and lignin of structural plant tissues), there are several mechanisms by which they may acquire nutrients from their food (Steedman and Anderson 1985). Many of these mechanisms rely on the fact that detritus and woody material are typically consumed along with a suite of microorganisms (fungi, bacteria, protozoa) that decompose the detritus within the aquatic environment. (1) Insects may absorb molecules liberated from the wood by the enzymes of microorganisms before the wood was ingested or whilst wood is in the gut, provided the enzymes remain active. In this situation, neither the wood nor the fungi and bacteria are digested. (2) Enzymes produced by the insect may digest the microorganisms that are consumed in association with the detritus, but the woody material itself is not digested. (3) Fungal enzymes that are consumed by the insect may somehow be sequestered or retained within the insect to digest the consumed wood, and the insect then absorbs the products of this digestion. (4) Symbiotic microorganisms (typically bacteria and protozoa) may be kept in the gut to digest wood. (5) Insects may produce their own cellulase to digest wood. While cellulase production may be common in terrestrial insects, it appears to be absent or rare among aquatic insects. In practice, many insects may use a combination of these mechanisms and it may be difficult to distinguish between them.

Mechanisms (1) and (2) are often difficult to separate, and some insects may use a combination of both. Acquired enzymes, as in mechanism (1), is common in insects that consume fungi in association with detritus, and the ingested fungal enzymes can remain active and continue to digest the plant tissues when inside the gut (Martin 1987, 1992). In the gut, ingested fungal enzymes can continue to digest cellulose in the detritus and increase the efficiency of food utilization, at least a little, in some larval stoneflies and caddisflies (Martin et al. 1981; Sinsabaugh et al. 1985). Enzymes are not sequestered by the insects, and a constant supply of detritus with associated microorganisms is required. In some other insects, however, degradative activity of consumed microorganisms may be halted by the gut milieu, for example by high pH (approx. pH 10) in the midgut of some craneflies *Tipula* (Tipulidae) (Martin et al. 1980; Sinsabaugh et al. 1985; Canhoto and Graça 2006). For mechanism (2), detritivores that are poorly adapted biochemically to exploit woody detritus itself may have enzymes (carbohydrases, lipases, proteinases) that are able to digest the microorganisms, as seen in detritivorous stoneflies and beetles (Martin et al. 1981). Larvae of the beetle *Lara avara* (Elmidae) consume wood near the surface of logs, but have no cellulase and no diverticula or fermentation chamber in the hindgut. Instead, they appear to use a combination of (1) and (2) to digest the products of the microbial assemblage, and the assemblage itself when it is consumed coincidentally with the detritus (Steedman and Anderson 1985).

Symbiotic microorganisms, mechanism (4), are common in terrestrial xylophages, such as termites, and similar processes occur in some aquatic insects. Craneflies in the genera *Tipula* and *Lipsothrix*

(Tipulidae), for example, live in galleries within submerged logs that are created as they consume the wood. They have a hindgut that includes a fermentation pouch (Figure 14.5) in which the symbionts, such as filamentous bacteria, are housed (Martin et al. 1980; Sinsabaugh et al. 1985; Dudley and Anderson 1987) and they digest a portion of the insect's food that is diverted into these caeca. The composition of the symbiotic bacterial assemblage is typically different from that found associated with the consumed detritus (Klug and Kotarski 1980). Cellulases remain bound in these bacteria, but the products of the enzymatic digestion of the wood are released into the insect gut. While bacteria in caeca may be capable of digesting woody material, the location of the caeca in the hindgut is posterior to the main sites of absorption (i.e. the midgut), and the hindgut may have only limited capacity to absorb the products of bacterial digestion. This has led to the suggestion that some xylophagous species may consume their own faeces and reap the benefits of bacterial digestion through re-ingestion (Anderson et al. 1978).

Unusually, symbiotic bacteria have been found in the posterior region of the midgut of a xylophagous chironomid, *Xylotopus par*, and a tipulid, *Epiphragma* (Kaufman et al. 1986). This is unusual for at least two reasons: first, because bacteria are usually found in the insect hindgut, except in some termites, and second, because the bacteria were beneath the peritrophic matrix and attached to the gut epithelium, even though this matrix is thought to exclude microbes from the gut tissue. While it is tempting to think that these bacteria play some nutritional roles, their location below the peritrophic matrix would preclude contact with ingested food. The function of this unusual relationship remains unknown.

14.5.4 Plant chemical compounds

Along with the nutritional challenges of a herbivorous diet (e.g. low nitrogen content), potential herbivores may be repelled by the chemical defenses of plants, which adversely affect insect behaviour and/or physiology by reducing food digestibility or by poisoning. Although chemical defenses have been well studied in terrestrial plant–insect interactions, they also occur in an aquatic context and

chemical defenses occur in macrophytes (Ostrofsky and Zettler 1986; Newman 1991; Newman et al. 1996; Prusak et al. 2005), as well as various algae and cyanobacteria (Fink 2007). Plant chemicals that have adverse effects on insects are assumed to be defensive. Alternatively, they may have, or once have had, other metabolic functions or may simply be metabolic waste products that coincidentally have adverse effects on consumers. Such substances are often called secondary plant compounds, allelochemicals, or noxious phytochemicals. They are a diverse group of chemicals, including phenolics (such as tannins), terpenoid compounds, alkaloids (as found in floating-leaved water lilies, Nymphaeacea), cyanogenic glycosides and glucosinolates (as found in watercress, *Nasturtium officinale*). In terrestrial systems, some very specific plant–insect interactions revolve around phytochemicals, such as the monarch butterflies and various milkweed bugs that are able to sequester the toxic cardiac glycosides from their food plant (milkweed plants) and reuse these chemicals for their own anti-predator defenses. Such specific interactions have not been documented in aquatic insects, and the chemicals of aquatic plants and algae appear to be broad-spectrum defenses against a wide range of enemies. Similarly, most aquatic insects that consume live tissues of macrophytes are polyphagous and will feed on several plant species. Many phytophagous insects actively avoid plants that contain allelochemicals (Newman 1991), suggesting the presence of chemosensory (taste) organs (Chapter 7). If allelochemicals are consumed, the midgut epithelium seems to be the main target organ for the toxic effects of plant chemicals, which makes sense given its role in absorption and uptake of nutrients.

Some aquatic insects are able to feed on macrophytes containing phytochemicals, but often suffer reduced growth rates, smaller size at maturity, or fewer individuals reaching reproductive maturity, compared to the same insect species fed plants without phytochemicals. For example, lepidopteran larvae of *Acentria ephemerella* (Crambidae) grew faster on *Potamogeton perfoliatus* than on *Myriophylum spicatum*, and this was attributed to the negative effects of the hydrolyzable tannins (polyphenols) in *M. spicatum*, not to any differences in plant nitrogen content (Choi et al. 2002). However, *Acentria* feeds naturally on *Myriophylum* (Gross et al. 2002) and the antimicrobial activity of polyphenols may protect these tannin-adapted larvae from bacterial and fungal pathogens, thus compensating for reduced growth (Walenciak et al. 2002). The physico-chemical gut conditions of *Acentria* appear able to hydrolyze and/or oxidize the ingested polyphenols rapidly, thus significantly reducing the concentration of toxic tannins in the gut and, presumably, reducing uptake of those toxins (Gross et al. 2008). The allelochemcials (including flavonoids) of *Elodea nuttallii*, however, have a much larger negative effect on growth of *Acentria* and larvae normally avoid this plant (Erhard et al. 2007).

Many algae and cyanobacteria also have noxious toxins and these are periodically of great concern to water quality managers, because they can be toxic or simply unpleasant tasting to humans, and bad news for aquaculture (Watson 2004; Watson and Ridal 2004). The potential allelopathic function of these chemicals, however, is poorly understood. In limnetic systems, grazing by zooplankton can be responsible for the release of volatile organic compounds and there is some evidence that these compounds have a repellant function and/or inhibit growth of zooplankton (review in Fink 2007). There is little information on whether aquatic insects are negatively affected by allelochemicals from algae.

Once plant tissues are dead, plant compounds can have no further benefit to the plant, but may have negative effects on consumers. Some compounds will be translocated within the plant before senescence or abscission; others leach rapidly from plant tissues once they enter the water. Insect shredders that normally avoid consuming live macrophyte tissues will consume the same tissues shortly after senescence, but before a microbial assemblage has colonized the tissues, or if the chemical defense system has been shut down experimentally (Suren and Lake 1989; Newman et al. 1996). Other compounds, such as tannins, may persist in plant tissues long after they die and enter the water, and their toxicity to shredders depends on age and state of decay of the plant tissue (i.e. concentration of toxin remaining), as much on the levels of detoxifying enzymes in the insect (David et al. 2000). For example, differences among species in their capacity to metabolize leaf litter toxicants may explain, in part, spatial distribution patterns among larval mosquito taxa in subalpine breeding sites (Tilquin et al. 2004).

References

Abbot, J. C. and Stewart, K. W. (1993). Male search behavior of the stonefly, *Pteronercella badia* (Hagen) (Plecoptera: Pteronarcyidae), in relation to drumming. *Journal of Insect Behavior*, 6, 467–81.

Abedi, Z. H. and Brown, A. W. A. (1961). Peritrophic membrane as vehicle for DDT and DDE excretion in *Aedes aegypti* larvae. *Annals of the Entomological Society of America*, 54, 539–42.

Abellán, P., Gómez-Zurita, J., Millán, A., Sánchez-Fernández, D., Velasco, J., Galián, J., and Ribera, I. (2007). Conservation genetics in hypersaline inland waters: mitochondrial diversity and phylogeography of an endangered Iberian beetle (Coleoptera: Hydraenidae). *Conservation Genetics*, 8, 79–88.

Addo-Bediako, A., Chown, S. L., and Gaston, K. J. (2000). Thermal tolerance, climatic variability and latitude. *Proceedings of the Royal Society, London B*, 267, 739–45.

Ahlroth, P., Alatalo, R. V., and Suhonen, J. (2010). Reduced dispersal propensity in the wingless waterstrider *Aquarius najas* in a highly fragmented landscape. *Oecologia*, 162, 323–30.

Aiken, R. B. (1982). Sound production and mating in a waterboatman *Palmacorixa nana* (Heteroptera: Corixidae). *Animal Behaviour*, 30, 54–61.

Aiken, R. B. (1985). Sound production by aquatic insects. *Biological Reviews*, 65, 163–211.

Aiken, R. B. (1992). The mating behaviour of a boreal water beetle, *Dytiscus alaskanus* (Coleoptera, Dytiscidae). *Ethology Ecology & Evolution*, 4, 245–54.

Aiken, R. B. and Khan, A. (1992). The adhesive strength of the palettes of males of a boreal water beetle, *Dytiscus alaskanus* J. Balfour Browne (Coleoptera, Dytiscidae). *Canadian Journal of Zoology*, 70, 1321–4.

Albariño, R. J. (2001). The food habits and mouthpart morphology of a south Andes population of *Klapopteryx kuscheli* (Plecoptera: Austroperlidae). *Aquatic Insects*, 23, 171–81.

Alcocer, J., Escobar, E. G., Lugo, A., Lozano, L. M., and Oseguera, L.A. (2001). Benthos of a seasonally-astatic, saline, soda lake in Mexico. *Hydrobiologia*, 466, 291–7.

Alcock, J. (1994). Postinsemination associations between males and females in insects: the mate-guarding hypothesis. *Annual Review of Entomology*, 39, 1–21.

Alexander, L. C., Hawthorne, D. J., Palmer, M. A., and Lamp, W. O. (2011). Loss of genetic diversity in the North American mayfly *Ephemerella invaria* associated with deforestation of headwater streams. *Freshwater Biology*, 56, 1456–67.

Allan, J. D. and Feifarek, B. P. (1989). Distances travelled by drifting mayfly nymphs: factors influencing return to the substrate. *Journal of the North American Benthological Society*, 8, 322–30.

Alstad, D. N. (1982). Current speed and filtration rate link caddisfly phylogeny and distributional patterns on a stream gradient. *Science*, 216, 533–4.

Altner, H. and Loftus, R. (1985). Ultrastructure and function of insect thermo- and hygroreceptors. *Annual Review of Entomology*, 30, 273–95.

Amano, H., Hayashi, K., and Kasuya, E. (2008). Avoidance of egg parasitism through submerged oviposition by tandem pairs in the water strider, *Aquarius palaeum insularis* (Heteroptera: Gerridae). *Ecological Entomology*, 33, 560–3.

Amédégnato, C. and Devriese, H. (2008). Global diversity of true and pygmy grasshoppers (Acridomorpha, Orthoptera) in freshwater. *Hydrobiologia*, 595, 535–43.

Andersen, N. M. (1976). A comparative study of locomotion on the water surface in semiaquatic bugs (Insecta, Hemiptera, Gerromorpha). *Videnskabelige Meddelelser fra Dansk Naturhistorisk Forening*, 139, 337–96.

Andersen, N. M. (1982). The semiaquatic bugs (Hemiptera, Gerromorpha). Phylogeny, adaptations, biogeography, and classification. *Entomonograph*, 3, 1–455.

Andersen, N. M. (1995). Cladistic inference and evolutionary scenarios: Locomotory structure, function, and performance in water striders. *Cladistics*, 11, 279–95.

Andersen, N. M. and Cheng, L. (2004). The marine insect *Halobates* (Heteroptera: Gerridae): Biology, adaptations, distribution, and phylogeny *Oceanography and Marine Biology: An Annual Review*, 42, 119–79.

Anderson, B. R. (1980). Scanning electron mocroscope study of the mesothoracic and metathoracic scolopophorous organs of *Lethocerus* (Belostomatidae, Heteroptera). *Journal of Morphology,* 163, 27–35.

Anderson, D. T. and Lawson-Kerr, C. (1977). The embryonic development of the marine caddis fly, *Philanisus plebius* Walker (Trichoptera: Chathamidae). *Biological Bulletin,* 153, 98–105.

Anderson, J. (1987). Reproductive strategies and gonotrophic cycles of black flies. In K. C. Kim and R. W. Merritt, eds. *Black Flies: Ecology, Population Management, and Annotated World List.* pp. 276–93. The Pennsylvania State University, University Park.

Anderson, N. H., Sedell, J. R., Roberts, L. M., and Triska, F. J. (1978). The role of aquatic invertebrates in processing of wood debris in coniferous forest streams. *American Midland Naturalist,* 100, 64–82.

Anderson, R. C. (2009). Do dragonflies migrate across the western Indian Ocean? *Journal of Tropical Ecology,* 25, 347–58.

Angelon, K. A. and Petranka, J. W. (2002). Chemicals of predatory mosquitofish (*Gambusia affinis*) influence selection of oviposition site by *Culex* mosquitoes. *Journal of Chemical Ecology,* 28, 797–806.

Anusa, A., Ndagurwa, H. G. T., and Magadza, C. H. D. (2012). The influence of pool size on species diversity and water chemistry in temporary rock pools on Domboshawa Mountain, northern Zimbabwe. *African Journal of Aquatic Sciences,* 37, 89–99.

Aoyagi, M. and Ishii, M. (1991). Host acceptance behavior of the Japanese aquatic wasp, *Agriotypus gracilus* (Hymenoptera, Ichneumonidae) toward the caddisfly host, *Goera japonica* (Trichoptera, Limnephilidae). *Journal of Ethology,* 9, 113–19.

Apodaca, C. K. and Chapman, L. J. (2004). Larval damselflies in extreme environments: behavioral and physiological response to hypoxic stress. *Journal of Insect Physiology,* 50, 767–75.

Archangelsky, M. and Durand, M. E. (1992). Description of the preimaginal stages of *Dibolocelus ovatus* (Gemminger and Harold, 1868) (Coleoptera, Hydrophilidae: Hydrophilinae). *Aquatic Insects,* 14, 107–12.

Archangelsky, M. and Fikáček, M. (2004). Descriptions of the egg case and larva of *Anacaena* and a review of the knowledge and relationships between larvae of Anacaenini (Coleoptera: Hydrophilidae: Hydrophilinae). *European Journal of Entomology,* 101, 629–36.

Arens, W. (1989). Comparative functional morphology of the mouthparts of stream animals feeding on epilithic algae. *Archiv für Hydrobiologie, Supplement,* 83, 253–354.

Arens, W. (1990). Wear and tear of mouthparts: a critical problem in stream animals feeding on epilithic algae. *Canadian Journal of Zoology,* 68, 1896–914.

Arnqvist, G., Jones, T. M., and Elgar, M. A. (2003). Reversal of sex roles in nuptial feeding. *Nature,* 424, 387.

Arnqvist, G., Jones, T. M., and Elgar, M. A. (2006). Sex-role reversed nuptial feeding reduces male kleptoparasitism of females in Zeus bugs. *Biology Letters,* 2, 491–3.

Arnqvist, G., Jones, T. M., and Elgar, M. A. (2007). The extraordinary mating system of Zeus bugs (Heteroptera: Veliidae: *Phoreticovelia* sp.). *Australian Journal of Zoology,* 55, 131–7.

Arnqvist, G. and Rowe, L. (1995). Sexual conflict and arms races between the sexes—A morphological adaptation for control of mating in a female insect. *Proceedings of the Royal Society, London B,* 261, 123–7.

Arvy, L. and Brittain, J. E. (1984). The structure, function and development of the eyes of Ephemeroptera. In V. Landa, T. Soldán, and M. Tonner, eds. *Proceedings of the 4th International Conference on Ephemeroptera,* pp. 173–9. Institute of Entomology, Czechoslovak Academy of Sciences, Ceské Budejovice.

Autrum, H. (1963). Anatomy and Physiology of sound receptors in invertebrates. In R. G. Busnel, ed. *Acoustic Behavior of Animals.* Elsevier, New York.

Baggiano, O., Schmidt, D. J., Sheldon, F., and Hughes, J. M. (2011). The role of altitude and associated habitat stability in determining patterns of population genetic structure in two species of *Atalophlebia* (Ephemeroptera: Leptophlebiidae). *Freshwater Biology,* 56, 230–49.

Bailey, P. C. E. (1985). 'A prey in the hand', multi-prey capture behaviour in a sit-and-wait predator, *Rantra dispar* (Heteroptera: Nepidae), the water stick insect. *Journal of Ethology,* 3, 105–12.

Baker, G. T. (2001). Distribution patterns and morphology of sensilla on the apical segment of the antennae and palpi of Hydradephaga (Coleoptera: Adephaga). *Microscopy Research and Technique,* 55, 330–8.

Baldwin, E. F., West, A. S., and Gomery, J. (1975). Dispersal pattern of black flies (Diptera: Simuliidae) tagged with [32]P. *Canadian Entomologist,* 107, 113–18.

Bale, J. S. (1987). Insect cold hardiness: freezing and supercooling: an ecological perspective. *Journal of Insect Physiology,* 33, 899–908.

Ball, S. L. (2002). Population variation and ecological correlates of tychoparthenogenesis in the mayfly, *Stenonema femoratum. Biological Journal of the Linnean Society,* 75, 101–23.

Banga, N., Albert, P. J., Kapoor, N. N., and McNeil, J. N. (2003). Structure, distribution, and innervation of sensilla on the ovipositor of the spruce budworm, *Choristoneura fumiferana,* and evidence of a gustatory function for type II sensilla. *Canadian Journal of Zoology,* 81, 2032–7.

Barber-James, H. M., Gattolliat, J. -L., Sartori, M., and Hubbard, M. D. (2008). Global diversity of mayflies

(Ephemeroptera, Insecta) in freshwater. *Hydrobiologia*, 595, 339–50.

Barbier, R. and Chauvin, G. (1974). The aquatic egg of *Nymphula nymphaeata* (Lepidoptera: Pyralidae). On the fine structure of the egg shell. *Cell and Tissue Research*, 149, 473–9.

Barmuta, L. A., McKenny, C. E. A., and Swain, R. (2001). The response of lotic mayfly *Nousia* sp. (Ephemeroptera: Leptophlebiidae) to moving water and light of different wavelengths. *Freshwater Biology*, 46, 567–73.

Barr, D. and Smith, B. P. (1980). Stable swimming by diagonal phase synchrony in arthropods. *Canadian Journal of Zoology*, 58, 782–95.

Batzer, D. P. and Wissinger, S. A. (1996). Ecology of insect communities in nontidal wetlands. *Annual Review of Entomology*, 41, 75–100.

Bäumer, C., Pirow, R., and Paul, R. J. (2000). Respiratory adaptations to running-water microhabitats in mayfly larvae *Epeorus sylvicola* and *Ecdyonurus torrentis*, Ephemeroptera. *Physiological and Biochemical Zoology*, 73, 77–85.

Bayly, I. A. E. and Williams, W. D. (1973). *Inland Waters and Their Ecology*, Longman Australia Pty Ltd, Hawthorn, Victoria.

Beam, B. D. and Wiggins, G. B. (1987). A comparative study of the biology of five species of *Neophylax* (Trichoptera: Limnephilidae) in southern Ontario, Canada. *Canadian Journal of Zoology*, 65, 1741–54.

Becker, G. (1987). Net-building behaviour, tolerance and development of two caddisfly species from the River Rhine (*Hydropsyche contubernalis* and *H. pellucidlua*) in relation to the oxygen content. *Oecologia*, 73, 242–50.

Becnel, J. J. and Dunkle, S. W. (1990). Evolution of micropyles in dragonfly eggs (Anisoptera). *Odonatologica*, 19, 235–41.

Behrend, K. (1971). Riechen in Wasserund in Luft bei *Dytiscus marginalis* L. *Zeitschrift fur Vergleichende Physiologie*, 75, 108–22.

Bendele, H. (1986). Mechanosensory cues control chasing behaviour of whirligig beetles (Coleoptera, Gyrinidae). *Journal of Comparative Physiology A*, 158, 405–11.

Benfield, E. F. (1972). A defensive secretion of *Dineutes discolor* (Coleoptera: Gyrinidae). *Annals of the Entomological Society of America*, 65, 1324–7.

Bennett, A. M. R. (2001). Phylogeny of Agriotypinae (Hymenoptera: Ichneumonidae) with comments on the subfamily relationships of the basal Ichneumonidae. *Systematic Entomology*, 26, 329–56.

Bennett, A. M. R. (2008). Global diversity of hymenopterans (Hymenoptera; Insecta) in freshwater. *Hydrobiologia*, 595, 529–34.

Bennett, R. R. (1967). Spectral sensitivity studies on the whirligig beetle, *Dineutes ciliatus*. *Journal of Insect Physiology*, 13, 621–33.

Bennett-Clark, H. C. (1997). Tymbal mechanics and the control of song frequency in the cicada *Cyclochila australasiae*. *Journal of Experimental Biology*, 200, 1681–94.

Bentley, M. D. and Day, J. F. (1989). Chemical ecology and behavioral aspects of mosquito oviposition. *Annual Review of Entomology*, 34, 401–21.

Berendonk, T. U. (1999). Influence of fish kairomones on the ovipositing behavior of *Chaoborus* imagines. *Limnology and Oceanography*, 44, 454–8.

Berg, M. B. (1995). Larval food and feeding behaviour. In P. D. Armitage, P. S. Cranston and L. C. V. Pinder, eds. *The Chironomidae: Biology and Ecology of Non-Biting Midges.* pp. 136–68. Chapman and Hall, London.

Bergman, E. A. and Hilsenhoff, W. F. (1978). Parthenogenesis in the mayfly genus *Baëtis* (Ephemeroptera: Baetidae). *Annals of the Entomological Society of America*, 71, 167–8.

Bergsten, J. and Miller, K. B. (2007). Phylogeny of diving beetles reveals a coevolutionary arms race between the sexes. *PLoS ONE*, 2, e522.

Bergsten, J., Töyrä, A., and Nilsson, A. N. (2001). Intraspecific variation and intersexual correlation in secondary sexual characters of three diving beetles (Coleoptera: Dytiscidae). *Biological Journal of the Linnean Society*, 73, 221–32.

Berry, R. P., Stange, G., and Warrant, E. J. (2007). Form vision in the insect dorsal ocelli: An anatomical and optical analysis of the dragonfly median ocellus. *Vision Research*, 47, 1394–409.

Berté, A. B. and Prichard, G. (1983). The structure and hydration dynamics of trichopteran (Insecta) egg masses. *Canadian Journal of Zoology*, 61, 378–84.

Beutel, R. G. and Gorb, S. N. (2001). Ultrastructure of attachment specializations of hexapods (Arthropoda): evolutionary patterns inferred from a revised ordinal phylogeny. *Journal of Zoological and Systematic Evolutionary Research*, 39, 177–207.

Bilton, D. T., Freeland, J. R., and Okamura, B. (2001). Dispersal in freshwater invertebrates. *Annual Review of Ecology and Systematics*, 32, 159–81.

Bilton, D. T., Thompson, A., and Foster, G. N. (2007). Inter- and intrasexual dimorphism in the diving beetle *Hydroporus memnonius* Nicolai (Coleoptera: Dytiscidae). *Biological Journal of the Linnean Society*, 94, 685–97.

Binckley, C. A. and Resetarits Jr, W. J. (2007). Effects of forest canopy on habitat selection in treefrogs and aquatic insects: implications for communities and metacommunities. *Oecologia*, 153, 951–8.

Bjostad, L. B., Jewett, S. K., and Brigham, D. L. (1996). Sex pheromone of caddisfly *Hesperophylax occidentalis* (Banks) (Trichoptera: Limnephilidae). *Journal of Chemical Ecology*, 22, 103–21.

Blakely, T. J., Harding, J. S., McIntosh, A. R., and Winterbourn, M. J. (2006). Barriers to the recovery of aquatic insect communities in urban streams. *Freshwater Biology*, 51, 1634–45.

Blaustein, L. (1998). Influence of the predatory backswimmer, *Notonecta maculata*, on invertebrate community structure. *Ecological Entomology*, 23, 246–52.

Blaustein, L. and Kotler, B. P. (1993). Oviposition habitat selection by the mosquito, *Culiseta longiareolata*: effects of conspecifics, food and green toad tadpoles. *Ecological Entomology*, 18, 104–108.

Blum, M. S. and Hilker, M. (2002). Chemical protection of insect eggs. In M. Hilker and T. Meiners, eds. *Chemoecology of Insect Eggs and Egg Deposition*. pp. 61–90. Blackwell Publishing, Oxford.

Boiteau, G., Vincent, C., Meloche, F., Leskey, T. C., and Colpitts, B. G. (2011). Evaluation of tag entanglement as a factor in harmonic radar studies of insect dispersal. *Environmental Entomology*, 40, 94–102.

Boix, D., Magnusson, A. K., Gascón, S., Sala, J., and Williams, D. D. (2011). Environmental influence on flight activity and arrival patterns of aerial colonizers of temporary ponds. *Wetlands*, 31, 1227–40.

Boix, D., Sala, J., Gascón, S., Martinoy, M., Gifre, J., Brucet, S., Badosa, A., López-Flores, R., and Quintana, X. D. (2007). Comparative biodiversity of crustaceans and aquatic insects from various water body types in coastal Mediterranean wetlands. *Hydrobiologia*, 584, 347–59.

Boling Jr, R. H., Goodman, E. D., Van Sickle, J. A., Zimmer, J. O., Cummins, K. W., Petersen Jr, R. C., and Reice, S. R. (1975). Toward a model of detritus processing in a woodland stream. *Ecology*, 56, 141–51.

Bouchard Jr, R. W., Carrillo, S. A., and Ferrington Jr, L. C. (2006a). Lower lethal temperature for adult male *Diamesa mendotae* Muttkowski (Diptera: Chironomidae), a winter-emerging Diamesinae. *Aquatic Insects*, 28, 57–66.

Bouchard Jr, R. W., Carrillo, S. A., Kells, S. A., and Ferrington Jr, L. C. (2006b). Freeze tolerance in larvae of the winter-active *Diamesa mendotae* Muttkowsko (Diptera: Chironomidae): a contrast to adult strategy for survival at low temperatures. *Hydrobiologia*, 568, 403–16.

Bouchard Jr, R. W., Schuetz, B. E., Ferrington Jr, L. C., and Kells, S. A. (2009). Cold hardiness in the adults of two winter stonefly species: *Allocapnia granulata* (Claassen, 1924) and *A. pygmaea* (Burmeister, 1839) (Plecoptera: Capniidae). *Aquatic Insects*, 31, 145–55.

Boulton, A. J. and Lake, P. S. (2008). Effects of drought on aquatic insects and its ecological consequences. In J. Lancaster and R. A. Briers, eds. *Aquatic Insects: Challenges to Populations*. pp. 81–102. CAB International, Wallingford.

Brackenbury, J. (1999a). Water skating in the larvae of *Dixella aestivalis* (Diptera) and *Hydrobius fuscipes* (Coleoptera). *Journal of Experimental Biology*, 202, 845–53.

Brackenbury, J. (1999b). Regulation of swimming in the *Culex pipiens* (Diptera, Culicidae) pupa: kinematics and locomotory trajectories. *Journal of Experimental Biology*, 202, 2521–9.

Brackenbury, J. (2000). Locomotory modes in the larva and pupa of *Chironomus plumosus* (Diptera, Chironomidae). *Journal of Insect Physiology*, 46, 1517–27.

Brackenbury, J. (2001a). The vortex wake of the free-swimming larva and pupa of *Culex pipiens* (Diptera). *Journal of Experimental Biology*, 204, 1855–67.

Brackenbury, J. (2001b). Locomotion through use of the mouth brushes in the larva of *Culex pipiens* (Diptera: Culicidae). *Proceedings of the Royal Society of London Series B*, 268, 101–106.

Brackenbury, J. (2002). Kinematics and hydrodynamics of an invertebrate undulatory swimmer: the damsel-fly larva. *Journal of Experimental Biology*, 205, 627–39.

Brackenbury, J. (2003). Swimming kinematics and wake elements in a worm-like insect: the larva of the midge *Chironomus plumosus* (Diptera). *Journal of Zoology*, 260, 195–201.

Brackenbury, J. (2004). Kinematics and hydrodynamics of swimming in the mayfly larva. *Journal of Experimental Biology*, 207, 913–22.

Bradley, T. J. (1985). The excretory system: structure and physiology. In G. A. Kerkut and L. I. Gilbert, eds. *Comprehensive Insect Physiology, Biochemistry and Pharmacology, Volume 4*. pp. 421–65. Pergamon Press, Oxford.

Braimah, S. A. (1987a). Mechanics of filter feeding in immature *Simulium bivittatum* Malloch (Diptera: Simuliidae) and *Isonychia campestris* McDunnough (Ephemeroptera: Oligoneuriidae). *Canadian Journal of Zoology*, 65, 504–13.

Braimah, S. A. (1987b). The influence of water velocity on particle capture by the labial fans of larvae of *Simulium bivattatum* Malloch (Diptera: Simuliidae). *Canadian Journal of Zoology*, 65, 2395–9.

Braimah, S. A. (1987c). Pattern of flow around fliter-feeding structures of immature *Simulium bivittatum* Malloch (Diptera: Simuliidae) and *Isonychia campestris* McDunnough (Ephemeroptera: Oligoneuriidae). *Canadian Journal of Zoology*, 65, 514–21.

Breidenbaugh, M. S., Clark, J. W., Brodeur, R. M., and de Szalay, F. A. (2009). Seasonal and diel patterns of biting midges (Ceratopogonidae) and mosquitoes (Culicidae) on the Parris Island Marine Corps Recruit Depot. *Journal of Vector Ecology*, 34, 129–40.

Briers, R. A., Gee, J. H. R., Cariss, H. M., and Geoghegan, R. (2004). Inter-population dispersal by adult stoneflies detected by stable isotope enrichment. *Freshwater Biology*, 49, 425–31.

Briscoe, A. D. and Chittka, L. (2001). The evolution of color vision in insects. *Annual Review of Entomology*, 46, 471–510.

Brittain, J. E. and Eikeland, T. (1988). Invertebrate drift—a review. *Hydrobiologia*, 166, 177–93.

Brodersen, K. P., Pedersen, O., Walker, I. R., and Jensen, M. T. (2008). Respiration of midges (Diptera Chironomidae) in British Columbian lakes: oxy-regulation, temperature and their role as palaeo-indicators. *Freshwater Biology*, 53, 593–602.

Brodin, T., Johansson, F., and Bergsten, J. (2006). Predator related oviposition site selection of aquatic beetles (*Hydroporus* spp.) and effects on offspring life-history. *Freshwater Biology*, 51, 1277–85.

Brodin, Y. and Andersson, M. H. (2009). The marine splash midge *Telmatogon japonicus* (Diptera; Chironomidae)—extreme and alien? *Biological Invasions*, 11, 1311–17.

Brown, D. S. (1961). The morphology and functioning of the mouthparts of *Chloëon dipterum* L. and *Baëtis rhodani* Pictet (Insecta, Ephemeroptera). *Proceedings of the Zoological Society of London*, 136, 147–76.

Brown, E. S. (1948). The ecology of Saldidae (Hemiptera-Heteroptera) inhabiting a salt marsh, with observations on the evolution of aquatic habits in insects. *Journal of Animal Ecology*, 17, 180–8.

Brown, S. A., Ruxton, G. D., Pickup, R. W., and Humphries, S. (2005). Seston capture by *Hydropsyche siltalai* and the accuracy of capture efficiency estimates. *Freshwater Biology*, 50, 113–26.

Bruce, T. J. A., Wadhams, L. J., and Woodcock, C. M. (2005). Insect host location: a volatile situation. *Trends in Plant Science*, 10, 269–74.

Brundin, L. (1967). Insects and the problem of austral disjunctive distribution. *Annual Review of Entomology*, 12, 149–68.

Buden, D. W. (2010). *Pantela flavescens* (Insecta: Odonata) rides west winds into Ngulu Atoll, Micronesia: Evidence of seasonality and wind-assisted dispersal. *Pacific Science*, 64, 141–3.

Burghause, F. (1976). Adaptationserscheinungen in den komplexaugen von *Gyrinus natator* L. (Coleoptera: Gyrinidae). *International Journal of Insect Morphology and Embryology*, 5, 335–48.

Burmester, T. (1999). Evolution and function of the insect hexamerins. *European Journal of Entomology*, 96, 213–25.

Buschbeck, E. K., Sbita, S. J., and Morgan, R. C. (2007). Scanning behavior by larvae of the predacious diving beetle, *Thermonectus marmoratus* (Coleoptera: Dytiscidae) enlarges visual field prior to prey capture. *Journal of Comparative Physiology A*, 193, 973–82.

Bush, J. W. M. and Hu, D. L. (2006). Walking on water: biolocomotion at the interface. *Annual Review of Fluid Mechanics*, 38, 339–69.

Bush, J. W. M., Hu, D. L., and Prakash, M. (2008). The integument of water-walking arthropods: form and function. *Advances in Insect Physiology*, 34, 117–92.

Butler, M. G. (1984). Life histories of aquatic insects. In V. H. Resh and D. M. Rosenberg, eds. *The Ecology of Aquatic Insects.* pp. 24–55. Praeger, New York.

Byrne, K. and Nichols, R. A. (1999). *Culex pipiens* in London Underground tunnels: differentiation between surface and subterranean populations. *Heredity*, 82, 7–15.

Cabrera-Walsh, G., Schooler, S., and Julien, M. (2011). Biology and preliminary host range of *Hydrotimetes natans* Kolbe (Coleoptera: Curculionidae), a natural enemy candidate for biological control of *Cabomba caroliniana* Gray (Cabombaceae) in Australia. *Australian Journal of Entomology*, 50, 200–206.

Cairns, A. and Wells, A. (2008). Contrasting modes of handling moss for feeding and case-building by the caddisfly *Scelotrichia willcairnsi* (Insecta: Trichoptera). *Journal of Natual History*, 42, 2609–15.

Caldwell, J. P., Thorp, J. H., and Jervey, T. O. (1980). Predator-prey relationships among larval dragonflies, salamanders and frogs. *Oecologia*, 46, 285–9.

Calosi, P., Bilton, D. T., and Spicer, J. I. (2007). The diving response of a diving beetle: effects of temperature and acidification. *Journal of Zoology*, 273, 289–97.

Cameron, G. N. (1976). Do tides affect coastal insect communities? *American Midland Naturalist*, 95, 279–87.

Campbell, R. W. and Dower, F. (2003). Role of lipids in the maintenance of neutral buoyancy by zooplankton. *Marine Ecology Progress Series*, 263, 93–9.

Candan, S., Suludere, Z., Koç, H., and Kuyucu, N. (2005). External morphology of eggs of *Tipula (Lunatipula) decolor, Tipula (Lunatipula) dedecor*, and *Tipula (Acutipula) latifurca* (Diptera: Tipulidae). *Annals of the Entomological Society of America*, 98, 346–50.

Canhoto, C. and Graça, M. A. S. (2006). Digestive tract and leaf processing capacity of the stream invertebrate *Tipula lateralis*. *Canadian Journal of Zoology*, 84, 1087–95.

Cantonati, M., Füreder, L., Gerecke, R., Jüttner, I., and Cox, E. J. (2012). Crenic habitats, hotspots for freshwater biodiversity conservation: toward an understanding of their ecology. *Freshwater Science*, 31, 463–80.

Cardinale, B. J., Palmer, M. A., and Collins, S. L. (2001). Species diversity enhances ecosystem functioning through interspecific facilitation. *Nature*, 415, 426–9.

Carling, P. A. (1992). The nature of the fluid boundary layer and the selection of parameters for benthic ecology. *Freshwater Biology*, 28, 273–84.

Carter, C. E. (1980). The life cycle of *Chironomus anthracinus* in Lough Neagh. *Holarctic Ecology*, 3, 214–17.

Casey, T. M. (1988). Thermoregulation and heat exchange. *Advances in Insect Physiology*, 20, 119–46.

Caudell, J. N. and Conover, M. R. (2006). Energy content and digestibility of brine shrimp (*Artemia franciscana*) and other prey items of eared grebes (*Podiceps nigricollis*) on the Great Salt Lake, Utah. *Biological Conservation*, 130, 251–4.

Caudill, C. C. (2003). Measuring dispersal in a metapopulation using stable isotope enrichment: high rates of sex-biased dispersal between patches in a mayfly metapopulation. *Oikos*, 101, 624–30.

Chaloner, D. T., Wipfli, M. S., and Caouette, J. P. (2002). Mass loss and macroinvertebrate colonisation of Pacific Salmon carcasses in south-eastern Alaskan streams. *Freshwater Biology*, 47, 263–73.

Chance, M. M. and Craig, D. A. (1986). Hydrodynamics and behaviour of Simuliidae larvae (Diptera). *Canadian Journal of Zoology*, 64, 1295–309.

Chapman, R. F. (1964). The structure and wear of the mandibles in some African grasshoppers. *Proceedings of the Zoological Society of London*, 142, 107–21.

Cheer, A. Y. L. and Koehl, M. A. R. (1987). Paddles and rakes: fluid flow through bristled appendages of small organisms. *Journal of Theoretical Biology*, 129, 17–39.

Chen, P. S. (1984). The functional morphology and biochemistry of insect male accessory glands and their secretions. *Annual Review of Entomology*, 29, 233–55.

Cheng, L. (ed.) (1976). *Marine Insects*, New Holland Publishers, Amsterdam.

Cheng, L. (1985). Biology of *Halobates* (Heteroptera: Gerridae). *Annual Review of Entomology*, 30, 111–35.

Cheng, L. and Collins, J. D. (1980). Observations on behavior, emergence and reproduction of the marine midges *Pontomyia* (Diptera: Chironomidae). *Marine Biology*, 58, 1–5.

Cheng, L. and Roussis, V. (1998). Sex attractant in the marine insect *Trochopus plumbeus* (Heteroptera: Veliidae): a preliminary report. *Marine Ecology Progress Series*, 170, 283–6.

Chesson, J. (1984). Effect of notonectids (Hemiptera: Notonectidae) on mosquitoes (Diptera: Culicidae): predation or selective oviposition? *Environmental Entomology*, 13, 531–8.

Chester, E. T. and Robson, B. J. (2011). Drought refuges, spatial scale and recolonisation by invertebrates in non-perennial streams. *Freshwater Biology*, 56, 2094–104.

Chin, K. S. and Taylor, P. D. (2009). Interactive effects of distance and matrix on the movements of a peatland dragonfly. *Ecography*, 32, 715–22.

Choi, C., Bareliss, C., Walenciak, O., and Gross, E. M. (2002). Impact of polyphenols on growth of the aquatic herbivore *Acentria ephemeralla*. *Journal of Chemical Ecology*, 28, 2245–56.

Chown, S. L., Gibbs, A. G., Hetz, S. K., Klok, C. J., Lighton, J. R. B., and Marais, E. (2006). Discontinuous gas exchange in insects: a clarification of hypotheses and approaches. *Physiological and Biochemical Zoology*, 79, 333–43.

Clegg, J. S. (2001). Cryptobiosis—a peculiar state of biological organization. *Comparative Biochemistry and Physiology B*, 128, 613–24.

Clements, A. N. (1992). *The Biology of Mosquitoes. Volume 1. Development, Nutrition and Reproduction*, Chapman & Hall, London.

Clements, A. N. (1999). *The Biology of Mosquitoes. Volume 2. Sensory Reception and Behaviour*, CABI, Wallingford.

Clifford, H. F. (1982). Life cycles of mayflies (Ephemeroptera), with special reference to voltinism. *Quaestiones Entomologicae*, 18, 15–90.

Cline, A. R., Shockley, F. W., and Puttler, B. (2002). Description of surface swimming by *Hypera eximia* LeConte (Coleoptera: Curculionidae): behavioral, morphological, and phylogenetic implications. *Annals of the Entomological Society of America*, 95, 637–45.

Clissold, F. J. (2008). The biomechanics of chewing and plant fracture: mechanisms and implications. *Advances in Insect Physiology*, 34, 317–72.

Cloarec, A. (1984). Development of the compound eyes of the water stick insect, *Ranatra linearis*. *Physiological Entomology*, 9, 253–62.

Cloarec, A. (1991). Handling time and multi-prey capture by a water bug. *Animal Behaviour*, 42, 607–13.

Cohen, A. C. (1995). Extra-oral digestion in predaceous terrestrial arthropods. *Annual Review of Entomology*, 40, 85–103.

Colless, D. H. (1986). The Australian Chaoboridae (Diptera). *Australian Journal of Zoology, Supplementary Series*, 124, 1–66.

Collier, K. J. and Smith, B. J. (1998). Dispersal of adult caddisflies (Trichoptera) into forests alongside three New Zealand streams. *Hydrobiologia*, 361, 53–65.

Collins, N. (1980). Population ecology of *Ephydra cinerea* Jones (Diptera: Ephydridae), the only benthic metazoan of the Great Salt Lake, USA *Hydrobiologia*, 68, 99–112.

Conn, D. B. and Quinn, C. M. (1995). Ultrastructure of the vitellogenic egg chambers of the caddisfly *Brachycentrus incanus* (Insecta: Trichoptera). *Invertebrate Biology*, 114, 334–43.

Conrad, K. F., Willson, K. H., Harvey, I. F., Thomas, C. J., and Sherratt, T. N. (1999). Dispersal characteristics of seven odonate species in an agricultural landscape. *Ecography*, 22, 524–31.

Contreras, H. L. and Bradley, T. J. (2011). The effect of ambient humidity and metabolic rate on the gas-exchange pattern of the semi-aquatic insect *Aquarius remigis*. *Journal of Experimental Biology*, 214, 1086–91.

Convey, P. and Block, W. (1996). Antarctic Diptera: Ecology, physiology and distribution. *European Journal of Entomology*, 93, 1–13.

Corallini, C. and Gaino, E. (2001). Peculiar digestion patterns of sponge-associated zoochlorellae in the caddisfly *Ceraclea fulva*. *Tissue & Cell*, 33, 402–407.

Corallini, C. and Gaino, E. (2003). The caddisfly *Ceraclea fulva* and the freshwater sponge *Ephydatia fluviatilis*: a successful relationship. *Tissue & Cell*, 35, 1–7.

Corbet, P. S. (1957). The life-history of the emperor dragon-fly *Anax imperator* Leach (Odonata: Aeshnidae). *Journal of Animal Ecology,* 26, 1–69.

Corbet, P. S. (1964). Temporal patterns of emergence in aquatic insects. *Canadian Entomologist,* 96, 264–79.

Corbet, P. S. (1967). Facultative autogeny in arctic mosquitoes. *Nature,* 215, 662–3.

Corbet, P. S. (2004). *Dragonflies: Behaviour and Ecology of Odonata,* Harley Books, Colchester, UK.

Córdoba-Aguilar, A. (ed.) (2008). *Dragonflies and Damselflies: Model Organisms for Ecological and Evolutionary Research,* Oxford University Press, Oxford, UK.

Córdoba-Aguilar, A. and Cordero-Rivera, A. (2008). Cryptic female choice and sexual conflict. In A. Córdoba-Aguilar, ed. *Dragonflies and Damselflies: Model Organisms for Ecological and Evolutionary Research.* pp. 189–202. Oxford University Press, Oxford.

Córdoba-Aguilar, A., Uhía, E., and Cordero-Rivera, A. (2003). Sperm compeition in Odonata (Insecta): the evolution of female sperm storage and rivals' sperm displacement. *Journal of Zoology,* 261, 381–98.

Coupland, J. B. (1991). Oviposition response of *Simulium reptans* (Diptera: Simuliidae) to the presence of conspecific eggs. *Ecological Entomology,* 16, 11–15.

Courtney, S. P. (1984). The evolution of egg clustering by butterflies and other insects. *American Naturalist,* 123, 127–281.

Coutant, C. C. (1982) Evidence for upstream dispersion of adult caddisflies (Trichoptea: Hydropsychidae) in the Columbia River. *Aquatic Insects,* 4, 61–66.

Cover, M. R. and Resh, V. (2008). Global diversity of dobsonflies, fishflies, and alderflies (Megaloptera; Insecta) and spongillaflies, nevrorthids, and osmylids (Neuroptera; Insecta) in freshwater. *Hydrobiologia,* 595, 409–17.

Cowley, D. R. (1978). Studies on the larvae of New Zealand Trichoptera. *New Zealand Journal of Zoology,* 5, 639–750.

Cranston, F. P. and Sprague, I. B. (1961). A morphological study of the head capsule of *Gerris remigis* Say. *Journal of Morphology,* 108, 287–98.

Cranston, P. S. (1995a). Biogeography. In P. D. Armitage, P. S. Cranston, and L. C. V. Pinder, eds. *The Chironomidae.* pp. 62–84. Chapman & Hall, London.

Cranston, P. S. (1995b). Systematics. In P. D. Armitage, P. S. Cranston, and L. C. V. Pinder, eds. *The Chironomidae.* pp. 31–61. Chapman & Hall, London.

Cranston, P. S., Edward, D. H. D., and Cook, L. G. (2002). New status, species, distribution records and phylogeny for Australian mandibulate Chironomidae (Diptera). *Australian Journal of Entomology,* 41, 357–66.

Cranston, P. S. and Nolte, U. (1996). *Fissimentum,* a new genus of drought-tolerant Chironomini (Diptera: Chironomidae) from teh Americas and Australia. *Entomological News,* 107, 1–15.

Crowl, T. A. and Alexander, J. E. (1989). Parental care and foraging ability in male waterbugs (*Belostoma flumineum*). *Canadian Journal of Zoology,* 67, 513–15.

Csabai, Z., Boda, P., Bernáth, B., Kriska, G., and Horváth, G. (2006). A 'polarisation sun-dial' dictates the optimal time of day for dispersal by flying aquatic insects. *Freshwater Biology,* 51, 1341–9.

Cuda, J. P., Charudattan, R., Grodowitz, M. J., Newman, R. M., Shearer, J. F., Tamayo, M. L., and Villegas, B. (2008). Recent advances in biological control of submersed aquatic weeds. *Journal of Aquatic Plant Management,* 46, 15–32.

Culp, J. M., Glozier, N. E., and Scrimgeour, G.J. (1991). Reduction of predation risk under the cover of darkness: Avoidance responses of mayfly larvae to a benthic fish. *Oecologia,* 86, 163,169.

Currie, D. C. and Adler, P. H. (2008). Global diversity of black flies (Diptera: Simuliidae) in freshwater. *Hydrobiologia,* 595, 469–75.

Currie, D. C. and Craig, D. A. (1987). Feeding strategies of larval blackflies. In K. C. Kim and R. W. Merritt, eds. *Black Flies. Ecology, Population Management, and Annotated World List.* pp. 155–70. The Pennsylvania State University, University Park.

Daborn, G. R. (1971). Survival and mortality of coenagrionid nymphs (Odonata: Zygoptera) from the ice of an aestival pond. *Canadian Journal of Zoology,* 49, 569–71.

Dahl, J. and Peckarsky, B. L. (2002). Induced morphological defenses in the wild: predator effects on a mayfly, *Drunella coloradensis. Ecology,* 83, 1620–34.

Dahmen, H. (1991). Eye specialisation in waterstriders: an adaptation to life in a flat world. *Journal of Comparative Physiology A,* 169, 623–32.

Dale, P. and Breitfuss, M. (2009). Ecology and management of mosquitoes. In N. Saintilan, ed. *Australian Saltmarsh Ecology.* pp. 167–78. CSIRO Publishing, Collingwood, Australia.

Dallas, H. F. and Rivers-Moore, N. A. (2012). Critical thermal maxima of aquatic macroinvertebrates: towards identifying bioindicators of thermal alteration. *Hydrobiologia,* 679, 61–76.

Danks, H. V. (1971). Overwintering of some north temperate arctic Chironomidae. II. Chironomid biology. *Canadian Entomologist,* 103, 1875–910.

Danks, H. V. (1987). *Insect Dormancy: An Ecological Perspective,* Ottawa Biological Survey, Ontario, Canada.

Danks, H. V. (2000). Dehydration in dormant insects. *Journal of Insect Physiology,* 46, 837–52.

Danks, H. V. (2005). How similar are daily and seasonal biological clocks? *Journal of Insect Physiology,* 51, 609–19.

Danks, H. V. (2007). How aquatic insects live in cold climates. *Canadian Entomologist,* 139, 443–71.

Danks, H. V. (2008). Aquatic insect adaptations to winter cold and ice. In J. Lancaster and R. A. Briers, eds. *Aquatic*

Insects: Challenges to Populations. pp. 1–19. CAB International, Wallingford.

Danks, H. V., Downes, J. A., Larson, D. J., and Scudder, G. G. E. (1997). Insects of the Yukon: Characteristics and History. In H. V. Danks and J. A. Downes, eds. *Insects of the Yukon.* pp. 963–1013. The Biological Survey of Canada (Terrestrial Arthropods), Ottawa.

David, J. -P., Rey, D., Pautou, M. -P., and Meyran, J. -C. (2000). Differential toxicity of leaf litter to dipteran larvae of mosquito developmental sites. *Journal of Invertebrate Pathology,* 75, 9–18.

Davis, C. C. (1964). A study of the hatching process in aquatic invertebrates. VII. Observations on hatching in *Notonecta melaena* Kirkaldy (Hemiptera, Notonectidae) and on *Ranatra absona* D. & DeC. (Hemiptera, Nepidae). VIII. The hatching process of *Amnicola* (?) *hydrobioides* (Ancey) (Prosobranchia, Hydrobiidae). *Hydrobiologia,* 23, 253–66.

Davis, C. C. (1965). A study of the hatching process in aquatic invertebrates. XII. The eclosion process in *Trichocorixa naias* (Kirkaldy) (Heteroptera, Corixidae). *Transactions of the American Microscopical Society,* 84, 60–5.

Day, J. C. (2010). *Simulium angustipes* (Diptera: Simuliidae) found in the tidal region of a stream in Cornwall. *British Simuliid Group Bulletin,* Issue 33, 8–10.

de Jong, H., Oosterbroek, P., Gelhaus, J., Reusch, H., and Young, C. (2008). Global diversity of craneflies (Insecta, Diptera: Tipulidea or Tipulidae *sensu lato*) in freshwater. *Hydrobiologia,* 595, 457–67.

de Moor, F. C. and Ivanov, V. D. (2008). Global diversity of caddisflies (Trichoptera: Insecta). *Hydrobiologia,* 595, 407.

de Sousa, W., Marques, M. I., Rosado-Neto, G. H., and Adis, J. (2007). Surface swimming behavior of the curculionid *Ochetina uniformis* Pascoe (Erirhininae, Stenopelmini) and *Ludovix fasciatus* (Gyllenhal) (Curculioninae, Erodiscini). *Revista Brasileira de Entomologia,* 51, 87–92.

Denis, C. and Le Lannic, J. (1977). Les modalités de la vitellogenèse chez let Trichoptères. *Annales de Zoologie Ecologie Animale,* 9, 627–35.

Denis, C. and Malicky, H. (1985). Etude du cycle biologique de deux Limnephilidae: *Limnephilus minos* et *Limnephilus germanus* (Trichoptera). *Annales de Limnologie—International Journal of Limnology,* 21, 71–6.

Denlinger, D. L. (2002). Regulation of diapause. *Annual Review of Entomology,* 47, 93–122.

Denny, M. W. (2004). Paradox lost: answers and questions about walking on water. *Journal of Experimental Biology,* 207, 1601–6.

Deutsch, W. G. (1984). Oviposition of Hydropsychidae (Trichoptera) in a large river. *Canadian Journal of Zoology,* 62, 1988–94.

Deutsch, W. G. (1985). Swimming modifications of adult female Hydropsychidae compared with other Trichoptera. *Freshwater Invertebrate Biology,* 4, 35–40.

Devarakonda, R., Barth, F. G., and Humphrey, J. A. C. (1996). Dynamics of arthropod filifrom hairs. IV Hair motion in air and water. *Philosophical Transactions of the Royal Society of London. B.,* 351, 933–46.

Dillon, R. J. and Dillon, V. M. (2004). The gut bacteria of insects: nonpathogenic interactions. *Annual Review of Entomology,* 49, 71–92.

Dimitriadis, S. and Cranston, P. S. (2007). From the mountains to the sea: assemblage structure and dynamics in Chironomidae (Insecta: Diptera) in the Clyde River estuarine gradient, New South Wales, south-eastern Australia. *Australian Journal of Entomology,* 46, 188–97.

Disney, R. H. L. (1971a). Notes on *Simulium ovazzae* Grenier and Mouchet (Diptera: Simuliidae) and river crabs (Malacostraca: Potamidae) and their association. *Journal of Natural History,* 5, 677–89.

Disney, R. H. L. (1971b). Association between blackflies (Simuliidae) and prawns (Atyidae), with a discussion of the phoretic habit in simuliids. *Journal of Animal Ecology,* 40, 83–92.

Ditsche-Kuru, P. and Koop, J. H. E. (2009). New insights into a life in current: do the gill lamellae of *Epeorus assimilis* and *Iron alpicola* larvae (Ephemeroptera: Heptageniidae) function as a sucker or as friction pads? *Aquatic Insects,* 31, 495–506.

Ditsche-Kuru, P., Koop, J. H. E., and Gorb, S. N. (2010). Underwater attachment in current: the role of setose attachment structures on the gills of the mayfly larvae *Epeorus assimilis* (Ephemeroptera, Heptageniidae). *Journal of Experimental Biology,* 213, 1950–9.

Dodds, G. S. and Hisaw, F. L. (1924). Ecological studies of aquatic insects. I. Adaptations of mayfly nymphs to swift streams. *Ecology,* 5, 137–48.

Dodson, S. I., Crowl, T. A., Peckarsky, B. L., Kats, L. B., Covich, A. P., and Culp, J. M. (1994). Non-visual communication in freshwater benthos: an overview. *Journal of the North American Benthological Society,* 13, 268–82.

Dominguez, E. and Cuezzo, M. G. (2002). Ephemeroptera egg chorion characters: A test of their importance in assessing phylogenetic relationships. *Journal of Morphology,* 253, 148–65.

Donoughe, S., Crall, J. D., Merz, R. A. and Combes, S. A. (2011). Resilin in dragonfly and damselfly wings and its implications for wing flexibility. *Journal of Morphology,* 272, 1409–21.

Douglas, A. E. (2009). The microbial dimension in insect nutritional ecology. *Functional Ecology,* 23, 38–47.

Dow, J. A. T. (1986). Insect midgut function. *Advances in Insect Physiology,* 19, 187–328.

Downes, B. J., Bellgrove, A., and Street, J. L. (2005). Drifting or walking? Colonisation routes used by different instars and species of lotic, macroinvertebrate filter feeders. *Marine and Freshwater Research,* 56, 815–24.

Downes, B. J. and Lancaster, J. (2010). Does dispersal control population densities in advection-dominated systems? A fresh look at critical assumptions and a direct test. *Journal of Animal Ecology*, 79, 235–48.

Downes, J. A. (1965). Adaptations of insects in the arctic. *Annual Review of Entomology*, 10, 257–74.

Downes, J. A. (1969). The swarming and mating flight of Diptera. *Annual Review of Entomology*, 14, 271–98.

Dudley, T. L. and Anderson, N. H. (1987). The biology and life cycles of *Lipsothrix* spp. (Diptera: Tipulidae) inhabiting wood in Western Oregon streams. *Freshwater Biology*, 17, 437–51.

Duffield, R. M., Blum, M. S., Wallace, J. B., Lloyd, H. A., and Regnier, F. E. (1977). Chemistry of the defensive secretion of the caddisfly *Pycnopsyche scabripennis* (Trichoptera: Limnephilidae). *Journal of Chemical Ecology*, 3, 649–56.

Eastham, L. E. S. (1934). Metachronal rhythms and gill movements of the nymph of *Caenis horaria* (Ephemeroptera) in relation to water flow. *Proceedings of the Royal Society of London. Series B, Containing Papers of a Biological Character*, 115, 30–48.

Eastham, L. E. S. (1936). The rhythmical movements of the gills of nymphal *Leptophlebia marginata* (Ephemeroptera) and the currents produced by them in water. *Journal of Experimental Biology*, 13, 443–9.

Eastham, L. E. S. (1937). The gill movements of nymphal *Ecdyonurus venosus* (Ephemeroptera) and the currents produced by them in water. *Journal of Experimental Biology*, 14, 219–28.

Eastham, L. E. S. (1939). Gill movements of nymphal *Ephemera danica* (Ephemeroptera) and the water currents caused by them. *Journal of Experimental Biology*, 16, 18–33.

Eaton, A. E. (1888). A revisional monograph of the recent Ephemeridae or mayflies. *Transactions of the Linnean Society, Series 2*, 3, 1–352.

Edgerly, J. S. and Marvier, M. A. (1992). To hatch or not to hatch? Egg hatch response to larval density and to larval contact in a treehole mosquito. *Ecological Entomology*, 17, 28–32.

Edington, J. M. and Hildrew, A. G. (1995). *Caseless caddis larvae of the British Isles*, Scientific Publication No. 53, Freshwater Biological Association, Ambleside, UK.

Edmunds Jr, G. F. and McCafferty, W. P. (1988). The mayfly subimago. *Annual Review of Entomology*, 33, 509–29.

Eisner, T. and Aneshansley, D. J. (2000). Chemical defence: Aquatic beetle (*Dineutes hornii*) vs. fish (*Micropterus salmoides*). *Proceedings of the National Academy of Sciences of the U.S.A.*, 97, 11313–18.

Eisner, T., Rossini, C., González, A., Iyengar, V. K., Siegler, M. V. S., and Smedley, S. R. (2002). Paternal investment in egg defence. In M. Hilker and T. Meiner, eds. *Chemoecology of Insect Eggs and Egg Deposition*. pp. 91–116. Blackwell Publishing, Oxford.

Eitam, A. and Blaustein, L. (2004). Oviposition habitat selection by mosquitoes in response to predator (*Notonecta maculata*) density. *Physiological Entomology*, 29, 188–91.

Elliott, J. M. (1971a). Upstream movements of benthic invertebrates in a Lake District stream. *Journal of Animal Ecology*, 40, 235–52.

Elliott, J. M. (1971b). The distances travelled by drifting invertebrates in a Lake District stream. *Oecologia*, 6, 350–79.

Elliott, J. M. (1977). *A Key to the Larvae and Adults of British Freshwater Megaloptera and Neuroptera*, Freshwater Biological Association, Scientific Publication No. 35, Ambleside.

Elliott, J. M. (1996). *British Freshwater Megaloptera and Neuroptera: A Key with Ecological Notes*, Freshwater Biological Association, Scientific Publication No. 54, Ambleside.

Encalada, A. C. and Peckarsky, B. L. (2006). Selective oviposition of the mayfly *Baetis bicaudatus*. *Oecologia*, 148, 526–37.

Endrass, U. (1976). Physiological adaptations of a marine insect. II. Characteristics of swimming and sinking egg masses. *Marine Biology*, 36, 47–60.

Engelhardt, C. H. M., Haase, P., and Pauls, S. U. (2011). From the western Alps across central Europe: Postglacial recolonisation of the tufa stream specialist *Rhyacophila pubescens* (Insecta,Trichoptera). *Frontiers in Zoology*, 8, 10.

Engster, M. S. (1976a). Studies on silk secretion in the Trichoptera (F. Limnephilidae). *Cell and Tissue Research*, 169, 77–92.

Engster, M. S. (1976b). Studies on silk secretion in Trichoptera (F. Limnephilidae). 1. Histology, histochemsitry, and ultrastructure of silk glands. *Journal of Morphology*, 150, 183–212.

Eplers, C. and Tomka, I. (1995). Food-filtering mechanisms of the larvae of *Oligoneuriella rhenana* Imhoff (Ephemeroptera: Oligoneuriidae). In L. D. Corkum and J. J. H. Ciborowski, eds. *Current Directions in Research of Ephemeroptera*. pp. 283–94. Canadian Scholars' Press Inc., Toronto.

Erezyilmaz, D. F. (2006). Imperfect eggs and oviform nymphs: a history of ideas about the origins of insect metamorphosis. *Integrative and Comparative Biology*, 46, 795–807.

Erhard, D., Pohnert, G., and Gross, E. M. (2007). Chemical defense in *Elodea nuttallii* reduces feeding and growth of aquatic herbivorous Lepidoptera. *Journal of Chemical Ecology*, 33, 1646–61.

Eriksen, C. H. (1986). Respiratory roles of caudal lamellae (gills) in a lestid damselfly (Odonata: Zygoptera). *Journal of the North American Benthological Society*, 5, 16–27.

Erman, N. A. (1981). Terrestrial feeding migration and life history of the stream-dwelling caddisfly, *Desmona bethula*

(Trichoptera: Limnephilidae). *Canadian Journal of Zoology,* 59, 1658–65.

Estevez, A. L., de Reyes, C. A., and Schnack, J. A. (2006). Successful hatching from eggs carried by females and naturally removed from incubant males in *Belostoma* spp. water bugs (Heteroptera: Belostomatidae). *Revista de Biologia Tropical,* 54, 515–17.

Evans, M. E. K. and Dennehy, J. J. (2005). Germ banking: Bet-hedging and variable release from egg and seed dormancy. *Quarterly Review of Biology,* 80, 431–51.

Evans, W. G. (1982). *Oscillatoria* sp. (Cyanobacteria) mat metabolites implicated in habitat selection in *Bembidion obtusidens* (Coleoptera: Carabidae). *Journal of Chemical Ecology,* 8, 671–8.

Ewing, A. W. (1989). *Arthropod Bioacoustics. Neurobiology and Behaviour,* Edinburgh University Press, Edinburgh.

Fairbairn, D. J. (1988). Adaptive significance of wing dimorphism in the absence of dispersal: a comparative study of wing morphs in the waterstrider, *Gerris remigis. Ecological Entomology,* 13, 273–81.

Feng, H. -Q., Wu, K. -M., Ni, Y. -X., Cheng, D. -F., and Guo, Y. -Y. (2006). Nocturnal migration of dragonflies over the Bohai Sea in northern China. *Ecological Entomology,* 31, 511–20.

Ferrari, J. and Vavre, F. (2011). Bacterial symbionts in insects or the story of communities affecting communities. *Philosophical Transactions of the Royal Society of London. B.,* 366, 1389–400.

Ferrington Jr, L. C. (2008a). Global diversity of non-biting midges (Chironomidae; Insecta-Diptera) in freshwater. *Hydrobiologia,* 595, 447–55.

Ferrington Jr, L. C. (2008b). Global diversity of scorpionflies and hangingflies (Mecoptera) in freshwater. *Hydrobiologia,* 595, 443–5.

Field, L. H. and Matheson, T. (1998). Chordotonal organs of insects. *Advances in Insect Physiology,* 27, 1–228.

Figuerola, J., Green, A. J., and Santamaría, L. (2003). Passive internal transport of aquatic organisms by waterfowl in Doñana, south-west Spain. *Global Ecology & Biogeography,* 12, 427–36.

Fingerut, J. T., Hart, D. D., and McNair, J. N. (2006). Silk filaments enhance the settlement of stream insect larvae. *Oecologia,* 150, 202–12.

Fingerut, J. T., Schamel, L., Faugno, A., Mestrinaro, M., and Habdas, P. (2009). Role of silk threads in the dispersal of larvae through stream pools. *Journal of Zoology,* 279, 137–43.

Fink, P. (2007). Ecological functions of volatile organic compounds in aquatic systems. *Marine and Freshwater Behaviour and Physiology,* 40, 155–68.

Fish, F. E. and Nicastro, A. J. (2003). Aquatic turning performance by the whirligig beetle: constraints on maneuverability by a rigid biological system. *Journal of Experimental Biology,* 206, 1649–56.

Flenner, I., Richter, O., and Suhling, F. (2010). Rising temperature and development in dragonfly populations at different latitudes. *Freshwater Biology,* 55, 397–410.

Flynn, M. R. and Bush, J. W. M. (2008). Underwater breathing: the mechanics of plastron respiration. *Journal of Fluid Mechanics,* 608, 275–96.

Fochetti, R. and Tierno de Figueroa, J. M. (2008). Global diversity of stoneflies (Plecoptera; Insecta) in freshwater. *Hydrobiologia,* 595, 365–77.

Fonseca, D. M. (1999). Fluid-mediated dispersal in streams: models of settlement from the drift. *Oecologia,* 121, 212–23.

Fonseca, D. M. and Hart, D. D. (1996). Density-dependent dispersal of black fly neonates is mediated by flow. *Oikos,* 75, 49–58.

Foote, B. A. (1995). Biology of shore flies. *Annual Review of Entomology,* 40, 417–42.

Formanowicz, D. R. J. (1982). Foraging tactics of larvae of *Dytiscus verticalis* (Coleoptera: Dytiscidae). An assessment of prey density. *Journal of Animal Ecology,* 51, 757–67.

Formanowicz, D. R. J. (1987). Foraging tactics of *Dytiscus verticalis* larvae (Coleoptera: Dytiscidae): prey detection, reactive distance and predator size. *Journal of the Kansas Entomological Society,* 60, 92–9.

Formanowicz, D. R. J. and Brodie, E. D. J. (1981). Prepupation behavior and pupation of the predaceous diving beetle *Dytiscus verticalis* Say (Coleoptera: Dytiscidae). *Journal of the New York Entomological Society,* 89, 152–7.

Forrester, G. E. (1994). Influences of predatory fish on the drift and local density of stream insects. *Ecology,* 75, 1208–18.

Fox, H. M. and Sidney, J. (1953). The influence of dissolved oxygen on the respiratory movements of caddis larvae. *Journal of Experimental Biology,* 30, 235–7.

Fox, H. M., Wingfield, C. A., and Simmonds, B. G. (1947). The oxygen consumption of ephemerid nymphs from flowing and from still waters in relation to the concentration of oxygen in the water. *Journal of Experimental Biology,* 14, 210–18.

Francis, T. B., Schindler, D. E., and Moore, J. W. (2006). Aquatic insects play a minor role in dispersing salmon-derived nutrients into riparian forests in southwestern Alaska. *Canadian Journal of Fisheries and Aquatic Sciences,* 63, 2543–52.

Frantsevich, L. and Wang, W. (2009). Gimbals in the insect leg. *Arthropod Structure & Development,* 38, 16–30.

Freilich, J. E. (1991). Movement patterns and ecology of *Pteronarcys* nymphs (Plecoptera): observations of marked individuals in a Rocky Mountain stream. *Freshwater Biology,* 25, 379–94.

Frenzel, P. (1990). The influence of chironomid larvae on sediment oxygen microprofiles. *Archiv für Hydrobiologie,* 119, 427–37.

Frisch, K. V. (1967). *The Dance Language and Orientation of Bees*, Belknap Press of Harvard University Press, Cambridge, Massachusetts.

Frutiger, A. (1987). Investigations on the life history of the stonefly *Dinocras cephalotes* Curt. (Plecoptera: Perlidae). *Aquatic Insects*, 9, 51–63.

Frutiger, A. (1998). Walking on suckers—new insights into the locomotory behavior of larval net-winged midges (Diptera: Blephariceridae). *Journal of the North American Benthological Society*, 17, 104–20.

Fry, K. A. and Craig, D. A. (1995). Larval blackfly feeding (Diptera: Simuliidae): use of endogenous glycoconjugates. *Canadian Journal of Zoology*, 73, 615–22.

Fry, K. M. and Craig, D. A. (1996). Endogenous glycoconjugates are not associated with filter feeding in mosquito larvae. *Canadian Journal of Zoology*, 73, 413–22.

Fu, X., Vencl, F. V., Nobuyoshi, O., Meyer-Rochow, V. B., Lei, C., and Zhang, Z. (2007). Structure and function of the eversible glands of the aquatic firefly *Luciola leii* (Coleoptera: Lampyridae). *Chemoecology*, 17, 117–24.

Funk, D. H., Jackson, J. K., and Sweeney, B. W. (2006). Taxonomy and genetics of the parthenogenetic mayfly *Centroptilium triangulifer* and its sexual sister *Centroptilium alamance* (Ephemeroptera: Baetidae). *Journal of the North American Benthological Society*, 25, 417–29.

Funk, D. H., Jackson, J. K., and Sweeney, B. W. (2008). A new parthenogenetic mayfly (Ephemeroptera: Ephemerellidae: *Eurylophella* Tiensuu) oviposits by abdominal bursting in the subimago. *Journal of the North American Benthological Society*, 27, 269–79.

Funk, D. H., Sweeney, B. W., and Jackson, J. K. (2010). Why stream mayflies can reproduce without males but remain bisexual: a case of lost genetic variation. *Journal of the North American Benthological Society*, 29, 1258–66.

Gaino, E. and Bongiovanni, E. (1992). Comparative morphology of epithemata (polar chorionic structures) in the eggs of *Ephemerella ignita* (Ephemeroptera: Ephemerellidae). *Transactions of the American Microscopical Society*, 111, 255–65.

Gaino, E. and Mazzini, M. (1987). Scanning electron microscopy of the egg attachment structures of *Electrogena zebrata* (Ephemeroptera: Heptageniidae). *Transactions of the American Microscopical Society*, 106, 114–19.

Gaino, E., Piersanti, S., and Rebora, M. (2009). The oviposition mechanism in *Habrophlebia eldae* (Ephemeroptera: Leptophlebiidae). *Aquatic Insects*, 31, Supplement 1, 515–22.

Gaino, E. and Rebora, M. (1999a). Flat-tipped sensillum in Baetidae (Ephemeroptera): a microcharacter for taxonomic and phylogenetic considerations. *Invertebrate Biology*, 118, 68–74.

Gaino, E. and Rebora, M. (1999b). Larval antennal sensilla in water-living insects. *Microscopy Research and Technique*, 47, 440–57.

Gaino, E. and Rebora, M. (2001). Apical antennal sensilla in nymphs of *Libellula depressa* (Odonata: Libellulidae). *Invertebrate Biology*, 120, 162–9.

Gaino, E. and Rebora, M. (2005). Egg envelopes of *Baetis rhodani* and *Cloeon dipterum* (Ephemeroptera, Baetidae): a comparative analysis between an oviparous and an ovoviviparous species. *Acta Zoologica*, 86, 63–9.

Galacatos, K., Cognato, A. I., and Sperling, F. A. H. (2002). Population genetic structure of two water strider species in the Ecuadorian Amazon. *Freshwater Biology*, 47, 391–9.

Gall, B. G., Hopkins, G. R., and Brodie Jr, E. D. (2011). Mechanics and ecological role of swimming behavior in the caddisfly larvae *Triaenodes tardus*. *Journal of Insect Behavior*, 24, 317–28.

Gallepp, G. (1974). Diel periodicity in the behaviour of the caddisfly, *Brachycentrus americanus* (Banks). *Freshwater Biology*, 4, 193–204.

Gambles, R. M. (1956). Eggs of *Lestinogomphus africanus* Fraser. *Nature*, 177, 663.

Ganesan, K., Mendki, M. J., Suryanarayana, M. V. S., Prakash, M., and Malhotra, R. C. (2006). Studies of *Aedes aegypti* (Diptera: Culicidae) ovipositional reponses to newly identified semiochemicals from conspecific eggs. *Australian Journal of Entomology*, 45, 75–80.

Gao, X. and Jiang, L. (2004). Water-repellent legs of water striders. *Nature*, 432, 36.

Garbuz, D. G., Zatsepina, O. G., Prxhiboro, A. A., Yushenova, I., Guzhova, I. V., and Evgen'ev, M. B. (2008). Larvae of related Diptera species from thermally contrasting habitats exhibit continous up-regulation of heat shock proteins and high thermotolerance. *Molecular Ecology*, 17, 4763–77.

Gardner, W. S., Nalepa, T. F., Slavenus, D. S., and Lairds, G. A. (1983). Patterns and rates of nitrogen release by benthic Chironomidae and Oligochaeta. *Canadian Journal of Fisheries and Aquatic Sciences*, 40, 259–66.

Garten Jr, C. T. and Gentry, J. B. (1976). Thermal tolerance of dragonfly nymphs. II. Comparison of nymphs from control and thermally altered environments. *Physiological Zoology*, 49, 206–13.

Gattolliat, J. L. and Sartori, M. (2000). Guloptiloides: an extraordinary new carnivorous genus of Baetidae (Ephemeroptera) *Aquatic Insects*, 22, 148–59.

Genkai-Kato, M., Nozaki, K., Mitsuhashi, H., Kohmatsu, Y., Miyasaka, H., and Nakanishi, M. (2000). Push-up response of stonefly larvae in low-oxygen conditions. *Ecological Research*, 15, 175–9.

Gibbs, D. G. (1967). The proventriculus of some trichopterous larvae. *Journal of Zoology*, 152, 245–56.

Gibbs, K. E. (1977). Evidence of obligatory partheogenesis and its possible effect on the emergence period of *Cloeon triangulifer* (Ephemeroptera: Baetidae). *Canadian Entomologist*, 109, 337–40.

Gibbs, K. E. (1979). Ovoviviparity and nymphal seasonal movements on *Callibaetis* (Ephemeroptera: Baetidae) in a pond in southwestern Quebec. *Canadian Entomologist*, 111, 927–32.

Giberson, D. J., MacInnis, A. J., and Blanchard, M. (1996). Seasonal frequency and positioning of parasitic midges (Chironomidae) on *Pteronarcys bilboa* nymphs (Plecoptera: Pteronarcyidae). *Journal of the North American Benthological Society*, 15, 529–36.

Gibson, C. M. and Hunter, M. S. (2010). Extraordinarily widespread and fantastically complex: comparative biology of endosymbiotic bacterial and fungal mutualists of insects. *Ecology Letters*, 13, 223–34.

Gibson, G. (1985). Swarming behaviour of the mosquito *Culex pipiens quinquefasciatus*: a quantitative anlaysis. *Physiological Entomology*, 10, 283–96.

Giguère, L. A. and Dill, L. M. (1979). The predatory responses of *Chaoborus* larvae to acoustic stimuli, and the acoustic characteristics of their prey. *Zeitschrift für Tierpsychologie*, 50, 113–23.

Giguère, L. A. and Northcote, T. G. (1987). Ingested prey increase risks of visual predation in transparent *Chaoborus* larvae. *Oecologia*, 73, 48–52.

Gilbert, C. (1994). Form and function of stemmata in larvae of holometabolous insects. *Annual Review of Entomology*, 39, 323–49.

Giller, P. S. (1986). The natural diet of the Notonectidae: field trials using electrophoresis. *Ecological Entomology*, 11, 163–72.

Gillooly, J. F. and Dodson, S. I. (2000). The relationship of egg size and incubation temperature to embryonic development time in univoltine and multivoltine aquatic insects. *Freshwater Biology*, 44, 595–604.

Gillott, C. (2002). Insect accessory reproductive glands: Key players in production and protection of eggs. In M. Hilker and T. Meiner, eds. *Chemoecology of Insect Eggs and Egg Deposition*. pp. 37–59. Blackwell Publishing, Oxford.

Glasgow, J. P. (1936). Internal anatomy of a caddis (*Hydropsyche colonica*). *Quarterly Review of the Microscopical Society*, 79, 151–79.

Goodwyn, P. P., Katsumata-Wada, A., and Okada, K. (2009). Morphology and neurophysiology of tarsal vibration receptors in the water strider *Aquarius paludum* (Heteroptera: Gerridae). *Journal of Insect Physiology*, 55, 855–61.

Gopfert, M. C., Briegel, H., and Robert, D. (1999). Mosquito hearing: Sound-induced antennal vibrations in male and female *Aedes aegypti*. *Journal of Experimental Biology*, 202, 2727–38.

Gorb, S. N. (1994). Central projections of ovipositor sense organs in the damselfly, *Sympecma annulata* (Zygoptera, Lestidae). *Journal of Morphology*, 220, 139–46.

Gorb, S. N. (1995). Design of the predatory legs of water bugs (Hemiptera: Nepidae, Naucoridae, Notonectidae, Gerridae). *Journal of Morphology*, 223, 289–302.

Gorb, S. N. (2008). Smooth attachment devices in insects: Functional morphology and biomechanics. *Advances in Insect Physiology*, 34, 117–92.

Gordon, N. D., McMahon, T. A., Finlayson, B. L., Gippel, C. J., and Nathan, R. J. (2004). *Stream Hydrology: An Introduction for Ecologists*, 2nd Edn, John Wiley and Sons, Chichester.

Gotceitas, V. (1985). Formation of aggregations by overwintering fifth instar *Dicosmoecus atripes* larvae (Trichoptera). *Oikos*, 44, 313–18.

Gotceitas, V. and Clifford, H. F. (1983). The life history of *Dicosmoecus atripes* (Hagen) (Limnephilidae: Trichoptera) in a Rocky Mountain stream in Alberta, Canada. *Canadian Journal of Zoology*, 61, 586–96.

Green, A. J. and Figuerola, J. (2005). Recent advances in the study of long-distance dispersal of aquatic invertebrates via birds. *Diversity and Distributions*, 11, 149–56.

Green, A. J. and Sánchez, M. I. (2006). Passive internal dispersal of insect larvae by migratory birds. *Biology Letters*, 2, 55–7.

Greeney, H. F. (2001). The insects of plant-held waters: a review and bibliography. *Journal of Tropical Ecology*, 17, 241–60.

Greenfield, M. D. (2002). *Signalers and Receivers. Mechanisms and Evolution of Arthropod Communication*, Oxford University Press, Oxford.

Grenier, S. (1970). Biologie d'*Argiotypus armatusi* Curtis (Hymenoptera: Agriotypidae), parasite de nymphes de Trichopteres. *Annals de Limnologie*, 6, 317–61.

Grey, J., Kelly, A., Ward, S., Sommerwerk, N., and Jones, R. I. (2004). Seasonal changes in the stable isotope values of lake-dwelling chironomid larvae in relation to feeding and life cycle variability. *Freshwater Biology*, 49, 681–9.

Grimaldi, D. and Engel, M. S. (2005). *Evolution of the Insects*, Cambridge University Press, New York.

Gross, E. M., Brune, A., and Walenciak, O. (2008). Gut pH, redox conditions and oxygen levels in an aquatic caterpillar: Potential effects on the fate of ingested tannins. *Journal of Insect Physiology*, 54, 462–71.

Gross, E. M., Feldbaum, C., and Choi, C. (2002). High abundance of herbivorous Lepidoptera larvae (*Acentria ephemerella* Denis & Schiffermüller) on submerged macrophytes in Lake Constance (Germany). *Archiv für Hydrobiologie*, 155, 1–21.

Guerra, P. A. (2011). Evaluating the life-history trade-off between dispersal capability and reproduction in wing dimorphic insects: a meta-analysis. *Biological Reviews*, 86, 813–35.

Gullefors, B. and Petersson, E. (1993). Sexual dimorphism in relation to swarming and pair formation patterns in leptocerid caddisflies (Trichoptera: Leptoceridae). *Journal of Insect Behavior*, 6, 563–77.

Gunter, G. and Christmas, J. Y. (1959). Corixid insects as part of the offshore fauna of the sea. *Ecology*, 40, 723–4.

Günther, K. (1912). Die Sehorgane der Larve und Imago von *Dytiscus marginalis*. *Zeitschrift für Wissenschaftliche Zoologie*, 100, 60–155.

Gupta, S., Gupta, A., and Meyer-Rochow, V. B. (2000). Post-embryonic development of the lateral eye of *Cloeon* sp. (Ephemeroptera: Baetidae) as revealed by scanning electron microscopy. *Entomologica Fennica*, 11, 89–96.

Guzik, M. T., Austin, A. D., Cooper, S. J. B., Harvey, M. S., Humphreys, W. F., Bradford, T., Eberhard, S. M., King, R. A., Leys, R., Muirhead, K. A., and Tomlinson, M. (2010). Is the Australian subterranean fauna uniquely diverse? *Invertebrate Systematics*, 24, 407–18.

Habeck, D. H. and Balciunas, J. K. (2005). Larvae of Nymphulinae (Lepidoptera: Pyralidae) associated with *Hydrilla verticillata* (Hydrocharitaceae) in North Queensland. *Australian Journal of Entomology*, 44, 353–63.

Hagberg, M. (1986). Ultrastructure and central projections of extraocular photoreceptors in caddisflies (Insecta: Trichoptera). *Cell and Tissue Research*, 245, 643–8.

Hagner-Holler, S., Pick, P., Girgenrath, S., Marden, J. H., and Burmester, T. (2007). Diversity of stonefly hexamerins and implications for the evolution of insect storage proteins. *Insect Biochemistry and Molecular Biology*, 37, 1064–74.

Hagner-Holler, S., Schoen, A., Erker, W., Marden, J. H., Rupprecht, R., Decker, H., and Burmester, T. (2004). A respiratory hemocyanin from an insect. *Proceedings of the National Academy of Sciences of the U.S.A.*, 101, 871–4.

Hahn, D. A. and Denlinger, D. L. (2007). Meeting the energetic demands of insect diapause: Nutrient storage and utilization. *Journal of Insect Physiology*, 53, 760–73.

Hallberg, E. and Hagberg, M. (1986). Ocellar fine structure in *Caenis robusta* (Ephemeroptera), *Trichostegia minor*, *Agrypnia varia*, and *Limnephilus clavicornis* (Trichoptera). *Protoplasma*, 135, 12–18.

Halse, S. A., Shiel, R. J., and Williams, W. D. (1998). Aquatic invertebrates of Lake Gregory, northwestern Australia, in relation to salinity and ionic composition. *Hydrobiologia*, 381, 15–29.

Hammer, U. T. (1978). Saline lakes of Saskatchewan. 3. Chemical characterization. *Internationale Revue der Gesamten Hydrobiologie*, 63, 311–35.

Hankin, M. A. (1921). The soaring flight of dragonflies. *Proceedings of the Cambridge Philosophical Society*, 20, 460–5.

Hansell, M. H. (1973). Improvements and termination of house building in the caddis larva *Lepidostoma hirtum* Curtis. *Behaviour*, 46, 141–53.

Harada, T. and Nishimoto, T. (2007). Feeding conditions modify the photoperiodically induced dispersal of the water strider, *Aquarius paludum* (Heteroptera: Gerridae). *European Journal of Entomology*, 104, 33–7.

Harding, J. S. (2006). 'Basking' behaviour in adults of *Spaniocercoides cowleyi* (Plecoptera: Nemouridae). *New Zealand Journal of Natural Science*, 31, 71–8.

Harker, J. E. (1997). The role of parthenogenesis in the biology of two species of mayfly (Ephemeroptera). *Freshwater Biology*, 37, 287–97.

Harker, J. E. (1999). The structure of the foregut and midgut of nymphs, subimagos and imagos of *Cloeon dipterum* (Ephemeroptera) and the functions of the gut of adult mayflies. *Journal of Zoology*, 248, 243–53.

Harlin, C. (2005). To have and have not: volatile secretions make a difference in gyrinid beetle predator defence. *Animal Behaviour*, 69, 579–85.

Harrison, J. F., Kaiser, A., and VandenBrooks, J. M. (2010). Atmospheric oxygen level and the evolution of insect body size. *Proceedings of the Royal Society, London B*, 277, 1937–46.

Harrison, R. G. (1980). Dispersal polymorphisms in insects. *Annual Review of Ecology and Systematics*, 11, 95–118.

Hart, D. D., Clark, B. D., and Jasentuliyana, A. (1996). Fine-scale field measurement of benthic flow environments inhabited by stream invertebrates. *Limnology and Oceanography*, 41, 297–308.

Hart, D. D. and Latta, S.C. (1986). Determinants of ingestion rates in filter-feeding larval black flies (Diptera: Simuliidae). *Freshwater Biology*, 16, 1–14.

Hart, D. D., Merz, R. A., Genovese, S. J., and Clark, B. D. (1991). Feeding postures of suspension-feeding larval blackflies: the conflicting demands of drag and food acquisition. *Oecologia*, 85, 457–63.

Hart, E. A. and Lovvorn, J. R. (2005). Patterns of macroinvertebrate abundance in inland saline wetlands: a trophic analysis. *Hydrobiologia*, 541, 45–54.

Hartland-Rowe, R. (1953). Feeding mechanism of an ephemeropteran nymph. *Nature*, 172, 1109–10.

Hartland-Rowe, R. (1958). The biology of a tropical mayfly *Povilla adusta* Navas (Ephemeroptera, Polymictarcidae) with special reference to the lunar rhythm of emergence. *Revue de Zoologie et de Botanique Africaines*, 58, 185–202.

Hayashi, F. (1993). Male mating costs in two insect species (*Protohermes*, Megaloptera) that produce large spermatophores. *Animal Behaviour*, 45, 343–9.

Hayashi, H. (1989). Microhabitat selection by the fishfly larva, *Parachauliodes japonicus*, in relation to its mode of respiration. *Freshwater Biology*, 21, 489–96.

Hayashi, H. (1996). Insemination through an externally attached spermatophore: bundled sperm and post-copulatory mate guarding by male fisherflies (Megaloptera: Corydalidae). *Journal of Insect Physiology*, 42, 859–66.

Hayashi, H. (1998). Sperm co-operation in the fishfly, *Parachauliodes japonicus*. *Functional Ecology*, 12, 347–50.

Hayashi, K. (1985). Alternative mating strategies in the water strider *Gerris elongatus* (Heteropetera, Gerridae). *Behavioral Ecology and Sociobiology*, 16, 301–306.

Hegedus, D., Erlandson, M., Gillott, C., and Toprak, U. (2009). New insights into peritrophic matrix, synthesis, architecture, and function. *Annual Review of Entomology*, 54, 285–302.

Heiman, D. R. and Knight, A. W. (1972). Upper lethal temperature relations of the nymphs of the stonefly, *Paragnetina media*. *Hydrobiologia*, 39, 479–93.

Heinis, F. and Crommentuijn, T. (1992). Behavioral responses to changing oxygen concentrations of deposit feeding chironomid larvae (Diptera) of littoral and profundal habitats. *Archiv für Hydrobiologie*, 124, 173–85.

Heinrich, B. and Casey, T. M. (1978). Heat transfer in dragonflies: 'Fliers' and 'perchers'. *Journal of Experimental Biology*, 74, 17–36.

Heinrich, B. and Vogt, F. D. (1980). Aggregation and foraging behavior of whirligig beetles (Gyrinidae). *Behavioral Ecology and Sociobiology*, 7, 179–86.

Heise, B. A. (1992). Sensitivity of mayfly nymphs to red light: implications for behavioural ecology. *Freshwater Biology*, 28, 331–6.

Henrikson, B. -I. and Stenson, J. A. E. (1993). Alarm substances in *Gyrinus aeratus* (Coleoptera, Gyrinidae). *Oecologia*, 93, 191–4.

Herbst, D. B. (1988). Comparative population ecology of *Ephydra hians* Say (Diptera: Ephydridae) at Mono Lake (California) and Abert Lake (Oregon). *Hydrobiologia*, 158, 145–66.

Herbst, D. B. (2001). Gradients of salinity stress, environmental stability and water chemistry as a templet for defining habitat types and physiological strategies in inland salt waters. *Hydrobiologia*, 466, 209–19.

Hershey, A. E., Pastor, J., Peterson, B. J., and Kling, G. W. (1993). Stable isotopes resolve the drift paradox for *Baetis* mayflies in an Arctic river. *Ecology*, 74, 2315–25.

Hickman, J. R. (1931). Contributions to the biology of the Haliplidae (Coleoptera). *Annals of the Entomological Society of America*, 24, 129–42.

Hildrew, A. G. and Wagner, R. (1992). The briefly colonial life of hatchlings of the net-spinning caddisfly *Plectrocnemia conspersa*. *Journal of the North American Benthological Society*, 11, 60–8.

Hinton, H. E. (1951). A new chironomid from Africa, the larva of which can be dehydrated without injury. *Proceedings of the Zoological Society of London*, 121, 371–80.

Hinton, H. E. (1955). On the structure, function, and distribution of the prolegs of Panorpoidea, with a criticism of the Berlese-Imms theory. *Transactions of the Royal Entomological Society*, 106, 455–545.

Hinton, H. E. (1960a). Cryptobiosis in the larva of *Polypedilum vanderplanki* Hint (Chironomidae). *Journal of Insect Physiology*, 5, 286–300.

Hinton, H. E. (1960b). A fly larva that tolerates dehydration and temperatures from -270° C to + 102 °C. *Nature*, 188, 366–7.

Hinton, H. E. (1961). The structure and function of the egg shell in the Nepidae (Hemiptera). *Journal of Insect Physiology*, 7, 224–57.

Hinton, H. E. (1964). The respiratory efficiency of the spiracular gill of *Simulium*. *Journal of Insect Physiology*, 10, 73–80.

Hinton, H. E. (1966). Respiratory adaptations of the pupae of beetles of the family Psephenidae. *Philosophical Transactions of the Royal Society of London. B.*, 251, 211–45.

Hinton, H. E. (1968a). Reversible suspension of metabolism and the origin of life. *Proceedings of the Royal Society, London B*, 171, 43–57.

Hinton, H. E. (1968b). Spiracular gills. *Advances in Insect Physiology*, 5, 65–162.

Hinton, H. E. (1968c). The structure and protective devices of the egg of the mosquito *Culex pipiens*. *Journal of Insect Physiology*, 14, 145–61.

Hinton, H. E. (1969). Respiratory systems of insect egg shells. *Annual Review of Entomology*, 14, 343–68.

Hinton, H. E. (1976a). Plastron respiration in bugs and beetles. *Journal of Insect Physiology*, 22, 1529–50.

Hinton, H. E. (1976b). The fine structure of the pupal plastron of simuliid flies. *Journal of Insect Physiology*, 22, 1061–70.

Hinton, H. E. (1981). *Biology of Insect Eggs, Vols 1–3*, Pergamon Press, Oxford.

Hirayama, H. and Kasuya, E. (2009). Oviposition depth in response to egg parasitism in the water strider: high-risk experience promotes deeper oviposition. *Animal Behaviour*, 78, 935–41.

Hirayama, H. and Kasuya, E. (2010). Cost of oviposition site selection in a water strider *Aquarius paludum insularis*: Egg mortality increases with oviposition depth. *Journal of Insect Physiology*, 56, 646–9.

Hix, R. L., Johnson, D. T., and Bernhardt, J. L. (2000). Swimming behavior of an aqutic weevil, *Lissorhoptrus oryzophilus* (Coleoptera: Curculionidae). *Florida Entomologist*, 83, 316–24.

Ho, J. -Z., Chiang, P. -H., Wu, C. -H., and Yang, P. -S. (2010). Life cycle of the aquatic firefly *Luciola ficta* (Coleoptera: Lampyridae). *Journal of Asia-Pacific Entomology*, 13, 189–96.

Hoback, W. W. and Stanley, D. W. (2001). Insects in hypoxia. *Journal of Insect Physiology*, 47, 533–42.

Hocking, B. (1952). Autolysis of flight muscles in a mosquito. *Nature*, 169, 1101.

Hoffman, A. and Hering, D. (2000). Wood-associated macroinvertebrate fauna in Central European streams. *International Review of Hydrobiology*, 85, 25–48.

Hoffmann, A. and Resh, V. H. (2003). Oviposition in three species of limnephilid caddisflies (Trichoptera): hierarchical influences in site selection. *Freshwater Biology*, 48, 1064–77.

Hoffmann, A. L., Olden, J. D., Monroe, J. B., Poff, N. L., Wellnitz, T., and Wiens, J. A. (2006). Current velocity and habitat patchiness shape stream herbivore movement. *Oikos*, 115, 358–68.

Hoffsten, P. -O. (2004). Site-occupancy in relation to flight-morphology in caddisflies. *Freshwater Biology*, 49, 810–17.

Honěk, A. (1993). Intraspecific variation in body size and fecundity in insects: a general relationship. *Oikos*, 66, 483–92.

Hopper, K. R. (1999). Risk-spreading and bet-hedging in insect population biology. *Annual Review of Entomology*, 44, 535–60.

Hora, S. L. (1930). Ecology, bionomics and evolution of the torrential fauna, with special reference to the organs of attachment. *Philosophical Transactions of the Royal Society of London Series B*, 218, 171–282.

Horridge, G. A., Marčelja, L., and Jahnke, R. (1982). Light guides in the dorsal eye of the male mayfly. *Proceedings of the Royal Society, London B*, 216, 25–51.

Horridge, G. A., Marčelja, L., Jahnke, R., and McIntyre, P. (1983). Daily changes in the compound eye of a beetle (*Macrogyrus*). *Proceedings of the Royal Society, London B*, 217, 265–85.

Horridge, G. A., Walcott, B., and Ioannides, A. C. (1970). The tiered retina of *Dytiscus*: a new type of compound eye. *Proceedings of the Royal Society, London B*, 175, 83–94.

Horváth, G., Blahó, M., Egri, A., Kriska, G., Seres, I., and Robertson, B. (2010). Reducing the maladaptive attractiveness of solar panels to polarotactic Insects. *Conservation Biology*, 24, 1644–53.

Horváth, G. and Kriska, G. (2008). Polarization vision in aquatic insects and ecological traps for polarotactic insects. In J. Lancaster and R. A. Briers, eds. *Aquatic Insects: Challenges to Populations*. pp. 204–29. CAB International, Wallingford.

Horváth, G. and Varjú, D. (2004). *Polarized Light in Animal Vision*, Springer-Verlag, Berlin.

Horváth, G. and Zeil, J. (1996). Kuwait oil lakes as insect traps. *Nature*, 379, 303–304.

Houlihan, D. F. (1969a). Respiratory physiology of the larva of *Donacia simplex*, a root-piercing beetle. *Journal of Insect Physiology*, 15, 1517–36.

Houlihan, D. F. (1969b). The structure and behaviour of *Notiphila riparia* and *Erioptera squalida*, two root-piercing insects. *Journal of Zoology, London*, 159, 249–67.

Hu, D. L. and Bush, J. W. M. (2005). Meniscus-climbing insects. *Nature*, 437, 733–6.

Hu, D. L. and Bush, J. W. M. (2010). The hydrodynamics of water-walking arthropods. *Journal of Fluid Mechanics*, 644, 5–33.

Hu, D. L., Chan, B., and Bush, J. W. M. (2003). The hydrodynamics of water strider locomotion. *Nature*, 424, 663–6.

Huang, D. and Cheng, L. (2011). The flightless marine midge *Pontomyia* (Diptera: Chironomidae): ecology, distribution, and molecular phylogeny. *Zoological Journal of the Linnean Society*, 162, 443–56.

Hughes, G. M. (1958). The co-ordination of insect movements III. Swimming in *Dytiscus*, *Hydrophilus*, and a dragonfly nymph. *Journal of Experimental Biology*, 35, 367–538.

Hughes, G. M. and Mill, P. J. (1966). Patterns of ventilation in dragonfly larvae. *Journal of Experimental Biology*, 44, 317–33.

Humpesch, U. H. (1980). Effect of temperature on the hatching time of parthenogenetic eggs of five *Ecdyonurus* spp. and two *Rhithrogena* spp. (Ephemeroptera) from Austrian streams and English lakes and rivers. *Journal of Animal Ecology*, 49, 927–37.

Humphrey, J. A. C. and Barth, F. G. (2008). Medium flow-sensing hairs: biomechanics and models. *Advances in Insect Physiology*, 34, 1–80.

Humphreys, W. F., Watts, C. H. S., Cooper, S. J. B., and Leijs, R. (2009). Groundwater estuaries of salt lakes: buried pools of endemic biodiversity on the western plateau, Australia. *Hydrobiologia*, 626, 79–95.

Hunter, D. M. (1977). Sugar-feeding in some Queensland blackflies (Diptera: Simuliidae). *Journal of Medical Entomology*, 14, 229–32.

Hynes, H. B. N. (1941). The taxonomy and ecology of the nymphs of British Plecoptera with notes on the adults and eggs. *Transactions of the Royal Entomological Society, London*, 91, 459–555.

Hynes, H. B. N. (1970). *The Ecology of Running Waters*, Liverpool University Press, Liverpool.

Hynes, H. B. N. (1976). Biology of the Plecoptera. *Annual Review of Entomology*, 21, 135–53.

Hynes, H. B. N. and Hynes, M. E. (1975). The life histories of many of the stoneflies (Plecoptera) of southeastern mainland Australia. *Australian Journal of Marine and Freshwater Research*, 26, 113–53.

Ichikawa, N. (1988). Male brooding behaviour of the giant water bug *Lethocerus deyrollei* Vuillefroy (Hemiptera: Belostomatidae). *Journal of Ethology*, 6, 121–7.

Ichikawa, N. (1991). Egg mass destroying and guarding behaviour of the giant water bug, *Lethocerus deyrollei* Vuilefroy (Heteroptera: Belostomatidae). *Journal of Ethology*, 9, 25–9.

Ide, F. P. (1937). The subimago of *Ephoron leukon* Will., and a discussion of the imago instar. (Ephem.). *Canadian Entomologist*, 69, 25–9.

Illies, J. (1963). The Plecoptera of the Aukland and Campbell Islands. *Records of the Dominion Museum*, 4, 255–65.

Illies, J. (1965). Phylogeny and zoogeography of Plecoptera. *Annual Review of Entomology*, 10, 117–40.

Inoda, T., Hirata, Y., and Kamimura, S. (2003). Asymmetric mandibles of water-scavenger larvae improve feeding effectiveness on right-handed snails. *The American Naturalist*, 162, 811–14.

Inoue, T. and Harada, T. (1997). Sensitive stages in the photoperiodic determination of wing forms and reproduction in the water strider *Aquarius paludum* (Fabricius). *Zoological Science*, 14, 21–7.

Irons III, J. G., Miller, K., and Oswood, M. W. (1993). Ecological adaptations of aquatic macroinvertebrates to overwintering in interior Alaska (U.S.A.) subarctic streams. *Canadian Journal of Zoology*, 71, 98–108.

Irving-Bell, R. J., Inynag, E. N., and Tamu, G. (1991). Survival of *Aedes vittatus* (Diptera, Culicidae) eggs in hot, dry rockpools. *Tropical Medicine and Parasitology*, 42, 63–6.

Ishihara, D., Yamashita, Y., Horie, T., Yoshida, S., and Niho, T. (2009). Passive maintenance of high angle of attack and its lift generation during flapping translation in crane fly wing. *Journal of Experimental Biology*, 212, 3882–91.

Ismail, I. A. H. (1962). Sense organs on the antennae of *Anopheles maculipennis troparvus* (v. Thiel), and their possible function in relation to the attraction of female mosquito to man. *Acta Tropica*, 19, 1–58.

Isobe, Y. and Uchida, S. (2009). Japanese species of the genus *Oyamia* (Plecoptera: Perlidae), with notes on *O. nigribasis* from Korea. *Aquatic Insects*, 31, Supplement 1, 231–44.

Ivanov, V. D. and Rupprecht, R. (1992). Substrate vibration for communication in adult *Agapetus fuscipes* (Trichoptera: Glossosomatidae). In C. Otto, ed. *Proceedings of the Seventh International Symposium on Trichoptera*. pp. 273–8. Backhuys Publishers, Leiden.

Jaag, O. and Ambühl, H. (1964). The effect of the current on the composition of biocoenoses in flowing water streams. In B. A. Southgate, ed. *Advances in Water Pollution Research Vol. 1*. pp. 31–44. Pergamon Press, Oxford.

Jäch, M. A. and Balke, M. (2008). Global diversity of water beetles (Coleoptera) in freshwater. *Hydrobiologia*, 595, 419–42.

Jackson, D. J. (1957). A note on the embryonic cuticle shed on hatching by the larvae of *Agabus bipustulatus* L. and *Dytiscus marginalis* L. (Coleoptera: Dytiscidae). *Proceedings of the Royal Entomological Society, London*, 32, 115–18.

Jackson, D. J. (1958). Egg-laying and egg-hatching in *Agabus bipustulatus* L., with notes on oviposition in other species of *Agabus* (Coleoptera: Dytiscidae). *Transactions of the Royal Entomological Society London*, 110, 53–80.

Jackson, D. J. (1960). Observations on egg-laying in *Ilybius fuliginosus* Fabricius and *I. ater* Degeer (Coleoptera, Dytiscidae), with an account of the female genitalia. *Transactions of the Royal Entomological Society, London*, 112, 37–52.

Jackson, D. J. (1961). Observations on the biology of *Caraphractus cinctus* Walker (Hymenoptera: Mymaridae), a parasitoid of the eggs of Dytiscidae (Coleoptera). 2. Immature stages and seasonal history with a review of mymarid larvae. *Parasitology*, 51, 269–94.

Jackson, D. J. (1966). Observations on the biology of *Caraphractus cinctus* Walker (Hymenoptera: Mymaridae), a parasitoid of the eggs of Dytiscidae (Coleoptera). III. The adult life and sex ratio. *Transactions of the Royal Entomological Society, London*, 118, 23–49.

Jackson, J. K. and Resh, V. H. (1991). Periodicity in mate attraction and flight activity of three species of caddisflies (Trichoptera). *Journal of the North American Benthological Society*, 10, 198–209.

Jackson, J. K., McElravy, E. P., and Resh, V. H. (1999). Long-term movements of self-marked caddisfly larvae (Trichoptera: Sericostomatidae) in a California coastal mountain stream. *Freshwater Biology*, 42, 525–36.

Jackson, J. K. and Sweeney, B. W. (1995). Egg and larval development times in 35 species of tropical stream insects from Costa Rica. *Journal of the North American Benthological Society*, 14, 115–30.

Jacobsen, D. (2000). Gill size of trichopteran larvae and oxygen supply in streams along a 4000-m gradient in altitude. *Journal of the North American Benthological Society*, 19, 329–43.

Jacobsen, D. and Brodersen, K. P. (2008). Are altitudinal limits of equatorial stream insects reflected in their respiratory performance? *Freshwater Biology*, 53, 2295–308.

Jansson, A. (1972). Mechanisms of sound production and morphology of the stridulatory apparatus in the genus *Cenocorixa* (Hemiptera, Corixidae). *Annales Zoologici Fennici*, 9, 120–9.

Jansson, A. (1973). Stridulation and its significance in the genus *Cenocorixa* (Hemiptera, Corixidae). *Behaviour*, 46, 1–36.

Jansson, A. and Scudder, G. G. E. (1974). The life cycle and sexual development of *Cenocorixa* species (Hemiptera, Corixidae) in the Pacific Northwest of North America. *Freshwater Biology*, 4, 73–92.

Jansson, A. and Vuoristo, T. (1979). Significance of stridulation in larval Hydropsychidae (Trichoptera). *Behaviour*, 71, 167–86.

Jensen, J. C. and Zacharuk, R. Y. (1991). The fine structure of uniporous and nonporous pegs on the distal antennal segment of the diving beetle *Graphoderus occidentalis* Horn (Coleoptera, Dytiscidae). *Canadian Journal of Zoology*, 69, 334–52.

Jensen, J. C. and Zacharuk, R. Y. (1992). The fine structure of the multiporous sensilla on the antenna of the diving beetle *Graphoderus occidentalis* Horn (Coleoptera, Dytiscidae). *Canadian Journal of Zoology*, 70, 825–32.

Jewett, S. G. (1963). A stonefly aquatic in the adult stage. *Science*, 139, 484–5.

Johansson, F. and Crowley, P. H. (2008). Larval cannibalism and population dynamics of dragonflies. In J. Lancaster and R. A. Briers, eds. *Aquatic Insects: Challenges to Populations.* pp. 36–54. CAB International, Wallingford.

Johansson, F., Söderquist, M., and Bokma, F. (2009). Insect wing shape evolution: independent effects of migratory and mate guarding flight on dragonfly wings. *Biological Journal of the Linnean Society*, 97, 362–72.

Johnsen, S. (2003). Hidden in plain sight: The ecology and physiology of organismal transparency. *Biological Bulletin*, 201, 301–18.

Johnstone, G. W. (1964). Stridulation by larval Hydropsychidae (Trichoptera). *Proceedings of the Royal Entomological Society, London (A)*, 39, 146–50.

Jónasson, P. M. and Kristiansen, J. (1967). Primary and secondary production in Lake Esrom. Growth of *Chironomus anthracinus* in relation to seasonal cycles of phytoplankton and dissolved oxygen. *Internationale Revue der Gesamten Hydrobiologie*, 52, 163–217.

Jones, R. E. (1975). Dehydration in an Australian rockpool chironomid larva, (*Paraborniella tonnoiri*). *Journal of Entomology Series A, General Entomology*, 49, 111–19.

Juday, C. (1921). Observations on the larvae of *Corethra punctipennis* Say. *Biological Bulletin Woods Hole*, 40, 271–86.

Junger, M. and Planas, D. (1994). Quantitative use of stable carbon isotope analysis to determine the trophic base of invertebrate communities in a boreal forest lotic system. *Canadian Journal of Fisheries and Aquatic Sciences*, 51, 52–61.

Kaitala, A. (1991). Phenotypic plasticity in reproductive behaviour of waterstriders: trade-offs between reproduction and longevity during food stress. *Functional Ecology*, 5, 12–18.

Kaitala, A. and Huldén, L. (1990). Significance of spring migration and flexibility in flight-muscle histolysis in waterstriders (Heteroptera, Gerridae). *Ecological Entomology*, 15, 409–18.

Kalkman, V. J., Clausnitzer, V., Dijkstra, K. -D. B., Orr, A. G., Paulson, D. R., and van Tol, J. (2008). Global diversity of dragonflies (Odonata) in freshwater. *Hydrobiologia*, 595, 351–63.

Kanou, M. and Shimazawa, T. (1983). The elicitation of the predatory labial strike of dragonfly larvae in response to a purely mechanical stimulus. *Journal of Experimental Biology*, 107, 391–404.

Kapoor, N. N. (1974). Some studies on the respiration of stonefly nymph, *Paragnetina media* (Walker). *Hydrobiologia*, 44, 37–41.

Kapoor, N. N. (1987). Fine structure of the coniform sensillar complex on the antennal flagellum of the stonefly nymph *Paragnetina media* (Plecoptera, Perlidae). *Canadian Journal of Zoology*, 65, 1827–32.

Kaufman, M. G., Pankratz, H. S., and Klug, M. J. (1986). Bacteria associated with the ectoperitrophic spaces in the midgut of the larvae of the midge *Xylopotus par* (Diptera: Chironomidae). *Applied Environmental Microbiology*, 51, 657–60.

Kaulenas, M. S. (1992). *Insect Accessory Reproductive Structures: Function, Structure and Development*, Springer Verlag, Berlin.

Keats, R. A. and Osher, L. J. (2007). The macroinvertebrates of *Ruppia* (Widgeon Grass) beds in a small Maine estuary. *Northeastern Naturalist*, 14, 481–91.

Kehl, S. and Dettner, K. (2009). Surviving submerged – setal tracheal gills for gas exchange in adult rheophilic diving beetles. *Journal of Morphology*, 270, 1348–1355.

Keilin, D. (1959). The problem of anabiosis or latent life: history and current concept. *Proceedings of the Royal Society, London B*, 150, 149–91.

Keiper, J. B. (2002). Biology and immature stages of coexisting Hydroptilidae (Trichoptera) from Northeastern Ohio lakes. *Annals of the Entomological Society of America*, 95, 608–16.

Keiper, J. B. and Foote, B. A. (2000). Biology and larval feeding habits of coexisting Hydroptilidae (Trichoptera) from a small woodland stream in Northeastern Ohio. *Annals of the Entomological Society of America*, 93, 225–34.

Keiper, J. B., Walton, W. E., and Foote, B. A. (2002). Biology and ecology of higher Diptera from freshwater wetlands. *Annual Review of Entomology*, 47, 207–32.

Kelly, L. C., Bilton, D. T., and Rundle, S. D. (2001). Population structure and dispersal in the Canary Island caddisfly *Mesophylax aspersus* (Trichoptera, Limnephilidae). *Heredity*, 86, 370–7.

Keltner, J. and McCafferty, W. P. (1986). Functional morphology of burrowing in the mayflies *Hexgenia limbata* and *Pentagenia vittigera*. *Zoological Journal of the Linnean Society*, 87, 139–62.

Kennedy, H. D. (1958). Biology and life history of a new species of mountain midge, *Deuterophlebia nielsoni*, from eastern California (Diptera: Deuterophlebiidae). *Transactions of the American Microscopical Society*, 77, 201–28.

Kesel, A. B. (2000). Aerodynamic characteristics of dragonfly wing sections compared with technical aerofoils. *Journal of Experimental Biology*, 203, 3125–35.

Khalifa, A. (1949). Spermatophore production in Trichoptera and some other insects. *Transactions of the Royal Entomological Society*, 100, 449–71.

Kiel, E. (2001). Behavioural response of blackfly larvae (Simuliidae, Diptera) to different current velocities. *Limnologica*, 31, 179–83.

Kight, S. L., Sprague, J., Kruse, K. C., and Johnson, L. (1995). Are egg-bearing male water bugs, *Belostoma flumineum* Say (Hemiptera: Belostomatidae), impaired swimmers? *Journal of the Kansas Entomological Society*, 68, 468–70.

Kikawada, T., Minakawa, N., Watanabe, M., and Okuda, T. (2005). Factors inducing successful anhydrobiosis in the African chironomid *Polypedilum vanderplanki*: significance of the larval tubular nest. *Integrative and Comparative Biology*, 45, 710–14.

King, I. M. (1976). Underwater sound production in *Micronecta batilla* Hale (Heteroptera: Corixidae). *Journal of the Australian Entomological Society*, 15, 35–43.

King, I. M. (1999). Species-specific sounds in water bugs of the genus *Micronecta*. Part 2, Chorusing. *Bioacoustics*, 10, 19–29.

King, P. E. and Fordy, M. R. (1984). Observations on *Aepophilus bonnairei* (Signoret) (Saldidae: Hemiptera) an intertidal insect of rocky shores. *Zoological Journal of the Linnean Society*, 80, 231–8.

King, P. E., Fordy, M. R., and Elliott, P. (1982). A comparison of the environmental adaptations of the intertidal carabids *Aepus robini* Laboulbéne and *Aepus marinus* Ström. *Journal of Natural History*, 16, 335–43.

King, R. A., Read, D. S., Traugott, M., and Symondson, W. O. C. (2008). Molecular analysis of predation: a review of best practice for DNA-based approaches. *Molecular Ecology*, 17, 947–63.

Kingsolver, J. and Huey, R. B. (2008). Size, temperature, and fitness: three rules. *Evolutionary Ecology Research*, 10, 251–68.

Kishi, M., Harada, T., and Fujisaki, K. (2007). Dispersal and reproductive responses of the water strider, *Aquarius paludum* (Hemiptera: Gerridae), to changing NaCl concentrations. *European Journal of Entomology*, 104, 377–83.

Kline, T. C. J., Goering, J. J., and Piorkowski, R. J. (1997). The effect of salmon carcasses on Alaskan freshwaters. In A. M. Milner and M. W. Oswood, eds. *Freshwaters of Alaska. Ecological Syntheses*. pp. 179–204. Springer-Verlag, New York.

Klug, M. J. and Kotarski, S. (1980). Bacteria associated with the gut tract of larval stages of the aquatic cranefly, *Tipula abdominalis* (Diptera: Tipulidae). *Applied Environmental Microbiology*, 40, 408–16.

Koehl, M. A. R. (1995). Fluid flow through hair-bearing appendages: feeding, smelling and swimming at low and intermediate Reynolds numbers. In C. P. Ellington and T. J. Pedley, eds. *Biological Fluid Dynamics*. pp. 157–82. The Company of Biologists Ltd, Cambridge.

Koehl, M. A. R. (1996). When does morphology matter? *Annual Review of Ecology and Systematics*, 27, 501–42.

Koehl, M. A. R. (2000). Consequences of size change during ontogeny and evolution. In J. H. Brown and G. B. West, eds. *Scaling in Biology*. pp. 67–86. Oxford University Press, New York.

Koehl, M. A. R. (2006). The fluid mechanics of arthropod sniffing in turbulent odor plumes. *Chemical Senses*, 31, 93–105.

Kohler, S. L. (1985). Identification of stream drift mechanisms: an experimental and observational approach. *Ecology*, 66, 1749–61.

Kohnert, S., Perry, S. F., and Schmitz, A. (2004). Morphometric analysis of the larval branchial chamber in the dragonfly *Aeshna cyanea* Müller (Insecta, Odonata, Anisoptera). *Journal of Morphology*, 261, 81–91.

Kolmes, S. A. (1983). Ecological and sensory aspects of prey capture by the whirligig beetle *Dineutes discolor* (Coleoptera: Gyrinidae). *Journal of the New York Entomological Society*, 91, 405–12.

Kolmes, S. A. (1985). Vibrational cues in the precopulatory behavior of whirligig beetles. *Journal of the New York Entomological Society*, 93, 1137–40.

Komnick, H. (1982). The rectum of larval dragonflies as jet-engine, respirator, fuel depot and ion pump. *Advances in Odonatology*, 1, 69–91.

Kon, M. and Hidaka, T. (1983). Chimney projecting behaviour of chironomid larvae (*Chironomus yoshimatsui*; Diptera, Chironomidae). *Journal of Ethology*, 1, 111–13.

Konopová, B. and Zrzavý, J. (2005). Ultrastructure, development, and homology of insect embryonic cuticles. *Journal of Morphology*, 264, 339–62.

Korboot, K. (1964). *Comparative Studies of the External and Internal Anatomy of Three Species of Caddis Flies (Trichoptera)*, University of Queensland Press, St. Lucia.

Koss, R. W. and Edmunds Jr, G. F. (1974). Ephemeroptera eggs and their contribution to phylogenetic studies of the order. *Zoological Journal of the Linnean Society*, 55, 267–349.

Košt'ál, V. (2006). Eco-physiological phases of insect diapause. *Journal of Insect Physiology*, 52, 113–27.

Kovac, D. and Maschwitz, U. (1989). Secretion-grooming in the water bug *Plea minutissima*: a chemical defence against microorganisms interfering with the hydrofuge porperties of the respiratory region. *Ecological Entomology*, 14, 403–11.

Kovac, D. and Maschwitz, U. (1991). The function of the metathoracic scent gland in corixid bugs (Hemiptera,

Corixidae): secretion-grooming on the water surface. *Journal of Natual History*, 25, 331–40.

Kovac, D. and Maschwitz, U. (1999). Protection of hydrofuge respiratory structures against detrimental microbiotic growth by terrestrial grooming in water beetles (Coleoptera: Hydrophilidae, Hydraenidae, Dryopidae, Elmidae, Curculionidae). *Entomologia Generalis*, 24, 277–92.

Kovats, Z. E., Ciborowski, J. J. H., and Corkum, L. D. (1996). Inland dispersal of adult aquatic insects. *Freshwater Biology*, 36, 265–76.

Kramer, M. G. and Marden, J. H. (1997). Almost airborne. *Nature*, 385, 403–404.

Kraus, J. M. and Vonesh, J. R. (2010). Feedbacks between community assembly and habitat selection shape variation in local colonization. *Journal of Animal Ecology*, 79, 795–802.

Kraus, W. F. (1989). Surface wave communication during courtship in the giant water bug, *Abedus indentatus* (Heteroptera: Belostomatidae). *Journal of the Kansas Entomological Society*, 62, 316–28.

Kriska, G., Csabai, Z., Boda, P., Malik, P., and Horváth, G. (2006). Why do red and dark-coloured cars lure aquatic insects? The attraction of aquatic insects to car paintwork explained by reflection-polarisation signals. *Proceedings of the Royal Society, London B*, 273, 1667–71.

Kriska, G., Horváth, G., and Andrikovics, S. (1998). Why do mayflies lay their eggs *en masse* on dry asphalt roads? Water-imitating polarized light reflected from asphalt attracts Ephemeroptera. *Journal of Experimental Biology*, 201, 2273–86.

Krosch, M. N., Baker, A. M., Mckie, B. G., Mather, P. B., and Cranston, P. S. (2009). Deeply divergent mitochondrial lineages reveal patterns of local endemism in chironomids of the Australian wet tropics. *Austral Ecology*, 34, 317–28.

Kruse, K. C. (1990). Male backspace availability in the giant waterbug (*Belostoma flumineum* Say). *Behavioral Ecology and Sociobiology*, 26, 281–9.

Kruse, K. C. and Leffler, T. R. (1984). Females of the giant water bug, *Belostoma flumineum* (Hemiptera, Belostomatidae), captured carrying eggs. *Annals of the Entomological Society of America*, 77, 20–20.

Kubow, K. B., Robinson, C. T., Shama, L. N. S., and Jokela, J. (2010). Spatial scaling in the phylogeography of an alpine caddisfly, *Allogamus uncatus*, within the central European Alps. *Journal of the North American Benthological Society*, 29, 1089–99.

Kutash, T. N. and Craig, D. A. (1998). Ontogenetic effects on locomotory gaits in nymphs of *Baetis tricaudatus* Dodds (Ephemeroptera: Baetidae). *Journal of the North American Benthological Society*, 17, 475–88.

Kuusela, K. and Huusko, A. (1996). Post-emergence migration of stoneflies (Plecoptera) into the nearby forest. *Ecological Entomology*, 21, 171–7.

Labandeira, C. C. (1997). Insect mouthparts: ascertaining the paleobiology of insect feeding strategies. *Annual Review of Ecology and Systematics*, 28, 153–93.

Labandeira, C. C., Johnson, K. R., and Wilf, P. (2002). Impact of the terminal Cretaceous event on plant-insect associations. *Proceedings of the National Academy of Sciences of the USA*, 99, 2061–6.

Labhart, T. and Nilsson, D. -E. (1995). The dorsal eye of the dragonfly *Sympetrum*: specializations for prey detection against the blue sky. *Journal of Comparative Physiology A*, 176, 437–53.

Lacoursière, J. O. (1992). A laboratory study of fluid flow and microhabitat selection by larvae of *Simulium vittatum* (Diptera: Simuliidae). *Canadian Journal of Zoology*, 70, 582–96.

Ladle, M. and Griffiths, B. S. (1980). A study on the faeces of some chalk stream invertebrates. *Hydrobiologia*, 74, 161–71.

Lancaster, J. (1990). Predation and drift of lotic macroinvertebrates during colonization. *Oecologia*, 85, 48–56.

Lancaster, J. (1999). Small scale movements of lotic macroinvertebrates with variations in flow. *Freshwater Biology*, 41, 605–19.

Lancaster, J. (2008). Movement and dispersion of insects in stream channels: what role does flow play? In J. Lancaster and R. A. Briers, eds. *Aquatic Insects: Challenges to Populations*. pp. 139–57. CAB International, Wallingford.

Lancaster, J., Bradley, D., Hogan, A., and Waldron, S. (2005). Intraguild omnivory in predatory stream insects. *Journal of Animal Ecology*, 74, 619–29.

Lancaster, J., Buffin-Bélanger, T., Reid, I., and Rice, S. (2006). Flow- and substratum-mediated movement by a stream insect. *Freshwater Biology*, 51, 1053–69.

Lancaster, J., Downes, B. J., and Arnold, A. (2010a). Oviposition site selectivity of some stream-dwelling caddisflies. *Hydrobiologia*, 652, 165–78.

Lancaster, J., Downes, B. J., and Arnold, A. (2010b). Environmental constraints on oviposition may limit density of a stream insect at multiple scales. *Oecologia*, 163, 373–84.

Lancaster, J., Downes, B. J., and Arnold, A. (2011). Lasting effects of maternal behaviour on the distribution of a dispersive stream insect. *Journal of Animal Ecology*, 80, 1061–9.

Lancaster, J., Downes, B. J., and Glaister, A. (2009). Interacting environmental gradients, trade-offs and reversals in the abundance–environment relationships of stream insects: when flow is unimportant. *Marine and Freshwater Research*, 60, 259–70.

Lancaster, J., Downes, B. J., and Reich, P. (2003). Linking landscape patterns of resource distribution with models of aggregation in ovipositing stream insects. *Journal of Animal Ecology*, 72, 969–78.

Lancaster, J., Hildrew, A. G., and Gjerlov, C. (1996). Invertebrate drift and longitudinal transport processes in streams. *Canadian Journal of Fisheries and Aquatic Sciences*, 53, 572–82.

Lancaster, J., Hildrew, A. G., and Townsend, C. R. (1988). Competition for space by predators in streams: field experiments on a net-spinning caddisfly. *Freshwater Biology*, 20, 185–93.

Lancaster, J. and Mole, A. (1999). Interactive effects of near-bed flow and substratum texture on the microdistribution of lotic macroinvertebrates. *Archiv fur Hydrobiologie*, 146, 83–100.

Land, M. F. (1969). Movements of the retinae of jumping spiders in response to visual stimuli. *Journal of Experimental Biology*, 51, 471–93.

Land, M. F., Gibson, G., Horwood, J., and Zeil, J. (1999). Fundamental differences in the optical structure of the eyes of nocturnal and diurnal mosquitoes. *Journal of Comparative Physiology A*, 185, 91–103.

Lang, H. H. (1980a). Surface wave discrimination between prey and nonprey by the back swimmer *Notonecta glauca* L. (Hemiptera, Heteroptera). *Behavioral Ecology and Sociobiology*, 6, 233–46.

Lang, H. H. (1980b). Surface wave sensitivity of the backswimmer *Notonecta glauca*. *Naturwissenschaften*, 67, 204–205.

Langton, P. H. (1995). The pupa and events leading to eclosion. In P. D. Armitage, P. S. Cranston, and L. C. V. Pinder, eds. *The Chironomidae: Biology and Ecology of Non-Biting Midges*. pp. 169–93. Chapman & Hall, London.

Larsson, M. C. and Hansson, B. S. (1998). Receptor neuron responses to potential sex pheromone components in the caddisfly *Rhyacophila nubila* (Trichotera: Rhyacophilidae). *Journal of Insect Physiology*, 44, 189–96.

Le Louarn, H. and Cloarec, A. (1997). Insect predation on pike fry. *Journal of Fish Biology*, 50, 366–70.

Lee, R. E. and Denlinger, D. L. (1991). *Insects at Low Temperature*, Chapman and Hall, London.

Lee, S. J., Hwang, J. M., and Bae, Y. J. (2008). Life history of a lowland burrowing mayfly, *Ephemera orientalis* (Ephemeroptera, Ephemeridae), in a Korean stream. *Hydrobiologia*, 596, 279–88.

Lehrian, S., Bálint, M., Haase, P., and Pauls, S.U. (2010). Genetic population structure of an autumn-emerging caddisfly with inherently low dispersal capacity and insights into its phylogeography. *Journal of the North American Benthological Society*, 29, 1100–18.

Leston, D. (1956). Systematics of the marine-bug. *Nature*, 178, 427–8.

Lewis, D. J., Reid, E. T., Crosskey, R. W., and Davies, J. B. (1960). Attachment of immature Simuliidae to other arthropods. *Nature*, 187, 618–19.

Lewis, S. M. and Cratsley, C. K. (2008). Flash signal evolution, mate choice, and predation in fireflies. *Annual Review of Entomology*, 53, 293–321.

Lichtwardt, R. W. (1986). *The Trichomycetes: Fungal Associates of Arthropods*, Springer-Verlag, New York.

Linley, J. R. and Lounibos, L. P. (1993). The eggs of *Anopheles* (*Nyssorhynchus*) *rangeli* and *Anopheles* (*Nyssorhynchus*) *dunhami* (Diptera: Culicidae). *Mosquito Systematics*, 25, 157–69.

Linley, J. R. and Turell, M. J. (1993). The eggs of *Aedes dentatus* and *Aedes fowleri* (Diptera, Culicidae). *Proceedings of the Entomological Society of Washington*, 95, 7–16.

Löfstedt, C., Bergmann, J., Francke, W., Jirle, E., Hansson, B. S., and Ivanov, V. D. (2008). Identification of a sex pheromone produced by sternal glands in females of the caddisfly *Molanna angustata* Curtis. *Journal of Chemical Ecology*, 34, 220–8.

Löfstedt, C., Hansson, B. S., Petersson, E., Valeur, P., and Richards, A. (1994). Pheromonal secretions from glands on the 5th abdominal sternite of hydropsychid and rhyacophilid caddisflies (Trichoptera). *Journal of Chemical Ecology*, 20, 153–70.

Lord, R. J. and Meier, P. G. (1977). Intraspecific variation in taxonomic characteristics of the mayfly *Potamanthus myops* (Walsh). *The Great Lakes Enotmologist*, 10, 51–8.

Loudon, C. and Alstad, D. N. (1990). Theoretical mechanics of particle capture: predictions for hydropsychid caddisfly distributional ecology. *The American Naturalist*, 135, 360–81.

Loudon, C. and Alstad, D. N. (1992). Architectural plasticity in net construction by individual caddisfly larvae (Trichoptera: Hydropsychidae). *Canadian Journal of Zoology*, 70, 1166–72.

Lowenberger, C. A. and Rau, M. E. (1994). Selective oviposition by *Aedes aegypti* (Diptera: Culicidae) in response to a larval parasite, *Plagiorchis elegans* (Trematoda: Plagiorchiidae). *Environmental Entomology*, 23, 1269–76.

Lucas, P. and Hunter, F. F. (1999). Phenotypic plasticity in the labral fan of simuliid larvae (Diptera): effect of seston load on primary-ray number. *Canadian Journal of Zoology*, 77, 1843–9.

Lyman, F. E. (1943). Swimming and burrowing activities of mayfly nymphs of the genus *Hexagenia*. *Annals of the Entomological Society of America*, 36, 250–6.

Lytle, D. A. (1999). Use of rainfall cues by *Abedus herberti* (Hempitera: Belostomatidae): a mechanism for avoiding flash floods. *Journal of Insect Behavior*, 12, 1–12.

Lytle, D. A. and White, N. J. (2007). Rainfall cues and flash-flood escape in desert stream insects. *Journal of Insect Behavior*, 20, 413–23.

Macneale, K. H., Peckarsky, B. L., and Likens, G. E. (2004). Contradictory results from different methods for measuring direction of insect flight. *Freshwater Biology*, 49, 1260–8.

Macneale, K. H., Peckarsky, B. L., and Likens, G. E. (2005). Stable isotopes identify dispersal patterns of stonefly populations living along stream corridors. *Freshwater Biology*, 50, 1117–30.

Maillard, Y. -P. and Sellier, R. (1970). La *pars stridens* des Hydrophilidae (Ins. Coléoptères); etude au microscope électronique á balayage. *Comptes Rendus de l'Academie des Sciences. Paris (D)*, 270, 2969–72.

Maksimovic, S., Cook, T. A., and Buschbeck, E. K. (2009). Spatial distribution of opsin-encoding mRNAs in the tiered larval retinas of the sunburst diving beetle *Thermonectus marmoratus* (Coleoptera: Dytiscidae). *Journal of Experimental Biology*, 212, 3781–94.

Maksimovic, S., Layne, J. E., and Buschbeck, E. K. (2011). Spectral sensitivity of the principal eyes of sunburst diving beetle, *Thermonectes marmoratus* (Coleoptera: Dytiscidae), larvae. *Journal of Experimental Biology*, 214, 3524–31.

Malmqvist, B. (2000). How does wing length relate to distribution patterns of stoneflies (Plecoptera) and mayflies (Ephemeroptera)? *Biological Conservation*, 93, 271–6.

Malmqvist, B. and Sjöström, P. (1980). Prey size and feeding patterns in *Dinocras cephalotes* (Plecoptera). *Oikos*, 35, 311–16.

Malmqvist, B., Sjöström, P., and Frick, K. (1991). The diet of two species of *Isoperla* (Plecoptera: Perlodidae) in relation to season, site, and sympatry. *Hydrobiologia*, 213, 191–203.

Malmqvist, B., Zhang, Y., and Adler, P. H. (1999). Diversity, distribution and larval habitats of North Swedish blackflies (Diptera: Simuliidae). *Freshwater Biology*, 42, 301–14.

Málnás, K., Polyák, L., Prill, E., Hegedüs, R., Kriska, G., Dévai, G., Horváth, G., and Lengyel, S. (2011). Bridges as optical barriers and population disruptors for the mayfly *Palingenia longicauda*: an overlooked threat to freshwater biodiversity? *Journal of Insect Conservation*, 15, 823–32.

Mandapaka, K., Morgan, R. C., and Buschbeck, E. K. (2007). Larval visual system of the diving beetle *Thermonectus marmoratus* (Coleoptera: Dytiscidae). *The Journal of Comparative Neurology*, 497, 166–81.

Mangan, B. P. (1992). Oviposition of the dobsonfly (*Corydalus cornutus*, Megaloptera) in a large river. *American Midland Naturalist*, 127, 348–54.

Mangan, B. P. (1994). Pupation ecology of the dobsonfly *Corydalus cornutus* (Corydalidae, Megaloptera) along a large river. *Journal of Freshwater Ecology*, 9, 57–62.

Marchant, R. and Hehir, G. (1999). Growth, production and mortality of two species of *Agapetus* (Trichoptera: Glossosomatidae) in the Acheron River, south-east Australia. *Freshwater Biology*, 42, 655–71.

Marden, J. H. (2000). Variability in the size, composition, and function of insect flight muscles. *Annual Review of Physiology*, 62, 157–78.

Marden, J. H. (2008). Dragonfly flight performance: a model system for biomechanics, physiological genetics, and animal competitive behaviour. In A. Córdoba-Aguilar, ed. *Dragonflies and damselflies: Model Organisms for Ecological and Evolutionary Research*. pp. 249–59. Oxford University Press, Oxford.

Marden, J. H. and Kramer, M. G. (1994). Surface-skimming stoneflies: a possible intermediate stage in insect flight evolution. *Science*, 266, 427–30.

Marden, J. H. and Kramer, M. G. (1995). Locomotor performance of insects with rudimentary wings. *Nature*, 377, 332–4.

Marden, J. H., Kramer, M. G., and Frisch, J. (1996). Age-related variation in body temperature, thermoregulation and activity in a thermally polymorphic dragonfly. *Journal of Experimental Biology*, 199, 529–35.

Marden, J. H., O'Donnell, B. C., Thomas, M. A., and Bye, J. Y. (2000). Surface-skimming stoneflies and mayflies: the taxonomic and mechanical diversity of two-dimensional aerodynamic locomotion. *Physiological and Biochemical Zoology*, 73, 751–64.

Marden, J. H. and Thomas, M. A. (2003). Rowing locomotion by a stonefly that possesses the ancestral pterygote condition of co-occurring wings and abdominal gills. *Biological Journal of the Linnean Society*, 79, 341–9.

Marko, M. D., Newman, R. M., and Gleason, F. K. (2005). Chemically mediated host-plant selection by the milfoil weevil: A freshwater insect-plant interaction. *Journal of Chemical Ecology*, 31, 2857–76.

Martin, I. D. and Mackay, R. J. (1982). Interpreting the diet of *Rhyacophila* larvae (Trichoptera) from gut analyses: an evaluation of techniques. *Canadian Journal of Zoology*, 60, 783–9.

Martin, I. D. and Mackay, R. J. (1983). Growth rates and prey selection of two congeneric predatory caddisflies (Trichoptera: Rhyacophilidae). *Canadian Journal of Zoology*, 61, 895–900.

Martin, M. M. (1987). *Invertebrate-Microbial Associations: Ingested Fungal Enzymes in Arthropod Biology*, Comstock Publishing Associates, Cornell University Press, Ithaca.

Martin, M. M. (1992). The evolution of insect-fungus associations: from contact to stable symbiosis. *American Zoologist*, 32, 593–605.

Martin, M. M., Martin, J. S., Kukor, J. J., and Merritt, R. W. (1980). The digestion of protein and carbohydrate by the stream detritivore, *Tipula abdominalis* (Diptera: Tipulidae). *Oecologia*, 46, 360–4.

Martin, M. M., Martin, J. S., Kukor, J. J., and Merritt, R. W. (1981). The digestive enzymes of detritus-feeding stonefly nymphs (Plecoptera; Pteronarcidae). *Canadian Journal of Zoology*, 59, 1947–51.

Massey, F. P. and Hartley, S. E. (2009). Physical defences wear you down: progressive and reversible impacts of silica on insect herbivores. *Journal of Animal Ecology*, 78, 281–91.

Masters, Z., Petersen, I., Hildrew, A. G., and Ormerod, S. J. (2007). Insect dispersal does not limit the biological recovery of streams from acidification. *Aquatic Conservation: Marine and Freshwater Ecosystems*, 17, 375–83.

Matthews, P. G. D. and Seymour, R. S. (2006). Diving insects boost their buoyancy bubbles. *Nature*, 441, 171.

Matthews, P. G. D. and Seymour, R. S. (2008). Haemoglobin as a buoyancy regulator and oxygen supply in the backswimmer (Notonectidae, *Anisops*). *Journal of Experimental Biology*, 211, 3790–9.

Matthews, P. G. D. and Seymour, R. S. (2010). Compressible gas gills of diving insects: Measurements and models. *Journal of Insect Physiology*, 56, 470–9.

Matthews, P. G. D. and White, C. R. (2011). Discontinuous gas exchange in insects: Is it all in their heads? *American Naturalist*, 177, 130–4.

Matushkina, N. and Gorb, S. (2002). Stylus of the odonate endophytic ovipositor: a mechanosensory organ controlling egg positioning. *Journal of Insect Physiology*, 48, 213–19.

Matushkina, N. and Gorb, S. N. (2007). Mechanical properties of the endophytic ovipositor in damselflies (Zygoptera, Odonata) and their oviposition substrates. *Zoology*, 110, 167–75.

Matushkina, N. A. and Lambret, P. F. (2011). Ovipositor morphology and egg laying behaviour in the dragonfly *Lestes macrostigma* (Zygoptera: Lestidae). *International Journal of Odonatology*, 14, 69–82.

May, B. M. (1970). Aquatic adaptation in the larva of *Desiantha ascita* (Coleoptera: Curculionidae). *New Zealand Entomologist*, 4, 7–11.

May, M. L. (1976). Thermoregulation and adaptation to temperature in dragonflies (Odonata: Anisoptera). *Ecological Monographs*, 46, 1–32.

May, M. L. (1979). Energy metabolism of dragonflies (Odonata: Anisoptera) at rest and during endothermic warmup. *Journal of Experimental Biology*, 83, 79–94.

May, M. L. and Matthews, J. H. (2008). Migration in Odonata: a case study of *Anax junius*. In A. Córdoba-Aguilar, ed. *Dragonflies and Damselflies: Model Organisms for Ecological and Evolutionary Research*. pp. 63–77. Oxford University Press, Oxford.

McCafferty, W. P. and Bloodgood, D. W. (1989). The female and male coupling apparatus in *Tortopus* mayflies. *Aquatic Insects*, 11, 141–6.

McCall, P. J. (2002). Chemoecology of oviposition in insects of medical and veterinary importance. In M. Hilker and T. Meiner, eds. *Chemoecology of Insect Eggs and Egg Deposition*. pp. 265–89. Blackwell Publishing, Oxford.

McCreadie, J. W., Beard, C. E., and Adler, P. H. (2005). Context-dependent symbiosis between black flies (Diptera: Simuliidae) and trichomycete fungi (Harpellales: Legeriomycetaceae). *Oikos*, 108, 363–70.

McIntosh, A. R. and Peckarsky, B. L. (1999). Criteria determining behavioural responses to multiple predators by a stream mayfly. *Oikos*, 85, 554–64.

McKie, B. G. and Cranston, P. S. (1998). Keystone coleopterans? Colonization by wood-feeding elmids of experimentally immersed woods in south-eastern Australia. *Marine and Freshwater Research*, 49, 79–88.

McKie, B. G., Cranston, P. S., and Pearson, R. G. (2004). Gondawanan mesotherms and cosmopolitan eurytherms: effects of temperature on the development and survival of Australian Chironomidae (Diptera) from tropical and temperate populations. *Marine and Freshwater Research*, 55, 759–68.

McLachlan, A. J. and Cantrell, M. A. (1976). Sediment development and its influence on the distribution and tube structure of *Chironomus plumosus* L. (Chironomidae, Diptera) in a new impoundment. *Freshwater Biology*, 6, 437–43.

McLellan, I. D. (1977). New alpine and southern Plecoptera from New Zeland, and a new classification of the Gripopterygidae. *New Zealand Journal of Zoology*, 4, 119–147.

McShaffrey, D. and McCafferty, W. P. (1986). Feeding behavior of *Stenacron interpunctatum* (Ephemeroptera: Heptageniidae). *Journal of the North American Benthological Society*, 5, 200–10.

McShaffrey, D. and McCafferty, W. P. (1988). Feeding behavior of *Rhithrogena pellucidula* (Ephemeroptera: Heptageniidae). *Journal of the North American Benthological Society*, 7, 87–99.

McShaffrey, D. and McCafferty, W. P. (1990). Feeding behavior and related functional morphology of the mayfly *Ephemerella needhami* (Ephemeroptera: Ephemerellidae). *Journal of Insect Behavior*, 3, 673–88.

Meinertzhagen, I. A., Menzel, R., and Kahle, G. (1983). The identification of spectral receptor types in the retina and lamina of the dragonfly *Sympetrum rubicundulum*. *Journal of Comparative Physiology A*, 151, 295–310.

Meissner, K. and Muotka, T. (2006). The role of trout in stream food webs: integrating evidence from field surveys and experiments *Journal of Animal Ecology*, 75, 421–33.

Melzer, R. R. and Paulus, H. F. (1991). Morphology of the visual system of *Chaoborus crystallinus* (Diptera, Chaoboridae). *Zoomorphology*, 110, 227–38.

Melzer, R. R., Paulus, H. F., and Kristensen, N. P. (1994). The larval eye of nannochoristid scorpionflies (Insecta, Mecoptera). *Acta Zoologica*, 75, 201–208.

Mendez, P. K. and Resh, V. H. (2008). Life history of the *Neophylax rickeri* (Trichoptera: Uenoidae) in two Northern California streams. *Annals of the Entomological Society of America*, 101, 573–84.

Mendki, M. J., Ganesan, K., Prakash, S., Suryanarayana, M. V. S., Malhotra, R. C., Rao, K. M., and Vaidyanathaswamy, R. (2000). Heneicosane: an oviposition-attractant pheromone of larval origin in *Aedes aegypti* mosquito. *Current Science*, 78, 1295–6.

Merrill, D. (1965). The stimulus for case-building activity in caddis-worms (Trichoptera). *Journal of Experimental Zoology*, 158, 123–30.

Merritt, R. W., Craig, D. A., Walker, E. D., Vanderploeg, H. A., and Wotton, R. S. (1992b). Interfacial feeding behavior and particle flow patterns of *Anopheles quadrimaculatus* larvae (Diptera: Culicidae). *Journal of Insect Behavior*, 5, 741–61.

Merritt, R. W., Dadd, R. H., and Walker, E. D. (1992a). Feeding behavior, natural food, and nutritional relationships of larval mosquitoes. *Annual Review of Entomology*, 37, 349–76.

Mey, W. and Speidel, W. (2008). Global diversity of butterflies (Lepidoptera) in freshwater. *Hydrobiologia*, 595, 521–8.

Meyer-Rochow, V. B. (1978). The dioptric system of the eye of *Cybister* (Dytiscidae: Coleoptera). *Proceedings of the Royal Society, London B*, 183, 159–78.

Miall, L. C. (1903). *The Natural History of Aquatic Insects*, Macmillan and Co. Ltd, London.

Mill, P. J. and Pickard, R. S. (1972). Anal valve movement and normal ventilation in aeshnid dragonfly larvae. *Journal of Experimental Biology*, 56, 537–43.

Miller, A. (1939). The egg and early development of the stonefly, *Pteronarcys proteus* Newman (Plecoptera). *Journal of Morphology*, 64, 555–609.

Miller, J. R., Huang, J., Vulule, J., and Walker, E. D. (2007). Life on the edge: African malaria mosquito (*Anopheles gambiae* s. l.) larvae are amphibious. *Naturwissenschaften*, 94, 195–9.

Miller, K. B. (2003). The phylogeny of diving beetles (Coleoptera: Dytiscidae) and the evolution of sexual conflict. *Biological Journal of the Linnean Society*, 79, 359–88.

Miller, P. L. (1994). The response of rectal pumping in some zygopteran larvae (Odonata) to oxygen and ion availability. *Journal of Insect Physiology*, 40, 333–9.

Miller, S. E. (1983). Late Quaternary insects of Rancho La Brea and McKittrick, California. *Quaternary Research*, 20, 90–104.

Mira, A. (2000). Exuviae eating: a nitrogen meal? *Journal of Insect Physiology*, 46, 605–10.

Mitchell, B. K., Itagaki, H., and Rivet, M. -P. (1999). Peripheral and central structures involved in insect gustation. *Microscopy Research and Technique*, 47, 401–15.

Mokany, A. and Shine, R. (2003). Oviposition site selection by mosquitoes is affected by cues from conspecific larvae and anuran tadpoles. *Austral Ecology*, 28, 33–7.

Molineri, C. and Emmerich, D. (2010). New species and new stage descriptions of *Campsurus major* species group (Polymictarcydiae: Campsurinae), with first report of silk-case construction in mayfly nymphs. *Aquatic Insects*, 32, 265–80.

Morales, M. E., Wesson, D. M., Sutherland, I. W., Impoinvil, D. E., Mbogo, C. M., Githure, J. I., and Beier, J. C. (2003). Determination of *Anopheles gambiae* larval DNA in the gut of insectivorous dragonfly (Libellulidae) nymphs by polymerase chain reaction. *Journal of the American Mosquito Control Association*, 19, 163–5.

Moreno, J. L., Angeler, D. G., and De las Heras, J. (2010). Seasonal dynamics of macroinvertebrate communities in a semiarid saline spring stream with contrasting environmental conditions. *Aquatic Ecology*, 44, 177–93.

Morgan, A. H. (1913). A contribution to the biology of mayflies. *Annals of the Entomological Society of America*, 6, 371–441.

Morgan, A. H. (1929). The mating flight and vestigial structures of the stump-legged mayfly, *Campsurus segnis* Needham. *Annals of the Entomological Society of America*, 22, 61–8.

Morgan, N. C. (1956). The biology of *Leptocerus aterrimus* Steph. with reference to its availability as a food for trout. *Journal of Animal Ecology*, 25, 349–65.

Muirhead-Thomson, R. C. (1956). Communal oviposition in *Simulium damnosum* Theobald (Diptera, Simuliidae). *Nature*, 178, 1297–9.

Müller, K. (1982). The colonization cycle of freshwater insects. *Oecologia*, 52, 202, 207.

Müller-Liebenau, I. (1973). Morphological characteristics used in revising the European species of the genus *Baetis* Leach. In W. L. Peters and J. G. Peters, eds. *Proceedings of the First International Conference of Ephemeroptera*. pp. 182–98. E.J. Brill, Leiden.

Muller-Liebenau, I. (1978). *Raptobaetopus*, eine neue carnivore Ephemeroptera f-Gattung aus Malaysia (Insecta, Ephemeroptera: Baetidae). *Archiv für Hydrobiologie*, 82, 465–81.

Munguía-Steyer, R., Favila, M. E., and Macías-Ordóñez, R. (2008). Brood pumping modulation and the benefits of paternal care in *Abedus breviceps* (Hemiptera: Belostomatidae). *Ethology*, 114, 693–700.

Muraji, M. and Nakasuji, F. (1988). Comparative studies on life history traits of three wing dimorphic water bugs,

Microvelia spp. Westwood (Heteroptera: Veliidae). *Reseaches on Population Ecology,* 30, 315–27.

Murphey, R. K. (1971). Sensory aspects of control of orientation to prey by waterstrider *Gerris remigis. Zeitschrift fur Vergleichende Physiologie,* 72, 168–85.

Murphey, R. K. and Mendenhall, B. (1973). Localization of receptors controlling orientation to prey by the backswimmer *Notonecta undulata. Journal of Comparative Physiology A,* 84, 19–30.

Múrria, C., Bonada, N., Ribera, C., and Prat, N. (2010). Homage to the Virgin of Ecology, or why an aquatic insect unadapted to desiccation may maintain populations in very small, temporary Mediterranean streams. *Hydrobiologia,* 653, 179–90.

Murvosh, C. M. (1971). Ecology of the water penny beetle *Psephenus herricki* (DeKay). *Ecological Monographs,* 41, 79–96.

Myers, E. M. (2003). The circadian control of eclosion. *Chronobiology International,* 20, 775–94.

Nachtigall, W. (1961). Funktionelle morphologie, kinematik und hydrmechanik des ruderapparates von *Gyrinus. Zeitschrift fur Vergleichende Physiologie,* 45, 193–226.

Nachtigall, W. (1985). Swimming in aquatic insects. In G.A. Kerkut and L.I. Gilbert, eds. *Comprehensive Insect Physiology Biochemistry and Pharmacology.* pp. 467–90. Pergamon Press, Oxford.

Nakahara, Y., Watanabe, M., Fujita, A., Kanamori, Y., Tanaka, D., Iwata, K., Furuki, T., Sakurai, M., Kikawada, T., and Okuda, T. (2008). Effects of dehydration rate on physiological responses and survival after rehydration in larvae of the anhydrobiotic chironomid. *Journal of Insect Physiology,* 54, 1220–5.

Nesterovitch, A. and Zwick, P. (2003). The development of *Nemurella pictetii* Klapálek (Plecoptera: Nemouridae) in two springstreams in central Europe. *Limnologica,* 33, 231–43.

Neumann, D. (1986). Diel eclosion rhythm of a sublittoral population of the marine insect *Pontomyia pacifica. Marine Biology,* 90, 461–5.

Newman, R. M. (1991). Herbivory and detritivory on freshwater macrophytes by invertebrates: a review. *Journal of the North American Benthological Society,* 10, 89–114.

Newman, R. M., Kerfoot, W. C., and Hanscome, Z. I. (1996). Watercress allelochemical defends high-nitrogen foliage against consumption: effects on freshwater invertebrate herbivores. *Ecology,* 77, 2312–23.

Nicola, S. J. (1968). Scavenging by *Alloperla* (Plecoptera: Chloroperlidae) nymphs on dead pink (*Oncorhynchus gorbuscha*) and chum (*O. keta*) embryos. *Canadian Journal of Zoology,* 46, 787–96.

Nilsson, A. N. (1996). Coleoptera Dytiscidae, diving water beetles. In A. Nilsson, ed. *Aquatic Insects of North Europe, Volume 1.* pp. 145–72. Apollo Books, Stenstrup.

Nondula, N., Marshall, D. J., Baxter, R., Sinclair, B. J., and Chown, S. L. (2004). Life history and osmoregulatory ability of *Telmatogeton amphibius* (Diptera, Chironomidae) at Marion Island. *Polar Biology,* 27, 629–35.

Notestine, M. K. (1994). Comparison of the respiratory currents produced by ephemeropteran nymphs with operculate gills. *Journal of the Australian Entomological Society,* 33, 399–403.

Nyhof, J. M. and McIver, S. B. (1987). Fine structure of ocelli of the larval black fly *Simulium vittatum* (Diptera: Simuliidae). *Canadian Journal of Zoology,* 65, 142–50.

O'Donnell, B. C. (2009). Early nymphal development in *Euphoron leukon* (Ephemeroptera: Polymitarcidae) with particular emphasis on mouthparts and abdominal gills. *Annals of the Entomological Society of America,* 102, 128–36.

O'Donnell, M. J. (1997). Mechanisms of excretion and ion transport in invertebrates. In W. H. Dantzler, ed. *Handbook of Physiology.* pp. 1207–89. Oxford University Press, Oxford.

Ohba, S. (2011). Field observation of predation on a turtle by a giant water bug. *Entomological Science,* 14, 364–5.

Ohba, S., Hidaka, K., and Sasaki, M. (2006). Notes on paternal care and sibling cannibalism in the giant water bug, *Lethocerus deyrolli* (Heteroptera, Belostomatidae). *Entomological Science,* 9, 1–5.

Ohba, S., Miyasaka, H., and Nakasuji, F. (2008b). The role of amphibian prey in the diet and growth of giant water bug nymphs in Japanese rice fields. *Population Ecology,* 50, 9–16.

Ohba, S. and Nakasuji, F. (2006). Dietary items of predacious aquatic bugs (Nepoidea: Heteroptera) in Japanese wetlands. *Limnology,* 7, 41–3.

Ohba, S. and Takagi, H. (2005). Food shortage affects flight migration of the giant water bug *Lethocerus deyrolli* in the prewintering season. *Limnology,* 6, 85–90.

Ohba, S., Tatsuta, H., and Nakasuji, F. (2008a). Variation in the geometry of foreleg claws in sympatric giant water bug species: an adaptive trait for catching prey? *Entomologia Experimentalis et Applicata,* 129, 223–7.

Olden, J. D., Hoffmann, A. L., Monroe, J. B., and Poff, N. L. (2004). Movement behaviour and dynamics of an aquatic insect in a stream benthic landscape. *Canadian Journal of Zoology,* 82, 1135–46.

Oldmeadow, D. F., Lancaster, J., and Rice, S. P. (2010). Drift and settlement of stream insects in a complex hydraulic environment. *Freshwater Biology,* 55, 1020–35.

Oliver, D. R. (1971). Life history of the Chironomidae. *Annual Review of Entomology,* 16, 211–30.

Olsson, T. I. (1981). Overwintering of benthic macroinvertebrates in ice and frozen sediment in a North Swedish river. *Holarctic Ecology,* 4, 161–6.

Ostrofsky, M. L. and Zettler, E. R. (1986). Chemical defences in aquatic plants. *Journal of Ecology,* 74, 279–87.

Otto, C. (1983). Behavioural and physiological adaptations to a variable habitat in two species of case-making caddis larvae using different food. *Oikos*, 41, 188–94.

Otto, C. (2000). Cost and benefit from shield cases in caddis larvae. *Hydrobiologia*, 436, 35–40.

Otto, C. and Johansson, A. (1995). Why do some caddis larvae in running waters construct heavy, bulky cases? *Animal Behaviour*, 49, 473–8.

Outomuro, D. and Ocharan, F. J. (2011). Wing pigmentation in *Calopteryx* damselflies: a role in thermoregulation? *Biological Journal of the Linnean Society*, 103, 36–44.

Pajunen, V. I. and Pajunen, I. (1991). Oviposition and egg cannibalism in rock-pool corixids (Hemiptera: Corixidae). *Oikos*, 60, 83–90.

Palmquist, K. R., Jenkins, J. J., and Jepson, P. C. (2008). Clutch morphology and the timing of exposure impact the susceptibility of aquatic insect eggs to esfenvalerate. *Environmental Toxicology and Chemistry*, 27, 1713–20.

Paulus, H. F. and Schmidt, M. (1978). Evolutionswege zum larvalauge der insekten: die stemmata der Trichoptera und Lepidoptera. *Zeitschrift fur Zoologische Systematik und Evolutionsforschung*, 16, 188–216.

Peck, S. B. (2006). Distribution and biology of the ectoparasitic beaver beetle *Platypsyllus castoris* Ritsema in North America (Coleoptera: Leiodidae: Platypsyllinae). *Insecta Mundi*, 20, 85–94.

Peckarsky, B. L., McIntosh, A. R., Taylor, B. W., and Dahl, J. (2002). Predator chemicals induce changes in mayfly life history traits: a whole stream manipulation. *Ecology*, 83, 612–18.

Peckarsky, B. L. and Penton, M. A. (1989). Early warning lowers risk of stonefly predation for a vulnerable mayfly. *Oikos*, 54, 301–309.

Peckarsky, B. L. and Wilcox, R. S. (1989). Stonefly nymphs use hydrodynamic cues to discriminate between prey. *Oecologia*, 79, 265–70.

Percival, E. and Whitehead, H. (1928). Observations on the ova and oviposition of certain Ephemeroptera and Plecoptera. *Proceedings of the Leeds Philosophical Society*, 1, 271–88.

Pereira, S. T., Secundino, N. F. C., Botelho, A. C. C., Pinheiro, V. C., Tadei, W. P., and Pimenta, P. F. P. (2006). Role of egg buster in hatching of *Aedes aegypti*: scanning electron microscopy study. *Journal of Medical Entomology*, 43, 68–72.

Peters, J. G. and Peters, W. L. (1986). Leg abscission and adult *Dolania* (Ephemeroptera: Behningiidae). *Florida Entomologist*, 69, 245–50.

Petersen, C. E. and Wiegert, R. G. (1982). Coprophagous nutrition in a population of *Paracoenia bisetosa* (Ephydridae) from Yellowstone National Park, USA. *Oikos*, 39, 251–5.

Petersen, I., Masters, Z., Ormerod, S. J., and Hildrew, A. G. (2006). Sex ratio and maturity indicate the local dispersal and mortality of adult stoneflies. *Freshwater Biology*, 51, 1543–51.

Petersen, L. B. -M. and Petersen, R. C. J. (1983). Anomalies in hydropsychid capture nets from polluted streams. *Freshwater Biology*, 13, 185–91.

Petersen, R. C., Petersen, L. B. -M., and Wallace, J. B. (1984). Influence of velocity and food availability on catchnet dimensions of *Neureclipsis bimaculata* (Trichoptera: Polycentropodidae). *Holarctic Ecology*, 7, 380–9.

Peterson, C. G., Vormittag, K. A., and Valett, H. M. (1998). Ingestion and digestion of epilithic algae by larval insects in a heavily grazed montane stream. *Freshwater Biology*, 40, 607–23.

Petersson, E. (1995). Male load-lifting capacity and mating success in the swarming caddis fly *Athripsodes cinereus*. *Physiological Entomology*, 20, 66–70.

Philipson, G. N. (1954). The effect of water flow and oxygen concentration on six species of caddis fly (Trichoptera) larvae. *Proceedings of the Zoological Society of London*, 124, 547–64.

Philipson, G. N. and Moorhouse, B. H. S. (1976). Respiratory behaviour of larvae of four species of the family Polycentropodidae (Trichoptera). *Freshwater Biology*, 6, 347–54.

Pickard, R. S. and Mill, P. J. (1972). Ventilatory muscle activity in intact preparations of aeshnid dragonfly larvae. *Journal of Experimental Biology*, 56, 527–36.

Pickard, R. S. and Mill, P. J. (1974). Ventilatory movements of the abdomen and branchial apparatus in dragonfly larvae (Odonoata: Anisoptera). *Journal of Zoology*, 174, 23–40.

Pickles, A. (1931). On the metamorphosis of the alimentary canal of certain Ephemeroptera. *Transactions of the Entomological Society of London*, 79, 263–74.

Piersanti, S., Rebora, M., Almaas, T. J., Salerno, G., and Gaino, E. (2011). Electrophysiological identification of thermo- and hygro-sensitive receptor neurons on the antennae of the dragonfly *Libellula depressa*. *Journal of Insect Physiology*, 57, 1391–8.

Piersanti, S., Rebora, M., and Gaino, E. (2010). A scanning electron microscope study of the antennal sensilla in adult Zygoptera. *Odonatologica*, 39, 235–41.

Piersanti, S., Rebora, M., Salerno, G., and Gaino, E. (2007). Behavioural strategies in the larval dragonfly *Libellula depressa* (Odonata: Libellulidae) in drying ponds. *Ethology Ecology & Evolution*, 19, 127–36.

Plague, G. R. and McArthur, J. V. (2003). Phenotypic plasticity of larval retreat design in a net-spinning caddisfly. *Behavioral Ecology*, 14, 221–6.

Poff, N. L., DeCino, R. D., and Ward, J. V. (1991). Size-dependent drift responses of mayflies to experimental hydrologic variation: active predator avoidance or passive hydrodynamic displacement? *Oecologia*, 88, 577–86.

Polhemus, J. T. (1994). Stridulatory mechanisms in aquatic and semiaquatic Heteroptera. *Journal of the New York Entomological Society,* 102, 270–4.

Polhemus, J. T. and Polhemus, D. A. (2008). Global diversity of true bugs (Heteroptera; Insecta) in freshwater. *Hydrobiologia,* 595, 379–91.

Prager, J. (1976). Das mesothorakale Tympanalorgan von-Corixa punctata Ill. (Heteroptera, Corixidae). *Journal of Comparative Physiology A,* 110, 33–50.

Prager, J. and Streng, R. (1982). The resonance properties of the physical gill of *Corixa punctata* and their significance in sound reception. *Journal of Comparative Physiology A,* 148, 323–5.

Prager, J. and Theiss, J. (1982). The effect of tympanal position on the sound-sensitivity of the mesothoracic scolopale organ of *Corixa. Journal of Insect Physiology,* 28, 447–52.

Pritchard, G. (1966). On the morphology of the compound eyes of dragonflies (Odonata: Anisoptera), with special reference to their role in prey capture. *Proceedings of the Royal Entomological Society, London A,* 41, 1–8.

Pritchard, G., Harder, L. D., and Mutch, R. A. (1996). Development of aquatic insect eggs in relation to temperature and strategies for dealing with different thermal environments. *Biological Journal of the Linnean Society,* 58, 221–44.

Pritchard, G., McKee, M. H., Pike, E. M., Scrimgeour, G. J., and Zloty, J. (1993). Did the first insects live in water or in air? *Biological Journal of the Linnean Society,* 49, 31–44.

Pritchard, G. and Leischner, T. G. (1973). The life history and feeding habits of *Sialis cornuta* Ross in a series of abandoned beaver ponds (Insecta; Megaloptera). *Canadian Journal of Zoology,* 51, 121–31.

Proctor, H. and Pritchard, G. (1989). Neglected predators: Water mites (Acari: Paraitengona: Hydrachnellae) in freshwater communities. *Journal of the North American Benthological Society,* 8, 100–11.

Prusak, A. C., O'Neal, J., and Kubanek, J. (2005). Prevalence of chemical defenses among water plants. *Journal of Chemical Ecology,* 31, 1145–60.

Pulikovsky, N. (1924). Metamorphosis of *Deuterophlebia* sp. (Diptera, Deuterophlebiidae Edw.). *Transactions of the Entomological Society of London,* 72, 45–62.

Purcell, A. H., Hoffmann, A., and Resh, V. H. (2008). Life history of a dipteran predator (Scathophagidae: *Acanthocnema*) of insect egg masses in a northern California stream. *Freshwater Biology,* 53, 2426–37.

Rahn, H. and Paganelli, C. V. (1968). Gas exchange in gas gills of diving insects. *Respiration Physiology,* 5, 145–64.

Ramdani, M., Flower, R. J., Elkhiati, N., Birks, H. H., Kraïem, M. M., and Fathi, A. A. (2001). Zooplankton (Cladocera, Ostracoda), Chironomidae and other benthic faunal remains in sediment cores from nine North African wetland lakes: the CASSARINA Project. *Aquatic Ecology,* 35, 389–403.

Rebora, M. and Gaino, E. (2008). The antennal sensilla of the nymph of *Ephemera danica.* In F. R. Hauer, J. A. Stanford, and R. L. Newell, eds. *International Advances in the Ecology, Zoogeography, and Systematics of Mayflies and Stoneflies.* pp. 301–12. University of California Press, Berkeley.

Rebora, M., Murányi, D., Piersantis, S., and Gaino, E. (2010). The lateral protrusions of the head of the stonefly larva *Leuctra* cf. *signifera* (Plecoptera: Leuctridae). *Aquatic Insects,* 32, 259–64.

Rebora, M., Piersanti, S., Almaas, T. J., and Gaino, E. (2007). Hygroreceptors in the larva of *Libellula depressa* (Odonata: Libellulidae). *Journal of Insect Physiology,* 53, 550–8.

Rebora, M., Piersanti, S., and Gaino, E. (2004). Visual and mechanical cues used for prey detection by the larva of *Libellula depressa* (Odonata Libellulidae). *Ethology, Ecology and Evolution,* 16, 133–44.

Rebora, M., Piersanti, S., Salerno, G., Conti, E., and Gaino, E. (2006). Water deprivation tolerance and humidity response in a larval dragonfly: a possible adaptation for survival in drying ponds. *Physiological Entomology,* 32, 121–6.

Reich, P. (2004). Patterns of composition and abundance in macroinvertebrate egg masses from temperate Australian streams. *Marine and Freshwater Research,* 55, 39–56.

Reich, P. and Downes, B. J. (2003a). The distribution of aquatic invertebrate egg masses in relation to physical characteristics of oviposition sites at two Victorian upland streams. *Freshwater Biology,* 48, 1497–513.

Reich, P. and Downes, B. J. (2003b). Experimental evidence for physical cues involved in oviposition site selection of lotic hydrobiosid caddisflies. *Oecologia,* 136, 465–75.

Reich, P., Hale, R., Downes, B. J., and Lancaster, J. (2011). Environmental cues or conspecific attraction as casues for egg mass aggregation in hydrobiosid caddisflies. *Hydrobiologia,* 661, 351–62.

Reilly, P. and McCarthy, T. K. (1990). Observations on the natural diet of *Cymatia bonsdorfi* (C. Sahlb.) (Heteroptera: Corixidae): an immunological analysis. *Hydrobiologia,* 196, 159–66.

Reinhardt, K. (1996). Negative effects of *Arrenurus* water mites on the flight distances of the damselfly *Nehalennia speciosa* (Odonata: Coenagrionidae). *Aquatic Insects,* 18, 233–40.

Reiskind, M. H., Greene, K. L., and Lounibos, L. P. (2009). Leaf species identity and combination affect performance and oviposition choice of two container mosquito species. *Ecological Entomology,* 34, 447–56.

Rejmankova, E., Higashi, R. M., Roberts, D. R., Lege, M., and Andre, R. G. (2000). The use of Solid Phase MicroExtraction (SPME) devices in analysis for potential mosquito oviposition chemicals from cyanobacterial mats. *Aquatic Ecology*, 34, 413–20.

Resetarits Jr., W. J. (2001). Colonization under threat of predation: avoidance of fish by an aquatic beetle, *Tropisternus lateralis* (Coleoptera: Hydrophilidae). *Oecologia*, 129, 155–60.

Resh, V. H. (1976). Biology and immature stages of caddisfly genus *Ceraclea* in Eastern North America (Trichoptera, Leptoceridae). *Annals of the Entomological Society of America*, 69, 1039–61.

Resh, V. H., Morse, J. C., and Wallace, J. B. (1976). Evolution of sponge feeding habit in caddisfly genus *Ceraclea* (Trichoptera, Leptoceridae). *Annals of the Entomological Society of America*, 69, 937–41.

Resh, V. H. and Rosenberg, D. M. (2010). Recent trends in life-history research on benthic macroinvertebrates. *Journal of the North American Benthological Society*, 29, 207–19.

Resh, V. H. and Wood, J. R. (1985). Site of sex pheromone production in three species of Trichoptera. *Aquatic Insects*, 7, 65–71.

Reynolds, J. D. and Scudder, G. G. E. (1987). Serological evidence of realized feeding niche in *Cenocorixa* species (Hemiptera: Corixidae) in sympatry and allopatry. *Canadian Journal of Zoology*, 65, 967–73.

Ribera, I. (2008). Habitat constraints and the generation of diversity in freshwater macroinvertebrates. In J. Lancaster and R.A. Briers, eds. *Aquatic Insects: Challenges to Populations*. pp. 289–311. CAB International, Wallingford.

Rice, S. P., Buffin-Bélanager, T., Lancaster, J., and Reid, I. (2008). Movements of a macroinvertebrate (*Potamophylax latipennis*) across a gravel-bed substrate: effects of local hydraulics and micro-topography under increasing discharge. In H. Habersack, T. Hoey, H. Piegay and M. Rinaldi, eds. *Gravel-bed Rivers: From Process Understanding to River Restoration*. pp. 637–60. Elsevier B.V., Amsterdam.

Richter, S. C. (2000). Larval caddisfly predation on the eggs and embryos of *Rana capito* and *Rana sphenocephala*. *Journal of Herpetology*, 34, 590–3.

Riley, J. R., Smith, A. D., Reynolds, D. R., Edwards, A. S., Osborne, J. L., Williams, I. H., Carreck, N. L., and Poppy, G. M. (1996). Tracking bees with harmonic radar. *Nature*, 379, 29–30.

Ro, A. -I. (1995). Pupil adjustments in the eye of the common backswimmer. *Journal of Experimental Biology*, 198, 71–7.

Ro, A. -I. and Nilsson, D. -E. (1994). Circadian and light-dependent control of the pupil mechanism in tipulid flies. *Journal of Insect Physiology*, 40, 883–91.

Roberts, D. M. (2001). Egg hatching of mosquitoes *Aedes caspius* and *Ae. vittatus* stimulated by water vibrations. *Medical and Veterinary Entomology*, 15, 215–18.

Roberts, D. R. (2004). Prolonged survival of eggs of the rock-pool mosquito, *Aedes vittatus*, in the extreme heat of the Arabian peninsula. *Journal of Arid Environments*, 57, 203–10.

Robinson, J. V., Shaffer, L. R., Hagemeier, D. D., and Smatresk, N. J. (1991). The ecological role of caudal lamellae loss in the larval damselfly, *Ischnura posita* (Hagen) (Odonata: Zygoptera). *Oecologia*, 87, 1–7.

Roff, D. A. (1990). The evolution of flightlessness in insects. *Ecological Monographs*, 60, 389–421.

Roland, J., McKinnon, G., Backhouse, C., and Taylor, P. D. (1996). Even smaller radar tags on insects. *Nature*, 381, 120.

Romey, W. L. and Galbraith, E. (2008). Optimal group positioning after a predator attack: the influence of speed, sex, and satiation within mobile whirligig swarms. *Behavioral Ecology*, 19, 338–43.

Room, P. M. (1990). Ecology of a simple plant-herbivore system: Biological control of *Salvinia*. *Trends in Ecology and Evolution*, 5, 74–9.

Rościszewska, E. (1991). Ultrastructural and histochemical studies of the egg capsules of *Perla marginata* (Panzer, 1799) and *Dinocras cephalotes* (Curtis, 1827) (Plecoptera, Perlidae). *International Journal of Insect Morphology & Embryology*, 20, 189–203.

Ross, D. H. and Merritt, R. W. (1987). Factors affecting larval blackfly distribution and population dynamics. In K. C. Kim and R. W. Merritt, eds. *Black Flies: Ecology, Population Management, and Annotated World List*. pp. 90–108. Pennsylvania State University Press, University Park, PA.

Rouquette, J. R. and Thompson, D. J. (2007). Patterns of movement and dispersal in an endangered damselfly and the consequences for its management. *Journal of Applied Ecolgy*, 44, 692–701.

Rubenstein, D. I. and Koehl, M. A. R. (1977). The mechanisms of filter feeding: some theoretical considerations. *The American Naturalist*, 111, 981–94.

Rubinoff, D. (2008). Phylogeography and ecology of an endemic radiation of Hawaiian aquatic case-bearing moths (*Hyposmocoma*: Cosmopterigidae). *Philosophical Transactions of the Royal Society, B.*, 363, 3459–65.

Rubinoff, D. and Schmitz, P. (2010). Multiple aquatic invasions by an endemic, terrestrial Hawaiian moth radiation. *Proceedings of the National Academy of Sciences of the USA*, 107, 5903–6.

Rudolph, P. (1967). Zum ortungsverfahren von *Gyrinus substriatus* Steph. *Zeitschrift für vergleichen Physiologie*, 56, 341–75.

Rueda, L. M. (2008). Global diversity of mosquitoes (Insecta: Diptera: Culicidae) in freshwater. *Hydrobiologia*, 595, 477–87.

Ruffieux, L., Elouard, J. M., and Sartori, M. (1998a). Flightlessness in mayflies and its relevance to hypotheses on the origin of insect flight. *Proceedings of the Royal Society of London B*, 265, 2135–40.

Ruffieux, L., Sartori, M., and L'Eplattenier, G. (1996). Palmen body: a reliable structure to estimate the number of instars in *Siphlonurus aestivalis* (Eaton) (Ephemeroptera: Siphlonuridae). *International Journal of Insect Morphology and Embryology*, 25, 341–4.

Rundle, S. D., Bilton, D. T., Abbott, J. C., and Foggo, A. (2007). Range size in North American *Ennallagma* damselflies correlates with wing size. *Freshwater Biology*, 52, 471–7.

Rűppell, G. (1989). Kinematic analysis of symmetrical flight manoeuvres of Odonata. *Journal of Experimental Biology*, 144, 13–42.

Rupprecht, R. (1972). Dialektbildung bei den Trommelsignalen von *Diura* (Plecoptera). *Oikos,* 23, 410–12.

Rupprecht, R. (1975). Die kommunikation von *Sialis* (Megaloptera) durch vibrationssignale. *Journal of Insect Physiology*, 21, 305–20.

Russell, B. M., Kay, B. H., and Shipton, W. (2001). Survival of *Aedes aegypti* (Diptera: Culicidae) eggs in surface and subterranean breeding sites during the Northern Queensland dry season. *Journal of Medical Entomology*, 38, 441–5.

Russell, R. W., May, M. L., Soltesz, K. L., and Fitzpatrick, J. W. (1998). Massive swarm migrations of dragonflies (Odonata) in eastern North America. *American Midland Naturalist*, 140, 325–42.

Rutherford, J. E. and Mackay, R. J. (1986). Patterns of pupal mortality in field populations of *Hydropsyche* and *Cheumatopsyche* (Trichoptera: Hydropsychidae). *Freshwater Biology*, 16, 337–50.

Ryker, L. C. (1976). Acoustic behavior of *Tropisternus ellipticus*, *T. columbianus*, and *T. lateralis limbalis* in Western Oregon (Coleoptera: Hydrophilidae). *The Coleopterists Bulletin*, 30, 147–56.

Sabando, M. C., Vila, I., Peñaloza, R., and Véliz, D. (2011). Contrasting population genetic structure of two widespread aquatic insects in the Chilean high-slope rivers. *Marine and Freshwater Research*, 62, 1–10.

Sahlen, G. (1990). Egg raft adhesion and chorion structure in *Culex pipiens* (Diptera, Culicidae). *International Journal of Insect Morphology & Embryology*, 19, 307–14.

Sakhuja, M., Williams, D. D., and Williams, N. E. (1983). The role of setae in the behaviour of larval *Phryganea cinerea* Walker (Trichoptera: Phryganeidae). *Canadian Journal of Zoology*, 61, 725–31.

Sakurai, M., Furuki, T., Akao, K., Tanaka, D., Nakahara, Y., Kikawada, T., Watanabe, M., and Okuda, T. (2008). Vitrification is essential for anhydrobiosis in an African chironomid, *Polypedilum vanderplanki*. *Proceedings of the National Academy of Sciences of the USA*, 105, 5093–8.

Salles, F. F., Pereira, S. M., and Serrão, J. E. (2003 (2005)). Redescription of *Camelobaetidius leentvaari* Demoulin, 1966 from Suriname and Brazil [Ephemeroptera, Baetidae]. *Ephemera*, 5, 69–75.

Salt, G. (1937). The egg-parasite of *Sialis lutaria*: A study of the influence of the host upon a dimorphic parasite. *Parasitology*, 29, 539–53.

Sánchez-Fernández, D., Calosi, P., Atfield, A., Arribas, P., Velasco, J., Spicer, J. I., Millán, A., and Bilton, D. T. (2010). Reduced salinities compromise the thermal tolerance of hypersaline specialist diving beetles. *Physiological Entomology*, 35, 265–73.

Sandberg, J. B. and Stewart, K. W. (2004). Capacity for extended egg diapause in six *Isogenoides* Klapálek species (Plecoptera: Perlodidae). *Transactions of the American Entomological Society*, 130, 411–23.

Sangpradub, N. and Giller, P. S. (1994). Gut morphology, feeding rate and gut clearance in five species of caddis larvae. *Hydrobiologia*, 287, 215–23.

Sartori, M., Keller, L., Thomas, A. G. B., and Passera, L. (1992). Flight energetics in relation to sexual differences in the mating behaviour of a mayfly, *Siphlonurus aestivalis*. *Oecologia*, 92, 172–6.

Saunders, D. S. (2010). Controversial aspects of photoperiodism in insects and mites. *Journal of Insect Physiology*, 56, 1491–502.

Saunders, D. S., Lewis, R. D., and Warman, G. R. (2004). Photoperiodic induction of diapause: opening the black box. *Physiological Entomology*, 29, 1–15.

Savage, A. A. (1989). *Adults of the British Aquatic Hemiptera Heteroptera*, Freshwater Biological Association, Ambleside.

Savolainen, E., Saura, A., and Hantula, J. (1993). Mode of swarming in relation to reproductive isolation in mayflies. *Evolution*, 47, 1796–804.

Sawchyn, W. W. and Gillott, C. (1974). The life histories of three species of *Lestes* (Odonata: Zygoptera) in Saskatchewan. *Canadian Entomologist*, 106, 1283–93.

Sawchyn, W. W. and Gillott, C. (1975). The biology of two related species of coenagrionid dragonflies (Odonata: Zygoptera) in Western Canada. *Canadian Entomologist*, 107, 119–28.

Schärer, M. T. and Epler, J. H. (2007). Long-range dispersal possibilities via sea turtle—a case for *Clunio* and *Pontomyia* (Diptera: Chironomidae) in Puerto Rico. *Entomological News*, 118, 273–7.

Schmidt, A. R., Perrichot, V., Svojtka, M., Anderson, K. B., Belete, K. H., Bussert, R., Dörfeldt, H., Jancke, S., Mohr, B., Mohrmann, E., Nascimbene, P. C., Nel, A., Nel, P., Ragazzi, E., Roghi, G., Saupe, E. E., Schmidt, K., Schneider, H., Selden, P. A., and Vávra, N. (2010). Cretaceous African life captured in amber. *Proceedings of the National Academy of Sciences of the USA*, 107, 7329–34.

Schneider, D. (1964). Insect antennae. *Annual Review of Entomology*, 9, 103–22.

Schuh, M. and Diesel, R. (1995). Breeding in a rockpool: Larvae of the semiterrestrial crab *Armases* [= *Sesarma*] *miersii* (Rathbun) (Decapoda: Grapsidae) develop in a highly variable environment. *Journal of Experimental Marine Biology and Ecology*, 185, 109–29.

Schultheis, A. S., Hendricks, A. C., and Weigt, L. A. (2002). Genetic evidence for 'leaky' cohorts in the semivoltine stonefly *Peltoperla tarteri* (Plecopetera: Peltoperlidae). *Freshwater Biology*, 47, 367–76.

Schwind, R. (1980). Geometrical optics of the *Notonecta* eye: Adaptations to optical environment and way of life. *Journal of Comparative Physiology A*, 140, 59–68.

Schwind, R. (1983). A polarization-sensitive response of the flying water bug *Notonecta glauca* to UV light. *Journal of Comparative Physiology A*, 150, 87–91.

Schwind, R. (1995). Spectral regions in which aquatic insects see reflected polarized light. *Journal of Comparative Physiology A*, 177, 439–48.

Scrimshaw, S. and Kerfoot, W. C. (1987). Chemical defenses of freshwater organisms: Beetles and bugs. In W. C. Kerfoot and A. Sih, eds. *Predation: Direct and Indirect Impacts on Aquatic Communities*. pp. 240–62. University Press of New England, Hanover.

Scudder, G. G. E. (1971). The postembryonic development of the indirect flight muscles in *Cenocorixa bifida* (Hung.) (Hemiptera: Corixidae). *Canadian Journal of Zoology*, 49, 1387–98.

Scudder, G. G. E. and Meredith, J. (1972). Temperature-induced development in the indirect flight muscle of adult *Cenocorixa* (Hemiptera: Corixidae). *Developmental Biology*, 29, 330–6.

Seenivasagan, T., Sharma, K. R., Sekhar, K., Ganesan, K., Prakash, S., and Vijayaraghavan, R. (2009). Electroantennogram, flight orientation, and oviposition responses of *Aedes aegypti* to the oviposition pheromone *n*-heneicosane. *Parasitology Research*, 104, 827–33.

Sei, M. (2004). Larval adaptation of the endangered maritime ringlet *Coenonympha tullia nipisiquit* McDonnough (Lepidoptera: Nymphalidae) to a saline wetland habitat. *Environmental Entomology*, 33, 1535–40.

Sensenig, A. T., Kiger, K. T., and Shultz, J. W. (2009). The rowing-to-flapping transition: ontogenetic changes in gill-plate kinematics in the nymphal mayfly *Centroptilum triangulifer* (Ephemeroptera, Baetidae). *Biological Journal of the Linnean Society*, 98, 540–55.

Sensenig, A. T., Kiger, K. T., and Shultz, J. W. (2010). Hydrodynamic pumping by serial gill arrays in the mayfly nymph *Centroptilum triangulifer*. *Journal of Experimental Biology*, 213, 3319–31.

Sephton, D. H. and Hynes, H. B. N. (1983). Food and mouthpart morphology of the nymphs of several Australian Plecoptera. *Australian Journal of Marine and Freshwater Research*, 34, 893–908.

Sérandour, J., Rey, D., and Raveton, M. (2006). Behavioural adaptation of *Coquillettidia* (*Coquillettidia*) *richiardii* larvae to underwater life: environmental cues governing plant-insect interaction. *Entomologia Experimentalis et Applicata*, 120, 195–200.

Sformo, T. and Doak, P. (2006). Thermal ecology of interior Alaska dragonflies. *Functional Ecology*, 20, 114–23.

Shama, L. N. S., Kubow, K. B., Jokela, J., and Robinson, C. T. (2011). Bottlenecks drive temporal and spatial genetic changes in alpine caddisfly metapopulations. *BMC Evolutionary Biology*, 11, 278.

Shashar, N., Hagan, R., Boal, J. G., and Hanlon, R. T. (2000). Cuttlefish use polarization sensitivity in predation on silvery fish. *Vision Research*, 40, 71–5.

Shashar, N., Hanlon, R. T., and Petz, A. D. (1998). Polarization vision helps detect transparent prey. *Nature*, 393, 222–3.

Sheppard, S. K. and Hardwood, J. P. (2005). Advances in molecular ecology: tracking trophic links through predator–prey food-webs. *Functional Ecology*, 19, 751–62.

Sherk, T. E. (1977). Development of the compound eyes of dragonflies (Odonata). I. Larval compound eyes. *Journal of Experimental Zoology*, 210, 391–416.

Sherk, T. E. (1978a). Development of the compound eyes of dragonflies (Odonoata) III. Adult compound eyes. *Journal of Experimental Zoology*, 203, 61–80.

Sherk, T. E. (1978b). Development of the compound eyes of dragonflies (Odonata) II. Development of the larval compound eyes. *Journal of Experimental Zoology*, 203, 47–60.

Short, A. E. Z. and Liebherr, J. K. (2007). Systematics and biology of the endemic water scavenger beetles of Hawaii (Coleoptera:Hydrophilidae, Hydrophilini). *Systematic Entomology*, 32, 601–24.

Sinsabaugh, R. L., Linkin, A. E., and Benfield, E. F. (1985). Cellulose digestion and assimilation by three leaf-shredding aquatic insects. *Ecology*, 66, 1464–71.

Smith, J. A. and Dartnell, A. J. (1980). Boundary layer control by water pennies (Coleoptera: Psephenidae). *Aquatic Insects*, 2, 65–72.

Smith, P. J. and Smith, B. J. (2009). Small-scale population-genetic differentiation in the New Zealand caddisfly *Orthopsyche fimbriata* and the crayfish *Paranephrops planifrons*. *New Zealand Journal of Marine and Freshwater Research*, 43, 723–34.

Smith, R. F., Alexander, L. C., and Lamp, W. O. (2009). Dispersal by terrestrial stages of stream insects in urban watersheds: a synthesis of current knowledge. *Journal of the North American Benthological Society*, 28, 1022–37.

Smith, R. L. (1973). Aspects of the biology of three species of the genus *Rhantus* (Coleoptera: Dytiscidae) with special reference to the acoustical behaviour of two. *Canadian Entomologist*, 105, 909–19.

Smith, R. L. (1976). Male brooding behaviour of the water bug *Aedes herberti* (Hemiptera: Belostomatidae). *Annals of the Entomological Society of America*, 69, 740–7.

Smith, R. L. and Larsen, E. (1993). Egg attendance and brooding by males of the giant water bug *Lethocerus medius* (Guerin) in the field (Heteroptera: Belostomatidae). *Journal of Insect Behavior*, 6, 93–106.

Snodgrass, R. E. (1935). *Principles of Insect Morphology*, Cornell University Press, Ithaca, New York.

Sode, A. and Wiberg-Larsen, P. (1993). Dispersal of adult Trichoptera at a Danish forest brook. *Freshwater Biology*, 30, 439–46.

Söderström, O. (1987). Upstream movements of invertebrates in running waters—a review. *Archiv für Hydrobiologie*, 111, 197–208.

Solarz, S. L. and Newman, R. M. (1996). Oviposition specificity and behavior of the watermilfoil specialist *Euhrychiopsis lecontei*. *Oecologia*, 106, 337–44.

Soong, K., Chen, G. -F., and Cao, J. -R. (1999). Life history studies of the flightless marine midges *Pontomyia* spp. (Diptera: Chironomidae). *Zoological Studies*, 38, 466–73.

Spangler, P. J. and Steiner Jr, W. E. (2005). A new aquatic beetle family, Meruidae, from Venezuela (Coleoptera: Adephaga). *Systematic Entomology*, 30, 339–57.

Spänhoff, B., Kock, C., Meyer, A., and Meyer, E. I. (2005). Do grazing caddisfly larvae of *Melampophylax mucoreus* (Limnephilidae) use their antennae for olfactory food detection? *Physiological Entomology*, 30, 134–43.

Spänhoff, B., Schulte, U., Alecke, C., Kaschek, N., and Meyer, E. I. (2003). Mouthparts, gut contents, and retreat-construction by the wood-dwelling larvae of *Lype phaeopa* (Trichoptera: Psychomyiidae). *European Journal of Entomology*, 100, 563–70.

Spence, J. R. and Andersen, N. M. (1994). Biology of water striders: interactions between systematics and ecology. *Annual Review of Entomology*, 39, 101–28.

Spence, J. R. and Wilcox, R. S. (1986). The mating system of two hybridizing species of water striders (Gerridae). II. Alternative tactics of males and females. *Behavioral Ecology and Sociobiology*, 19, 87–95.

Spence, J. R., Hughes Spence, D., and Scudder, G. G. E. (1980). Submergence behavior in *Gerris*: underwater basking. *American Midland Naturalist*, 103, 385–91.

Spitzer, K. and Danks, H. V. (2006). Insect biodiversity of boreal peat bogs. *Annual Review of Entomology*, 51, 137–61.

Srygley, R. B. and Dudley, R. (2008). Optimal strategies for insects migrating in the flight boundary layer: mechanisms and consequences. *Integrative and Comparative Biology*, 48, 119–33.

Staddon, B. W. (1955). The excretion and storage of ammonia by the aquatic larvae of *Sialis lutaria* (Neuroptera). *Journal of Experimental Biology*, 32, 84–94.

Staddon, B. W. (1959). Nitrogen excretion in nymphs of *Aeshna cyanea* (Müll.) (Odonata, Anisoptera). *Journal of Experimental Biology*, 36, 566–74.

Stamp, N. E. (1980). Egg deposition patterns in butterflies: Why do some species cluster their eggs rather than deposit them singly? *The American Naturalist*, 115, 367–80.

Standley, L. J., Sweeney, B. W., and Funk, D. H. (1994). Maternal transfer of chlordane and its metabolites to the eggs of a stream mayfly *Centroptilium triangulifer*. *Environmental Science and Technology*, 28, 2105–11.

Stanford, J. A. and Ward, J. V. (1988). The hyporheic habitat of river ecosystems. *Nature*, 335, 64–6.

Stange, G. (1981). The ocellar component of flight equilibrium control in dragonflies. *Journal of Comparative Physiology A*, 141, 335–47.

Stange, G., Stowe, S., Chahl, J. S., and Massaro, A. (2002). Anisotropic imaging in the dragonfly median ocellus: a matched filter for horizon detection. *Journal of Comparative Physiology A*, 188, 455–67.

Statzner, B. (1988). Growth and Reynolds number of lotic macroinvertebrates: a problem for adaptation of shape to drag. *Oikos*, 51, 84–7.

Statzner, B. (2008). How views about flow adaptations of benthic stream invertebrates changed over the last century. *International Review of Hydrobiology*, 93, 593–605.

Statzner, B. and Holm, T. F. (1989). Morphological adaptation of shape to flow: microcurrents around lotic macroinvertebrates with known Reynolds numbers at quasi-natural flow conditions. *Oecologia*, 78, 145–57.

Stav, G., Blaustein, L., and Margalit, J. (1999). Experimental evidence for predation risk sensitive oviposition by a mosquito, *Culiseta longiareolata*. *Ecological Entomology*, 24, 202–207.

Stecher, N., Morgan, R. C., and Buschbeck, E. K. (2010). Retinal ultrastructure may mediate polarization sensitivity in larvae of the Sunburst diving beetle, *Thermonectus marmoratus* (Coleoptera: Dytiscidae). *Zoomorphology*, 129, 141–52.

Steedman, R. J. and Anderson, N. H. (1985). Life history and ecological role of the xylophagous aquatic beetle, *Lara avara* LeConte (Dryopoidea: Elmidae). *Freshwater Biology*, 15, 535–46.

Steinmann, P. (1907). Die Tierwelt der Gebirgsbäche. Eine faunistisch-biologische Studie. *Annales de Biologie Lacustre*, 2, 30–150.

Steinman, A. D., McIntire, C. D., Gregory, S. V., Lamberti, G. A., and Ashkenas, L. R. (1987). Effects of herbivore type and density on taxonomic structure and physiognomy of algal assemblages in laboratory streams. *Journal of the North American Benthological Society*, 6, 175–88.

Sternberg, K. (1994). Temperature stratification in bog ponds. *Archiv für Hydrobiologie*, 129, 373–82.

Stewart, C. C. and Felgenhauer, B. E. (2003). Structure and function of the mouthparts and salivary gland complex of the giant waterbug, *Belostoma lutarium* (Stål) (Hemiptera: Belostomatidae). *Annals of the Entomological Society of America*, 96, 870–82.

Stewart, K. W. and Stark, B. P. (1988). *Nymphs of North American Stonefly Genera (Plecoptera)*, The Thomas Say Foundation, Ecological Society of America, Washington DC.

Stewart, K. W. and Maketon, M. (1991). Structures used by nearctic stoneflies (Plecoptera) for drumming, and their relationship to behavioral pattern diversity. *Aquatic Insects*, 13, 33–53.

Stewart, K. W. and Ricker, W. E. (1997). Stoneflies (Plecoptera) of the Yukon. In H. V. Danks and J. A. Downes, eds. *Insects of the Yukon*. pp. 201–22. Biological Survey of Canada, Ottawa.

Stewart, K. W. and Sandberg, J. B. (2006). Vibrational communication and mate searching behavior in stoneflies. In S. Drosopoulos and M. Claridge, eds. *Insect Sounds and Communication: Physiology, Behavior and Evolution.*, pp. 179–86. CRC Press, Taylor Francis Group, Boca Raton.

Stewart, K. W., Szczytko, S. W., and Stark, B. P. (1982). Drumming behavior of four species of North American Pteronarcyidae (Plecoptera): dialects in Colorado and Alaska *Pteronarcella badia*. *Annals of the Entomological Society of America*, 75, 530–3.

Stief, P., Nazarova, L., and de Beer, D. (2005). Chimney construction by *Chironomus riparius* larvae in response to hypoxia: microbial implications for freshwater sediments. *Journal of the North American Benthological Society*, 24, 858–71.

Stocks, I. C. (2010a). Comparative and functional morphology of wing coupling structures in Trichoptera: Annulipalpia. *Journal of Morphology*, 271, 152–68.

Stocks, I. C. (2010b). Comparative and functional morphology of wing coupling structures in Trichoptera: Integripalpia. *Annales Zoologici Fennici*, 47, 351–86.

Stuart, A. E. (2002). The cocoon-spinning behaviour of *Austrosimulium australense* (Diptera: Simuliidae) with a discussion of phylogenetic implications. *New Zealand Journal of Zoology*, 29, 5–14.

Stuart, A. E. and Hunter, F. F. (1995). A re-description of the cocoon-spinning behaviour of *Simulium vittatum* (Diptera Simuliidae). *Ethology Ecology & Evolution*, 7, 363–77.

Štys, P. and Soldán, T. (1980). Retention of tracheal gills in adult Ephemeroptera and other insects. *Acta Universitatis Carolinae—Biologica*, 1978, 409–35.

Suemoto, T., Kawai, K., and Imabayashi, H. (2004). A comparison of desiccation tolerance among 12 species of chironomid larvae. *Hydrobiologia*, 515, 107–14.

Sueur, J., Mackie, D., and Windmill, J. F. C. (2011). So small, so loud: Extremely high sound pressure level from a pygmy aquatic insect (Corixidae, Micronectinae). *PLoS ONE*, 6(6), e21089.

Sullivan, R. T. (1981). Insect swarming and mating. *Florida Entomologist*, 64, 44–65.

Suren, A. M. and Lake, P. S. (1989). Edibility of fresh and decomposing macrophytes to three species of freshwater invertebrate herbivores. *Hydrobiologia*, 178, 165–78.

Suren, A. M. and Winterbourn, M. J. (1991). Consumption of aquatic bryophytes by alpine stream invertebrates in New Zealand. *New Zealand Journal of Marine and Freshwater Research*, 25, 331–44.

Svensson, B. (1977). Life cycle, energy fluctuations and sexual differentiation in *Ephemera danica* (Ephemeroptera), a stream-living mayfly. *Oikos*, 29, 78–86.

Sweeney, B. W. and Vannote, R. L. (1982). Population synchrony in mayflies: A predator satiation hypothesis. *Evolution*, 36, 810–21.

Sweeney, B. W. and Vannote, R. L. (1987). Geographic parthenogenesis in the stream mayfly *Eurylophella funeralis* in eastern North America. *Holarctic Ecology*, 10, 52–9.

Symondson, W. O. C. (2002). Molecular identification of prey in predator diets. *Molecular Ecology*, 11, 627–41.

Szczytko, S. W. and Stewart, K. W. (1979). Drumming behavior of four Western nearctic *Isoperla* (Plecoptera) species. *Annals of the Entomological Society of America*, 72, 781–6.

Tachet, H. (1977). Vibrations and predatory behavior of *Plectrocnemia* larvae (Trichoptera). *Journal of Comparative Ethology*, 45, 61–74.

Tachet, H., Pierrot, J. P., Roux, C., and Bournaud, M. (1992). Net-building behaviour of six *Hydropsyche* species (Trichoptera) in relation to current velocity and distribution along the Rhône River. *Journal of the North American Benthological Society*, 11, 350–65.

Tallamy, D. W. (2001). Evolution of exclusive paternal care in arthropods. *Annual Review of Entomology*, 46, 139–65.

Taylor, J. M. and Kennedy, J. H. (2006). Life history and secondary production of *Caenis latipennis* (Ephemeroptera: Caenidae) in Honey Creek, Oklahoma. *Annals of the Entomological Society of America*, 99, 821–30.

Taylor, G. K. and Krapp, H. G. (2008). Sensory systems and flight stability: What do insects measure and why? *Advances in Insect Physiology*, 34, 231–316.

Telfer, W. H. and Kunkel, J. G. (1991). The function and evolution of insect storage hexamers. *Annual Review of Entomology*, 36, 205–28.

Tennessen, K. J. and Painter, M. K. (1994). Forced ejection of fecal pellets by nymphs of *Erithemis simplicicollis*. *Argia*, 6, 15.

Teraguchi, S. (1975a). Correction of negative buoyancy in the phantom larva, *Chaoborus americanus*. *Journal of Insect Physiology*, 21, 1659–70.

Teraguchi, S. (1975b). Detection of negative buoyancy in the phantom larva, *Chaoborus americanus*. *Journal of Insect Physiology*, 21, 1265–9.

Theiss, J. (1982). Generation and radiation of sounds by stridulating water insects as exemplified by the corixids. *Behavioral Ecology and Sociobiology*, 10, 225–35.

Theiss, J., Prager, J., and Streng, R. (1983). Underwater stridulation by corixids: stridulatory signals and sound producing mechanism in *Corixa dentipes* and *Corixa punctata*. *Journal of Insect Physiology*, 29, 761–71.

Thomas, M. A., Walsh, K. A., Wolf, M. R., McPheron, B. A., and Marden, J. H. (2000). Molecular phylogenetic analysis of evolutionary trends in stonefly wing structure and locomotor behavior. *Proceedings of the National Academy of Sciences of the United States of America*, 97, 13178–83.

Thorpe, W. H. (1950). Plastron respiration in aquatic insects. *Biological Reviews of the Cambridge Philosophical Society*, 25, 344–90.

Thorpe, W. H. and Crisp, D. J. (1947a). Studies on plastron respiration. III. The orientation responses of *Aphelocheirus* (Hemiptera, Aphelocheiridae (Naucoridae)) in relation to plastron respiration; together with an account of specialized pressure receptors in aquatic insects. *Journal of Experimental Biology*, 24, 310–28.

Thorpe, W. H. and Crisp, D. J. (1947b). Studies on plastron respiration. I. The biology of *Aphelocheirus* (Hemiptera, Aphelocheiridae (Naucoridae)) and the mechanism of plastron retention. *Journal of Experimental Biology*, 24, 227–69.

Thorpe, W. H. and Crisp, D. J. (1947c). Studies of plastron respiration. II. The respiratory efficiency of the plastron in *Aphelocheirus*. *Journal of Experimental Biology*, 24, 270–303.

Thorpe, W. H. and Crisp, D. J. (1949). Studies on plastron respiration. IV. Plastron respiration in the Coleoptera. *Journal of Experimental Biology*, 26, 219–60.

Tichy, H. and Kallina, W. (2010). Insect hygroreceptor responses to continuous changes in humidity and air pressure. *Journal of Neurophysiology*, 103, 3274–86.

Tichy, H. and Loftus, R. (1996). Hygroreceptors in insects and a spider: Humidity transduction models. *Naturwissenschaften*, 83, 255–63.

Tierno de Figueroa, J. M., Luzón-Ortega, J. M., and López-Rodriguez, M. J. (2006). Mating balls in stoneflies (Insecta, Plecoptera). *Zoologica baetica*, 17, 93–6.

Tilquin, M., Meyran, J. -C., and Marigo, G. (2004). Comparative capability to detoxify vegetable allelochemicals by larval mosquitoes. *Journal of Chemical Ecology*, 30, 1381–91.

Timms, B. V. (1983). Study of benthic communities in some shallow saline lakes of western Victoria, Australia. *Hydrobiologia*, 105, 165–77.

Timms, B. V. (1998). Further studies on the saline lakes of the eastern Paroo, inland New South Wales, Australia. *Hydrobiologia*, 381, 31–42.

Timms, B. V. (2001). A study of the Werewilka Inlet of the saline Lake Wyara, Australia—a harbour of biodiversity for a sea of simplicity. *Hydrobiologia*, 466, 245–54.

Timms, B. V. (2009). A study of the salt lakes and salt springs of Eyre Peninsula, South Australia. *Hydrobiologia*, 626, 41–51.

Tindall, A. R. (1964). The skeleton and musculature of the larval thorax of *Triaenodes bicolor* Curtis (Trichoptera: Limnephilidae). *Transactions of the Royal Entomological Society*, 116, 151–210.

Tojo, K., Sekiné, K., and Matsumoto, A. (2006). Reproductive mode of the geographic parthenogenetic mayfly *Ephoron shigae*, with findings from some new localities (Insecta: Ephemeroptera, Polymitarcyidae). *Limnology*, 7, 31–9.

Tokeshi, M. (1993). On the evolution of commensalism in the Chironomidae. *Freshwater Biology*, 29, 481–9.

Tokeshi, M. (1995). Species interactions and community structure. In P. D. Armitage, P. S. Cranston, and L. C. V. Pinder, eds. *The Chironomidae: Biology and Ecology of Non-Biting Midges.* pp. 297–335. Chapman & Hall, London.

Tokeshi, M. and Reinhardt, K. (1996). Reproductive behaviour in *Chironomus anthracinus* (Diptera: Culicidae), with a consideration of the evolution of swarming. *Journal of Zoology*, 240, 103–12.

Townsend, C. R. and Hildrew, A. G. (1979). Form and function of the prey catching net of *Plectrocnemia conspersa* larvae (Trichoptera). *Oikos*, 33, 412–18.

Tozer, W. (1979). Underwater behavioural thermoregulation in the adult stonefly, *Zapada cinctipes*. *Nature*, 281, 566–7.

Tozer, W. E., Resh, V. H., and Solem, J. O. (1981). Bionomics and adult behavior of a lentic caddisfly, *Nectopsyche albida* (Walker). *American Midland Naturalist*, 106, 133–44.

Trexler, J. D., Apperson, C. S., and Schal, C. (1998). Laboratory and field evaluations of oviposition responses of *Aedes albopictus* and *Aedes triseriatus* (Diptera: Culicidae) to oak infusions. *Journal of Medical Entomology*, 35, 967–76.

Trueman, J. W. H. (1990). Eggshells of Australian Gomphidae: plastron respiration in eggs of stream-dwelling Odonata (Anisoptera). *Odonatologica*, 19, 395–401.

Truman, J. W. and Riddiford, L. M. (2002). Endocrine insights into the evolution of metamorphosis in insects. *Annual Review of Entomology*, 47, 467–500.

Tsoukatou, M., Cheng, L., Vagias, C., and Roussis, V. (2001). Chemical composition and behavioural responses of the marine insect *Halobates hawaiiensis* (Heteroptera: Gerridae). *Zeitschrift für Naturforschung C*, 56, 597–602.

Tsubaki, Y., Kato, C., and Shintani, S. (2006). On the respiratory mechanism during underwater oviposition in a damselfly *Calopteryx cornelia* Selys. *Journal of Insect Physiology*, 52, 499–505.

Tucker, V. A. (1969). Wave-making by whirligig beetles (Gyrinidae). *Science*, 166, 897–9.

Turnbull, D. K. R. and Barmuta, L. A. (2006). Different light conditions influence the activity of *Eusthenia costalis* Banks (Plecoptera: Eustheniidae). *Aquatic Insects*, 28, 211–17.

Ubero-Pascal, N. and Piug, M. A. (2009). New type of egg attachment structure in Ephemeroptera and comparative analysis of chorion structure morphology in three species of Ephemerellidae. *Acta Zoologica*, 90, 87–98.

Upchurch, P. (2008). Gondwanan break-up: legacies of a lost world? *Trends in Ecology and Evolution*, 23, 229–36.

Usherwood, J. and Ellington, C. P. (2002). The aerodynamics of revolving wings II. Propeller force coefficients from mayfly to quail. *Journal of Experimental Biology*, 205, 1565–76.

Usherwood, J. and Lehman, F. -O. (2008). Phasing of dragonfly wings can improve aerodynamic efficiency by removing swirl. *Journal of the Royal Society Interface*, 5, 1303–7.

Van Dam, A. R. and Walton, W. E. (2008). The effect of predatory fish exudates on the ovipositional behavior of three mosquito species: *Culex quinquefasciatus*, *Aedes aegypti* and *Culex tarsalis*. *Medical and Veterinary Entomology*, 22, 399–404.

van der Geest, H. G. (2007). Behavioural responses of caddisfly larvae (*Hydropsyche angustipennis*) to hypoxia. *Contributions to Zoology*, 76, 255–60.

van Someren, V. D. and McMahon, J. (1950). Phoretic association between *Afronurus* and *Simulium* species, and the discovery of the early stages of *Simulium neavei* on freshwater crabs. *Nature*, 166, 350–1.

Vargas, A., Mittal, R., and Dong, H. (2008). A computational study of the aerodynamic performance of a dragonfly wing section in gliding flight. *Bioinspiration & Biomimetics*, 3, 026004.

Velasco, J. and Millan, A. (1998). Feeding habits of two large insects from a desert stream: *Abedus herberti* (Hemiptera: Belostomatidae) and *Thermonectus marmoratus* (Coleoptera: Dytiscidae). *Aquatic Insects*, 20, 85–96.

Verberk, W. C. E. P. and Bilton, D. T. (2011). Can oxygen set thermal limits in an insect and drive gigantism? *PLoS ONE*, 6, e22610.

Verrier, M. -L. (1956). *Biologie des Ephémères*, Librairie Armand Colin, Paris.

Verschuren, D., Cocquyt, C., Tibby, J., Roberts, C. N., and Leavitt, P. R. (1999). Long-term dynamics of algal and invertebrate communities in a small, fluctuating tropical soda lake. *Limnology and Oceanography*, 44, 1216–31.

Vickers, N. J. (2000). Mechanisms of animal navigation in odor plumes. *Biological Bulletin*, 198, 203–12.

Vinson, M. R. and Hawkins, C. P. (2003). Broad-scale geographical patterns in local stream insect genera richness. *Ecography*, 26, 751–67.

Virant-Doberlet, M. and Čokl, A. (2004). Vibrational communication in insects. *Neotropical Entomology*, 33, 121–34.

Vitek, C. J. and Livdahl, T. (2009). Hatch plasticity in response to varied inundation frequency in *Aedes albopictus*. *Journal of Medical Entomology*, 46, 766–71.

Vogel, S. (1994). *Life in Moving Fluids: The Physical Biology of Flow, 2nd edn*, Princeton University Press, Princeton NJ.

Vogel, S. (2006). Living in a physical world VIII Gravity and life in water. *Journal of Bioscience*, 31, 309–22.

Voise, J. and Casas, J. (2010). The management of fluid and wave resistances by whirligig beetles. *Journal of the Royal Society Interface*, 7, 343–52.

Vulinec, K. and Miller, M. C. (1989). Aggregation and predator avoidance in whirligig beetles (Coleoptera: Gyrinidae). *Journal of the New York Entomological Society*, 97, 438–47.

Waage, J. K. (1979). Dual function of the damselfly penis: sperm removal and transfer. *Science*, 203, 916–18.

Waage, J. K. (1986). Evidence for widespread sperm displacement ability among Zygoptera (Odonata) and the means for predicting its presence. *Biological Journal of the Linnean Society*, 28, 285–300.

Wagner, F. (2003). Flight behaviour of merolimnic insects from the Leutra River (Thuringia, Germany). *Aquatic Insects*, 25, 51–62.

Wagner, R., Barták, M., Borkent, A., Courtney, G., Goddeeris, B., Haenni, J. -P., Knutson, L., Pont, A., Rotheray, G. E., Rozkošný, R., Sinclair, B., Woodley, N., Zatwarnicki, T., and Zwick, P. (2008). Global diversity of dipteran families (Insecta Diptera) in freshwater (excluding Simuliidae, Culicidae, Chironomidae, Tipulidae and Tabanidae). *Hydrobiologia*, 595, 489–519.

Wakeling, J. M. and Ellington, C. P. (1997a). Dragonfly flight II. Velocities, accelerations and kinematics of flapping flight. *Journal of Experimental Biology*, 200, 557–82.

Wakeling, J. M. and Ellington, C. P. (1997b). Dragonfly flight III. Lift and power requirements. *Journal of Experimental Biology*, 200, 583–600.

Wakeling, J. M. and Ellington, C. P. (1997c). Dragonfly flight I. Gliding flight and steady-state aerodynamic forces. *Journal of Experimental Biology*, 200, 543–56.

Walcott, B. (1969). Movement of retinual cells in insect eyes on light adaptation. *Nature*, 223, 971–2.

Walcott, B. (1971). Cell movement on light adaptation in the retina of *Lethocerus* (Belostomatidae, Hemiptera). *Journal of Comparative Physiology*, 74, 1–16.

Walcott, B. and Horridge, G. A. (1971). The compound eye of *Archichauliodes* (Megaloptera). *Proceedings of the Royal Society of London. Series B,* 179, 65–72.

Waldbauer, G. P. (1968). The consumption and utilization of food by insects. *Advances in Insect Physiology,* 5, 229–88.

Waldbauer, G. P. and Friedman, S. (1991). Self-selection of optimal diets by insects. *Annual Review of Entomology,* 36, 43–63.

Walenciak, O., Zwisler, W., and Gross, E. M. (2002). Influence of *Myriophyllum spictatum*-derived tannins on gut microbiota of its herbivore. *Journal of Chemical Ecology,* 28, 2045–56.

Wallace, J. B. and Malas, D. (1976). The fine structure and capture nets of larval Philopotamidae (Trichoptera): with special emphasis on *Dolophilodea distinctus. Canadian Journal of Zoology,* 54, 1788–802.

Wallace, J. B. and Merritt, R. W. (1980). Filter-feeding ecology of aquatic insects. *Annual Review of Entomology,* 25, 103–32.

Wallace, J. B. and O'Hop, J. (1979). Fine particle suspension-feeding capabilities of *Isonychia* spp. (Ephemeroptera: Siphlonuridae). *Annals of the Entomological Society of America,* 72, 353–7.

Walshe, B. M. (1947). On the function of haemoglobin in *Chironomus* after oxygen lack. *Journal of Experimental Biology,* 24, 329–42.

Walshe, B. M. (1948). The function of haemoglobin in *Tanytarsus* (Chironomidae). *Journal of Experimental Biology,* 24, 343–51.

Walshe, B. M. (1950). The function of haemoglobin in *Chironomus plumosus* under natural conditions. *Journal of Experimental Biology,* 27, 73–95.

Walshe, B. M. (1951). The feeding habit of certain chironomid larvae (subfamily Tendipedinae). *Proceedings of the Zoological Society of London,* 121, 63–79.

Walter, J. K., Bilby, R. E., and Fransen, B. R. (2006). Effects of Pacific salmon spawning and carcass availability on the caddisfly *Ecclisomyia conspersa* (Trichoptera: Limnephilidae). *Freshwater Biology,* 51, 1211–18.

Walters Jr, K. R., Sformo, T., Barnes, B. M., and Duman, J. G. (2009). Freeze tolerance in an arctic Alaskan stonefly. *Journal of Experimental Biology,* 212, 305–12.

Walther, A. C., Benard, M. F., Boris, L. P., Enstice, N., Tindauer-Thompson, A., and Wan, J. (2008). Attachment of the freshwater limpet *Laevapex fuscus* to the hemelytra of the water bug *Belostoma flumineum. Journal of Freshwater Ecology,* 23, 337–9.

Waringer, J. A. (1993). The drag coefficient of cased caddis larvae from running waters: experimental determination and ecological applications. *Freshwater Biology,* 29, 419–27.

Watanabe, M. (2006). Anhydrobiosis in invertebrates. *Applied Entomology and Zoology,* 41, 15–31.

Watanabe, M., Kikawada, T., Minagawa, M., Yukuhiro, F., and Okuda, T. (2002). Mechanism allowing an insect to survive complete dehydration and extreme temperatures. *Journal of Experimental Biology,* 205, 2799–802.

Waterman, T. H. (2006). Reviving a neglected celestial underwater polarization compass for aquatic animals. *Biological Review,* 81, 111–15.

Waters, T. F. (1972). The drift of stream insects. *Annual Review of Entomology,* 17, 253–72.

Watson, S. B. (2004). Aquatic taste and odor: A primary signal of drinking-water integrity. *Journal of Toxicology and Environmental Health, A,* 67, 1779–95.

Watson, S. B. and Ridal, J. (2004). Periphyton: A primary source of widespread and severe taste and odour. *Water Science Technology,* 49, 33–9.

Watts, C. H. S. and Humphreys, W. F. (2009). Fourteen new Dytiscidae (Coleoptera) of the genera *Limbodessus* Guignot, *Paroster* Sharp, and *Exocelina* Broun from underground waters in Australia. *Transactions of the Royal Society of South Australia,* 133, 62–107.

Weaver III, J. S. (1992). Remarks on the evolution of Trichoptera: a critique of Wiggins and Wichard's classification. *Cladistics,* 8, 171–80.

Weber, R. E. and Vinogradov, S. N. (2001). Nonvertebrate hemoglobins: functions and molecular adaptations. *Physiological Reviews,* 81, 569–628.

Weber, W. and Grossmann, M. (1988). Ultrastructure of the chromatophore system on the tracheal bladders of the phantom larva of *Chaoborus crystallinus* (Insecta, Diptera). *Zoomorphology,* 108, 167–71.

Webster, D. R. and Weissburg, M. J. (2009). The hydrodynamics of chemical cues among aquatic organisms. *Annual Review of Fluid Mechanics,* 41, 73–90.

Wehner, R. (1987). 'Matched filters'—neural models of the external world. *Journal of Comparative Physiology A,* 161, 511–31.

Weiss, M. R. (2006). Defecation behavior and ecology of insects. *Annual Review of Entomology,* 51, 635–61.

Weissenberger, J., Spatz, H. C., Emanns, A., and Schwoerbel, J. (1991). Measurement of lift and drag forces in the mN range experienced by benthic arthropods at flow velocities below 1.2 m s^{-1}. *Freshwater Biology,* 25, 21–31.

Wellnitz, T. A. and Ward, J. V. (1998). Does light intensity modify the effect mayfly grazers have on periphyton? *Freshwater Biology,* 39, 135–49.

Wells, A. (1984). Comparative studies of antennal features of adult Hydroptilidae (Trichoptera). In J. C. Morse, ed. *Fourth International Symposium on Trichoptera.* pp. 423–40. Junk, The Hague.

Wells, A. (1985). Larvae and pupae of Australian Hydroptilidae (Trichoptera), with observations on general biology and relationships. *Australian Journal of Zoology Supplementary Series,* 113, 1–69.

Wells, A. (1992). The first parasitic Trichoptera. *Ecological Entomology*, 17, 299–302.

Wells, A. (2005). Parasitism by hydroptilid caddisflies (Trichoptera) and seven new species of Hydroptilidae from northern Queensland. *Australian Journal of Entomology*, 44, 385–91.

Wells, R. M. G., Hudson, M. J., and Brittain, T. (1981). Function of the hemoglobin and the gas bubble in the backswimmer *Anisops assimilis* (Hemiptera: Notonectidae). *Journal of Comparative Physiology*, 142, 515–22.

Wesenberg-Lund, C. (1943). *Biologie der Süsswasserinsekten*, Gyldendalske, Copenhagen and Springer, Berlin.

Whetmore, S. H., Mackay, R. J., and Newbury, R. W. (1990). Characterising the hydraulic habitat of *Brachycentrus occidentalis*, a filter-feeding caddisfly. *Journal of the North American Benthological Society*, 9, 157–69.

Whitaker Jr, J. O. (2006). Ectoparasites of North American aquatic rodents and comparison to European forms. *Acarina*, 14, 137–45.

Wichard, W. and Komnick, H. (1974). Structure and function of the respiratory epithelium in the tracheal gills of stonefly larvae. *Journal of Insect Physiology*, 20, 2397–406.

Wiese, K. (1972). Das mechanorezeptorische Beuteortungssystem von *Notonecta*. *Journal of Comparative Physiology*, 78, 83–102.

Wiese, K. (1974). The mechanoreceptive system of prey localization in *Notonecta*. II. Principle of prey localization. *Journal of Comparative Physiology*, 92, 317–25.

Wiggins, G. B. (2004). *Caddisflies: the Underwater Architects*, University of Toronto Press, Toronto.

Wiggins, G. B. and Wichard, W. (1989). Phylogeny of pupation in Trichoptera, with proposals on the origin and higher classification of the order. *Journal of the North American Benthological Society*, 8, 260–76.

Wikelski, M., Moskowitz, D., Adelman, J. S., Cochran, J., Wilcove, D. S., and May, M. L. (2006). Simple rules guide dragonfly migration. *Biology Letters*, 2, 325–9.

Wilcox, R. S. (1972). Communication by surface waves: mating behavior in a water strider. *Journal of Comparative Physiology*, 80, 255–66.

Wilcox, R. S. (1979). Discrimination in *Gerris remigis*: Role of surface wave signal. *Science*, 206, 1325–7.

Wilcox, R. S. and Spence, J. R. (1986). The mating system of two hybridizing species of water striders (Gerridae): I. Ripple signal functions. *Behavioral Ecology and Sociobiology*, 19, 79–85.

Wilcox, R. S. and Di Stefano, J. (1991). Vibratory signals enhance mate-guarding in a water strider (Hemiptera: Gerridae). *Journal of Insect Behavior*, 4, 44–50.

Wiles, P. R. (1982). A note on the watermite *Hydrodroma despiciens* feeding on chironomid egg masses. *Freshwater Biology*, 12, 83–7.

Williams, C. J. (1982). The drift of some chironomid egg masses (Diptera: Chironomidae). *Freshwater Biology*, 12, 573–8.

Williams, D. D. (1997). Temporary ponds and their invertebrate communities. *Aquatic Conservation: Marine and Freshwater Ecosystems*, 7, 105–17.

Williams, D. D. (2003). The brackishwater hyporheic zone: invertebrate community structure across a novel ecotone. *Hydrobiologia*, 510, 153–73.

Williams, D. D. (2006). *The Biology of Temporary Waters*, Oxford University Press, Oxford.

Williams, D. D., Heeg, N., and Magnusson, A. K. (2007). Habitat background selection by colonizing intermittent pond invertebrates. *Hydrobiologia*, 592, 487–98.

Williams, D. D., Tavares, A. F., and Bryant, E. (1987). Respiratory device or camouflage?—A case for the caddisfly. *Oikos*, 50, 42–52.

Williams, D. D. and Williams, N. E. (1993). The upstream/downstream movement paradox of lotic invertebrates: quantitative evidence from a Welsh mountain stream. *Freshwater Biology*, 30, 199–218.

Williams, D. D. and Williams, N. E. (1998). Aquatic insects in an estuarine environment: densities, distribution and salinity tolerance. *Freshwater Biology*, 39, 411–21.

Williams, D. S. (1980). Organisation of the compound eye of a tipulid fly during the day and night. *Zoomorphologie*, 95, 85–104.

Willkommen, J. and Hörnschemeyer, T. (2007). The homology of wing base sclerites and flight muscles in Ephemeroptera and Neoptera and the morphology of the pterothorax of *Habroleptoides confusa* (Insecta: Ephemeroptera: Leptophlebiidae). *Arthropod Structure & Development*, 36, 253–69.

Wilson, C. B. (1923). Life history of the scavenger water beetle, *Hydrous* (*Hydrophilus*) *traingularis*, and its economic relation to fish feeding. *Bulletin of the United States Bureau of Fisheries*, 39, 9–38.

Wingfield, C. A. (1939). The function of the gills of mayfly nymphs from different habitats. *Journal of Experimental Biology*, 16, 363–73.

Winterbourn, M. J. (1966). The ecology and life history of *Zelandoperla maculata* (Hare) and *Auklandobius trivacuatus* (Tillyard). *New Zealand Journal of Science*, 9, 312–23.

Winterbourn, M. J. (1974). The life histories, trophic relations and production of *Stenoperla prasina* (Plecoptera) and *Deleatidium* sp. (Ephemeroptera) in a New Zealand river. *Freshwater Biology*, 4, 507–24.

Winterbourn, M. J. (2005). Dispersal, feeding and parasitism of adult stoneflies (Plecoptera) at a New Zealand forest stream. *Aquatic Insects*, 27, 155–66.

Winterbourn, M. J. and Anderson, N. H. (1980). The life history of *Philanisus plebeius* Walker (Trichoptera:

Chathamiidae), a caddisfly whose eggs were found in a starfish. *Ecological Entomology*, 5, 293–303.

Winterbourn, M. J. and Crowe, A. L. M. (2001). Flight activity of insects along a mountain stream: is directional flight adaptive? *Freshwater Biology*, 46, 1479–89.

Wishart, M. J. and Hughes, J. M. (2001). Exploring patterns of population subdivision in the net-winged midge, *Elporia barnardi* (Diptera: Blephariceridae), in mountain streams of the southwestern Cape, South Africa. *Freshwater Biology*, 46, 479–90.

Wissinger, S. A., Whissel, J. C., Eldermire, C., and Brown, W. S. (2006). Predator defense along a permanence gradient: roles of case structure, behavior, and developmental phenology in caddisflies. *Oecologia*, 147, 667–78.

Wootton, R. J. (1988). The historical ecology of aquatic insects: An overview. *Palaeogeography, Palaeoclimatology, Palaeoecology*, 62, 477–92.

Wootton, R. J. (1990). The mechanical design of insect wings. *American Scientist*, 263, 114–20.

Wootton, R. J. (1992). Functional morphology of insect wings. *Annual Review of Entomology*, 37, 113–40.

Wootton, R. J. (1999). Invertebrate paraxial locomotory appendages: design, deformation and control. *Journal of Experimental Biology*, 202, 3333–45.

Wootton, R. J., Kukalova-Peck, J., Newman, D. J. S., and Muzón, J. (1998). Smart engineering in the mid-Carboniferous: How well could Palaeozoic dragonflies fly? *Science*, 282, 749–51.

Wotton, R. S. (1976). Evidence that blackfly larvae can feed on particles of colloidal size. *Nature*, 261, 697.

Wotton, R. S. (1980). Coprophagy as an economic feeding tactic in blackfly larvae. *Oikos*, 34, 282–6.

Wotton, R. S., Malmqvist, B., and Leonardsson, K. (2003). Expanding traditional views on suspension feeders—quantifying their role as ecosystem engineers. *Oikos*, 101, 441–3.

Wright, J. C., Westh, P., and Ramløv, H. (1992). Cryptobiosis in Tardigrada. *Biological Reviews*, 67, 1–29.

Wright, J. F., Hiley, P. D., and Berrie, A. D. (1981). A 9-year study of the life cycle of *Ephemera danica* Müll. (Ephemeridae: Ephemeroptera) in the River Lambourn, England. *Ecological Entomology*, 6, 321–31.

Yager, D. D. (1999). Structure, development, and evolution of insect auditory systems. *Microscopy Research and Technique*, 47, 380–400.

Yang, E. -C. and Osorio, D. (1991). Spectral sensitivities of photoreceptors and lamina monopolar cells in the dragonfly, *Hemicordulia tau*. *Journal of Comparative Physiology A*, 169, 663–9.

Ybarrondo, B. A. (1995). Habitat selection and thermal preference in two species of water scavenger beetles (Coleoptera: Hydrophilidae). *Physiological Zoology*, 68, 749–71.

Yee, D. A., Taylor, S., and Vamosi, S. M. (2009). Beetle and plant density as cues initiating dispersal in two species of adult predaceous diving beetles. *Oecologia*, 160, 25–36.

Young, D. and Bennett-Clark, H. C. (1995). The role of the tymbal in cicada sound production. *Journal of Experimental Biology*, 198, 1011–1019.

Young, E. C. (1965). Flight muscle polymorphism in British Corixidae: ecological observations. *Journal of Animal Ecology*, 34, 353–90.

Yule, C. and Jardel, J. -P. (1985). Observations on the eggs of species of *Dinotoperla* (Plecoptera: Gripopterygidae). *Aquatic Insects*, 7, 77–85.

Yuval, B. (2006). Mating systems of blood-feeding flies. *Annual Review of Entomology*, 51, 413–40.

Zacharuk, R. Y. and Shields, V. D. (1991). Sensilla of immature insects. *Annual Review of Entomology*, 36, 331–54.

Zahiri, N., Rau, M. E., and Davis, D. J. (1997a). Oviposition responses of *Aedes aegypti* and *Ae. atropalpus* (Diptera: Culicidae) females to waters from conspecific and heterospecific normal larvae and from larvae infected with *Plagiorchis elegans* (Trematoda: Plagiorchiidae). *Journal of Medical Entomology*, 34, 565–8.

Zahiri, N., Rau, M. E., and Lewis, D. J. (1997b). Intensity and site of *Plagiorchis elegans* (Trematoda: Plagiorchiidae) infections in *Aedes aegypti* (Diptera: Culicidae) larvae affect the attractiveness of their waters to ovipositing conspecific females. *Environmental Entomology*, 26, 920–3.

Zaidi, R. H., Jaal, Z., Hawkes, N. J., Hemingway, J., and Symondson, W. O. C. (1999). Can the detection of prey DNA amongst gut contents of invertebrate predators provide a new technique for quantifying predation in the field? *Molecular Ecology*, 8, 2081–8.

Zera, A. J. (1984). Differences in survivorship, development rate and fertility between the longwinged and wingless morphs of the waterstrider *Limnoporus canaliculatus*. *Evolution*, 38, 1023–32.

Zera, A. J. and Denno, R. F. (1997). Physiology and ecology of dispersal polymorphism in insects. *Annual Review of Entomology*, 42, 207–30.

Zhang, Y. and Malmqvist, B. (1996). Relationships between labral fan morphology, body size and habitat in North Swedish blackfly larvae (Diptera: Simuliidae). *Biological Journal of the Linnean Society*, 59, 261–80.

Zhang, Y. and Malmqvist, B. (1997). Phenotypic plasticity in a suspension-feeding insect, *Simulium lundstromi* (Diptera: Simuliidae), in response to current velocity. *Oikos*, 78, 503–10.

Zwick, P. (1977). Australian Blephariceridae (Diptera). *Australian Journal of Zoology, Supplementary Series*, 46, 1–121.

Zwick, P. (1982). A revision of the Oriental stonefly genus *Phanoperla* (Plecoptera: Perlidae). *Systematic Entomology*, 7, 87–126.

Zwick, P. (1996). Variable egg development of *Dinocras* spp. (Plecoptera, Perlidae) and the stonefly seed bank theory. *Freshwater Biology*, 35, 81–100.

Zwick, P. (2000). Phylogenetic system and zoogeography of the Plecoptera. *Annual Review of Entomology*, 45, 709–46.

Zwick, P. and Zwick, H. (2008). *Scirtes hemisphaericus* uses macrophyte snorkels to pupate underwater. With notes on pupae of additional European genera of Scirtidae (Coleoptera). *Aquatic Insects*, 30, 83–95.

Index

aerodynamics, 140–143
aggregation and sexual communication, 165–168
algae, 26, 236
  toxins in, 237
alimentary system
  epithelium, 227
  foregut, 226–227, 228
  gut morphology and diet, 227
  gut structure and function of non-feeding insects, 230–232
  hindgut, 226–227, 229–230
  ileum, 232
  Malphighian tubules, 232
  midgut, 226–227, 228–229
  preoral cavity, 226, 227–228
  rectum, 232
  structure of, 226–230
allelochemicals, 117–118
  allomones, 118
  kairomones, 118
  synomones, 118
allochthonous inputs in streams, 26–27
ametaboly, 4, 190
anhydrobiosis, 64
Anisoptera, 10–11, 22, 49
  aspect ratios of wings, 145
  ballistic ejection of faecal pellets, 233
  basal triangle-supratriangle complex, 143–144
  basking, 57
  body structure, 11
  colours of, 11
  copulation, 169–170
  dispersal in the terrestrial environment, 149
  flight capabilities, 143
  flight directions and distances, 147
  flying in tandem, 145
  genitalia, 169–170
  gliding, 145
  hunting in flight, 11

internal tracheal gills, 49
labial mask for capturing prey, 11, 214–215
larvae, 11
migration, 150
mouthparts, 214
oviposition, 179
parasites of, 144
rectal gills, 50
removal of sperm from a previous mating, 169
reproduction, 11
sperm competition, 170
stroking of wings, 144
tandem formation, 169
thermo-hygroreception, 117
thermoregulation, 58
tracheal gills, 49
vision, 89
wheel formation, 169
wing morphology and flight capability, 142, 143–145
antennae, 4, 167, 216
applied ecology, 155, 209
apolysis, 194, 201, 202
aquatic habitats, 24–27, 28–34
  artificial human-made environments, 34
  early freshwater feeding habitats, 26–27
  evolution of, 25
  flowing surface water, 29–30
  groundwater, 32–33
  marine environments, 33–34
  running water, 26
  semi-aquatic environments, 29
  standing surface water, 30–31
  still water, 26
  wetlands, springs, pools, puddles and phytotelmata, 31–32
aquatic insects, 9–19
  adaptations to maintain attachment in flowing water, 74–76
  adaptations to saline waters, 63

benthic insects, 29, 30
'blind' species, 83, 94
body shape, 71
complete dependence on atmospheric air, 42–43
creation of thrust, 78
detection of motion, 108
developmental stages, 155
diets of, 209
disjunctive distribution, 28
dispersal, 81–82, 137–153
dormancy or diapause, 54
dorso-ventral flattening, 73
ecological roles of, 24
environments inhabited by, 28–34
filter-feeding, 79–80
food of, 211–213
friction pads, 75–76
functional feeding groups, 211
halobionts, 31
hyporhetic insects, 29
migration, 54
movement, 81–82
origins of, 24–26
oviparous, 180
ovoviviparous, 180
oxygen conformers, 51
paddles and sieves, 78–79
polyphagous, 211
responses to abiotic extremes, 54
rowing and flapping appendages, 78–79
tarsal claws or hooks, 77
tolerance of salinity, 31
transition from terrestrial to aquatic environments, 157
upstream movement, 146
use of silk or other sticky secretions to maintain position, 77–78
viviparous, 180
water loss and gain, 62
water-surface dwellers, 30
aquatic plants, gas stores in, 43

Aquatic Entomology. First Edition. Jill Lancaster & Barbara J. Downes.
© Jill Lancaster & Barbara J. Downes 2013. Published 2013 by Oxford University Press.

9 780199 573226